The
Developmental
Biology of Reproduction

THE DEVELOPMENTAL BIOLOGY OF REPRODUCTION

The Thirty-Third Symposium of
The Society for Developmental Biology

Athens, Georgia, June 9-12, 1974

EXECUTIVE COMMITTEE:

1973-1974

ELIZABETH D. HAY, Harvard Medical School, *President*
THOMAS J. KING, National Cancer Institute, *Past-President*
DONALD D. BROWN, Carnegie Institution of Washington, *President-Designate*
JAMES A. WESTON, University of Oregon, *Secretary*
JOHN PAPACONSTANTINOU, Oak Ridge National Laboratory, *Treasurer*
PHILIP GRANT, University of Oregon, *Member-at-Large*

1974-1975

DONALD D. BROWN, Carnegie Institution of Washington, *President*
ELIZABETH D. HAY, Harvard Medical School, *Past-President*
WILLIAM J. RUTTER, University of California, *President-Designate*
JAMES A. WESTON, University of Oregon, *Secretary*
JOHN PAPACONSTANTINOU, Oak Ridge National Laboratory, *Treasurer*
VIRGINIA WALBOT, Washington University, *Member-at-Large*

Business Manager
CLAUDIA FORET
P.O. BOX 43
Eliot, Maine 03903

The
Developmental
Biology of Reproduction

Edited by

Clement L. Markert

Department of Biology
Yale University
New Haven, Connecticut

John Papaconstantinou

Biology Division
Oak Ridge National Laboratory
Oak Ridge, Tennessee

1975

ACADEMIC PRESS, INC. *New York   San Francisco   London*

A Subsidiary of Harcourt Brace Jovanovich, Publishers

ACADEMIC PRESS, INC.
111 Fifth Avenue, New York, New York 10003

United Kingdom Edition published by
ACADEMIC PRESS, INC. (LONDON) LTD.
24/28 Oval Road, London NW1

Library of Congress Cataloging in Publication Data

Society for Developmental Biology.
 The developmental biology of reproduction.

 (Symposium of the Society for Developmental Biology; 33d)
 Includes bibliographical references and index.
 1. Developmental biology—Congresses. 2. Reproduction—Congresses. I. Markert, Clement Law-
rence, (date) II. Papaconstantinou, John.
III. Title. IV. Series: Society for Developmental Biology. Symposium; 33d. [DNLM; 1. Reproduction—Congresses. W1 SO851M no. 33 1974 / QH471
S678d 1974]
QH 511.S6 no. 33 [QH491] 574.3'08s [599'.03'2]
ISBN 0–12–612979–7 75-14132

Contents

LIST OF CONTRIBUTORS vii
PREFACE ix
ACKNOWLEDGMENTS xi

I. Gametogenesis

Germinal Plasm and Determination of the Primordial Germ Cells   3
  L. Dennis Smith and Marilyn A. Williams

Gametogenesis in the Male: Prospects for Its Control   . . .   25
  Don W. Fawcett

Developmental Mechanisms in Volvox Reproduction   . . .   55
  Gary Kochert

II. Parthenogenesis

Teratogenesis and Spontaneous Parthenogenesis in Mice   . .   93
  Leroy C. Stevens

Induced Parthenogenesis in the Mouse   107
  Andrzej K. Tarkowski

III. Early Development

Genetic and Biochemical Activities in Preimplantation Embryos 133
  Cole Manes

The Regulation of Enzyme Synthesis in the Embryogenesis
and Germination of Cotton   165
  Virginia Walbot, Barry Harris, and Leon Dure, III

Chromosomal Basis for Hermaphrodism in Mice   189
  Wesley K. Whitten

Analysis of Determination and Differentiation
in the Early Mammalian Embryo Using
Intra- and Interspecific Chimeras 207
   R. L. Gardner

IV.   Hormonal Controls in Reproduction

Estrogen Feed-back and Gonadotrophin Secretion 239
   Béla Flerkó

Ontogeny of Development of the Hypothalamic
Regulation of Gonadotrophin Secretion:
Effects of Perinatal Sex Steroid Exposure 255
   Charles A. Barraclough and Judith L. Turgeon

V.   Implantation

The Relationship between the Early Mouse
Embryo and Its Environment 277
   Michael I. Sherman and David S. Salomon

A Determinant Role for Progesterone in the
Development of Uterine Sensitivity to
Decidualization and Ovo-Implantation 311
   Stanley R. Glasser and James H. Clark

Subject Index 347

List of Contributors

Boldface Denotes Chairmen

Charles Barraclough, Department of Physiology, University of Maryland Medical School, Baltimore, Maryland

Edgar J. Boell, Department of Biology, Yale University, New Haven, Connecticut

James H. Clark, Department of Cell Biology and Center for Population Research, Baylor College of Medicine, Houston, Texas

Stuart J. Coward, Department of Zoology, University of Georgia, Athens, Georgia

Leon S. Dure, III, Department of Biochemistry, University of Georgia, Athens, Georgia

Don W. Fawcett, Department of Anatomy, Harvard Medical School, Boston, Massachusetts

Béla Flerkó, Institute of Anatomy, Medical University of Pecs, Hungary

R. L. Gardner, Department of Zoology, Oxford University, Oxford, England

Stanley R. Glasser, Department of Cell Biology and Center for Population Research, Baylor College of Medicine, Houston, Texas

Barry Harris, Department of Biochemistry, University of Georgia, Athens, Georgia

Gary Kochert, Department of Botany, University of Georgia, Athens, Georgia

Cole Manes, Department of Pediatrics, University of Colorado Medical School, Denver, Colorado

Clement L. Markert (Program Chairman), Yale University, New Haven, Connecticut

Margaret W. Orsini, Department of Anatomy University of Wisconsin, School of Medicine Madison, Wisconsin

David M. Phillips, Rockefeller University, New York

Davis S. Salomon, Department of Cell Biology, Roche Institute of Molecular Biology, Nutley, New Jersey

Michael I. Sherman, Department of Cell Biology, Roche Institute of Molecular Biology, Nutley, New Jersey

L. Dennis Smith, Department of Biological Sciences, Purdue University, West Lafayette, Indiana

Leroy Stevens, The Jackson Laboratory, Bar Harbor, Maine

Steven Subtelny, Department of Biology, Rice University, Houston, Texas

Andrzej K. Tarkowski, Department of Embryology, Institute of Zoology, University of Warsaw, Warsaw, Poland

Judith Turgeon, Department of Physiology, University of Maryland Medical School, Baltimore, Maryland

Virginia Walbot, Department of Biochemistry, University of Georgia, Athens, Georgia

Wesley K. Whitten, The Jackson Laboratory, Bar Harbor, Maine

Marilyn R. Williams, Department of Biological Sciences, Purdue University, West Lafayette, Indiana

Preface

Recognizing that the continued growth of human populations threatens civilized existence, the Society for Developmental Biology selected the subject of Reproductive Biology as the focus for its 33rd Symposium. Reproductive biology encompasses a very large part of biology; if broadly defined – virtually all of it. This Symposium sought to center attention on basic aspects of reproduction in both plants and animals in the hope of stimulating research that might provide the necessary foundation for effective, practical control of human reproduction. Five areas were selected for emphasis: the formation of eggs and sperm, the activation of the egg to develop into an embryo, the genetic and biochemical events underlying the early development of the embryo, and then, on a physiological level of investigation, the hormonal controls operating in the reproductive process, and the general control of implantation and growth of the mammalian embryo in the uterus. Thirteen reports were given by distinguished researchers in each of these areas and, in accord with past practice of the Society, the international resources of science were used by the inclusion in the Symposium of three investigators from abroad.

In addition to the major scientific reports, one afternoon was devoted to a roundtable discussion of the social and biological implications of controls of reproduction. This roundtable was chaired by Clement L. Markert, with panelists Harriet Presser of the International Institute for the Study of Human Reproduction, Columbia University, and Sheldon Segal of the Population Council, Rockefeller University. The roundtable discussion was recorded on videotape and subsequently edited to produce a 45 minute, 16 mm film for general use by educational institutions and other groups interested in problems of population and human reproduction. One evening of the Symposium was devoted to workshops on a variety of topics related to research in reproduction. The 300 attendees at the Symposium were well accommodated by the host institution, The University of Georgia at Athens. The Society deeply appreciates the excellent organizational support by the local arrangements committee chaired by Stuart J. Coward, and is grateful for the financial support of the National Institute of Child Health and Human Development and the National Science Foundation. This financial support made it possible to bring before the Society a group of outstanding investigators to share their

PREFACE

wisdom and recent research results with those attending the Symposium and now, with the published volume, with a much larger section of the scientific community. All biologists interested in a broad understanding of problems of reproduction will find this Symposium interesting and important for their own work. We hope that the Symposium will contribute, at least in some small degree, to a solution of the most important problem facing human beings today – the control of their own reproduction.

Clement L. Markert

Acknowledgments

The 33rd Symposium was held at Athens, Georgia, June 9-12, 1974. The Society gratefully acknowledges the efficiency of the host committee, the hospitality of the University of Georgia and the support from The National Institute of Child Health and Human Development and The National Science Foundation.

VIDEOTAPE AVAILABLE ON IMPLICATIONS
OF CONTROL OF REPRODUCTION

The round-table discussion on *"Social Implications of Control of Repro-duction"* held at our 33rd Symposium in June was videotaped at the Georgia Center for Continuing Education. A 48 minute edited version is now available in color for classroom or other use. The discussion, features Harriet Presser of the International Institute for the Study of Human Reproduction (Columbia University), Sheldon Segal of the Population Council (Rockefeller University), and Clement L. Markert (Yale University).

Methods of birth control are discussed frankly, including the role of abortion as a "back-up" measure. The effect of population control in advanced countries on the balance of world politics is considered and the question of the overproduction of racial groups, such as Blacks, is handled very well. The issue of birth control in the younger group of the population by reversible methods versus permanent methods for older individuals is dealt with in depth. Dr. Markert raised one controversial issue after another for discussion by his distinguished panelists.

The tape is recommended for viewing in college courses introducing reproductive biology to beginning students as well as for showing in more advanced courses on reproduction and developmental biology. It is easily understood by the layman, as well. A small charge for rental will be made and advance booking is recommended. Write Dr. E. D. Hay, Dept. Anatomy, Harvard Medical School, 25 Shattuck St., Boston, MA 02115.

I. Gametogenesis

Germinal Plasm and Determination of the Primordial Germ Cells

L. Dennis Smith and Marilyn A. Williams

Department of Biological Sciences
Purdue University
West Lafayette, Indiana 47907

I. Introduction 3
II. Fine Structural Studies on Germ Plasm 4
    A. Identification of Electron Dense Bodies within Germ Plasm .. 4
    B. Origin and Continuity of Polar Granules in Drosophila 5
    C. Origin and Continuity of Germinal Granules in Amphibians .. 7
III. Experimental Studies on the Role of Germ Plasm in the Formation of Primordial Germ Cells 13
    A. Destruction or Deletion of Germ Plasm 13
    B. The Effect of UV Irradiation on Germinal Granules 14
    C. Commentary on UV Studies 17
    D. Definitive Demonstration of Germ Cell Determinants 19
IV. Nature and Action of Germ Cell Determinants 19
    References 21

I. INTRODUCTION

It appears to be reasonably well established that the functional gametes of many organisms are derived solely from primordial germ cells already present in early embryogenesis. Direct evidence to support this statement has existed for some time in those invertebrate organisms displaying chromosome diminution or elimination (reviews by Wilson, 1925; Gurdon and Woodland, 1968). Equally compelling evidence exists in at least one group of vertebrates, the amphibians, and is based on grafting experiments between genetically marked embryos at the gastrula (Smith, 1964) or neurula stages (Blackler, 1962, 1966, 1970; Blackler and Gecking, 1972a, b). Supportive evidence also has been obtained in the chick (Willier, 1937; Simon, 1957; Reynaud, 1969) and mouse (Mintz, 1960; see also Blackler, 1965).

The idea that the factor(s) responsible for this early germ line determination reside(s) in the cytoplasm traces back to the turn of the century. It is based partly on the demonstrated presence of deeply staining granular inclusions in the posterior pole plasm of insect eggs, and their incorporation into the early germ cells (review by Hegner, 1914; Wilson, 1925). Hegner (1914) originally considered these polar granules to be the "germ cell determinants", since their destruction resulted in embryos lacking germ cells. The fate of polar granules throughout most stages of the life cycle now has been followed, both in the light microscope (Counce, 1963) and electron microscope (see review by Mahowald, 1971b). Bounoure (1934) was the first to demonstrate the existence of a similar material in vertebrates. He was able to follow the embryonic history of primordial germ cells in the frog *Rana temporaria* by virtue of a specially staining "germ plasm", first identified in the vegetal hemisphere of fertilized but uncleaved eggs. This report has since been substantiated and extended to include a number of anuran amphibians (review by Blackler, 1966). In addition, considerable experimental evidence has implicated the stainable germ plasm in the formation of primordial germ cells (see Smith, 1966; Blackler, 1966, 1970).

Recent emphasis on the germ plasm in vertebrates, largely amphibians, has centered on additional descriptive studies, albeit at the fine structural level. These observations correspond to the earlier work done with *Drosophila*. On the other hand, recent work with *Drosophila* has been concerned with an experimental demonstration of the determinative nature of pole plasm in germ cell formation, considerably extending earlier amphibian work. In this paper a review of this recent evidence is presented both for amphibians and insects, in an attempt to present the "current status" of research concerned with the nature, origin, and function of germ plasm in germ cell determination.

II. FINE STRUCTURAL STUDIES ON GERM PLASM

A. *Identification of Electron Dense Bodies Within Germ Plasm*

The fine structure of polar granules within the posterior pole plasm in insects was first described by Mahowald (1962) in eggs and embryos of *Drosophila melanogaster*. Subsequent work was extended to other *Drosophila* species and included a description of polar granule morphology during several developmental stages (Mahowald, 1968; 1971a). While the size of the electron dense granules differed in each species (less than 0.5 μm to over 1 μm in diameter), and at different developmental stages, the basic fine structure always consisted of a densely staining fibril, 10 to 15 nm in diameter, which formed an interwoven mesh.

That a structure similar to polar granules might exist within the germinal plasm of amphibian eggs was first indicated by Balinsky (1966). He suggested that granular electron dense bodies observed amidst groups of mitochondria at the vegetal hemisphere of a South African frog egg "may represent the basophilic areas of cytoplasm found at the vegetal pole of amphibian eggs and known as germinal plasm". Such electron dense bodies were "rediscovered" in the vegetal hemisphere of *Rana pipiens* eggs five years later (Kessel, 1971; Mahowald and Hennen 1971; Williams and Smith, 1971).

Within the germ plasm region of the vegetal hemisphere of a fertilized uncleaved egg, one always sees distinctive electron dense bodies associated with large aggregations of mitochondria (Fig. 1). These bodies, whose diameters average about 0.2 μm, are not enclosed within a limiting membrane, and are usually seen in close proximity to mitochondria. Particles about the size of ribonsomes are also present at the periphery of the dense bodies, and numerous glycogen particles are evident within the immediate area. Similar electron dense bodies, although more variable in size and shape, subsequently have been identified within the germ plasm or vegetal hemisphere of fertilized *Xenopus laevis* eggs (Czolowska, 1972; Kalt, 1973; Ikenishi et al., 1974). Likewise, they have been observed within the vegetal hemisphere of fertilized *Engystomops* eggs (Williams and Smith, unpublished data), and near the marginal zone of fertilized *Axolotl* eggs (Williams and Smith, 1971). The latter observation represents the first demonstration of a morphological entity in urodele amphibians associated with a region which eventually contains primordial germ cells (see Smith, 1964). Finally, similar structures recently have been observed within primordial germ cells located in the primitive gut of 8-9 day old mouse embryos (Spiegelman and Bennett, 1973), and 10 day old rat embryos (Eddy, 1974).

Both in *Rana pipiens* and *Xenopus laevis*, the germinal granules consist of electron dense foci between 10-20 nm in diameter, with finer components ranging from 2 to 8 nm in diameter (Mahowald and Hennen, 1971; Williams and Smith, 1971; Kalt, 1973; Ikenishi et al., 1974). Thus the germinal granules are essentially identical in appearance and fine structural detail to *Drosophila* polar granules. This similarity, and observations that the electron dense bodies are apparently a unique feature of the stainable germ plasm led to their designation as germinal granules (Williams and Smith, 1971).

B. Origin and Continuity of Polar Granules in Drosophila

Mahowald's studies on *Drosophila* indicated that polar granules are very polymorphic during embryogenesis, progressing through cycles of fragmentation and fusion (review by Maholwald, 1971b). In the mature oocyte and freshly fertilized egg, polar granules frequently were attached to each other and to mitochondria. Shortly after fertilization, the granules became detached from mitochondria, fragmented into smaller spherical or rodlike bodies and now contained ribosomes closely associated with their periphery. After pole cells formed, polar granules again fused into large structures which, in one case, *D. immigrans*, resulted in rodlike sheets. Ribosomes no longer were present at the periphery. Finally, as the pole cells migrated to the embryonic

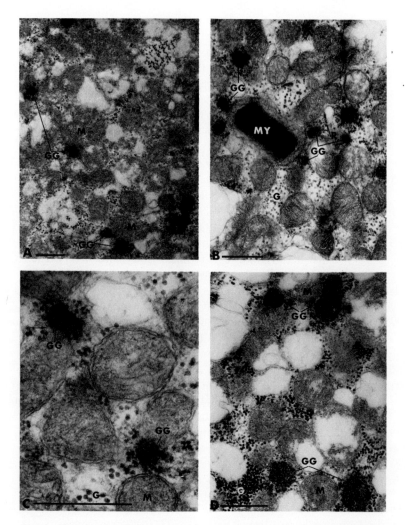

Fig. 1. Ultrastructure of germinal plasm region in fertilized *Rana pipiens* eggs: (A) 1.5 hours after fertilization; (B) 2-cell stage; (C) 4-cell stage; (D) 16-cell stage. These areas contain germinal granules (GG), mitochondria (M), mitochondria-containing yolk (MY), ribosomes (R), and glycogen (G). Arrow in C points to fibril connecting germinal granule and a ribosome. Scale represents 0.5 μ. From Williams and Smith (1971).

gonad, polar granules again fragmented, this time into amorphous fibrillar structures, which became associated with the nuclear membrane. This material is similar in fine structure to the "nuage" material observed during vertebrate gametogenesis (Eddy and Ito, 1971; Fawcett, 1972, Eddy, 1974).

Fibrous bodies still were found attached to the nuclear membrane in primordial germ cells and oogonia, but disappeared in oocytes long before polar granules reformed (midway through vitellogenesis). However, the nurse cells retained these structures until the end of vitellogenesis. Since the fibrous material could represent precursors to typical polar granules, these observations suggested the hypothesis that polar granule material is continuous throughout the life cycle of Drosophila (Mahowald, 1971a, b).

As Mahowald points out, doubt concerning the continuity of polar granules throughout the life cycle can be raised at two periods; the time pole cells differentiate into germ cells, and during oogenesis. The first period raises a question because it is possible that the fibrous bodies attached to the nuclear membrane during pole cell migration are newly formed, and not derived from polar granule fragmentation. Moreover, the fine structure of the two components is distinctly different. However, the main supporting evidence for derivation of fibrous bodies from polar granules comes from observations on the unique rodlike granule in D. immigrans. At the time fibrous bodies first appeared, the possible transition from rod to loosely woven fibrous structure has been described (Mahowald, 1971a).

C. Origin and Continuity of Germinal Granules in Amphibians

It is difficult to obtain extensive knowledge of the fate of germinal granules in amphibians, principally because primordial germ cells (identified in the light microscope by germ plasm) are embedded within masses of somatic cells and are difficult to find in thin sections. Accordingly, only a few selected developmental stages have been examined. In Rana pipiens, no change in germinal granule morphology is apparent during the first three cleavage divisions. By the 16 cell stage, some granules have increased in size (Fig. 1), suggesting coalescence of two or more dense bodies (Williams and Smith, 1971; also Mahowald and Hennen, 1971). Typical germinal granules, with associated ribosomes, are still present in blastula stage germ cells, both in Rana pipiens (Mahowald and Hennen, 1971) and Xenopus laevis (see Fig. 5), and have been observed in late gastrula stage embryos of Xenopus laevis (see Fig. 6). However, by the time primordial germ cells are localized in the genital ridges, germinal granules no longer are found (Mahowald and Hennen, 1971). Instead, one sees accumulations of irregularly shaped fibrous material, devoid of any associated ribosomes, adjacent to the nucleus and attached to mitochondria (Fig. 2). As in Drosophila, this material also is similar to the so-called nuage mentioned earlier. Thus, based upon the analogy with polar granule fragmentation in Drosophila, and light microscopic observations that germ plasm in amphibians associates with the nucleus during germ cell formation (Blackler, 1958), it has been suggested that the dense masses seen in amphibian primordial germ cells are derived from the original germinal granules (Mahowald and Hennen, 1971). However, since intervening "transition" stages have not been examined, this suggestion remains unsubstantiated, even though identity has been claimed (Kalt, 1973).

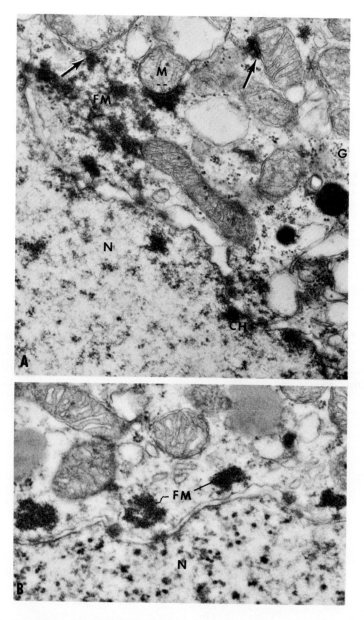

Fig. 2. Portions of primordial germ cells of a stage 25 *Rana pipiens* larva showing the irregularly shaped fibrous material (FM and arrows) attached to mitochondria (M) in Fig. 2A and associated with the nucleus (N) in Fig. 2B. Chromatin (CH) is also shown in Fig. 2A. Ribosomes are absent from the periphery of the fibrous material. Fig. 2A: x30,000; Fig. 2B: x43,000. From Mahowald and Hennen (1971).

The subsequent fate of these dense fibrous components has not been systematically followed throughout amphibian oogenesis. In *Rana* species, Kessel (1969), Massover (1968) and Eddy and Ito (1971) all have described dense material with a fibrogranular structure in oogonia and small oocytes. This material was found adjacent to the nuclear envelope and associated with aggregates of mitochondria, but is reported to disappear by the time yolk accumulation begins in large oocytes (Kessel, 1969; Eddy and Ito, 1971). Kessel (1969) also reported the existence of large "nucleolus-like" bodies in the cytoplasm of oocytes less than 200 μ in diameter. Studies on *Xenopus* have been more complete in that primordial germ cells, gonia, and oocyte and spermatocyte stages were examined (Al-Mukhtar and Webb, 1971; Reed and Stanley, 1972; Kalt, 1973; Coggins, 1973). Generally, the descriptions are in agreement. Primordial germ cells contain dense material (nuage) adjacent to the nuclear envelope, nuage-mitochondria aggregates, and large "nucleolus-like" bodies. These structures all persist into meiotic prophase, when the "nucleolus-like" bodies apparently disappear (Kalt, 1973), while the amount of nuage material is reported to increase in previtellogenic diplotene oocytes, and then also to disappear during later growth (Kalt, 1973).

It is not yet clear just when during oogenesis, dense nuage-like material no longer is detectable. We have observed such material in *Rana* oocytes as large as 500 μm in diameter (about one-third maximal size) but it is difficult to obtain a complete stage series in these animals because of the lengthy time required to reach sexual maturity; most oocytes in gravid females have already completed their growth. In *Xenopus*, however, oogenesis is a continuous but asynchronous process such that oocytes at all stages usually are present within the ovary at any given time. We have observed nuage-mitochondria aggregates in vitellogenic *Xenopus* oocytes (Fig. 3A) as large as about 600 μm in diameter (corresponding to late lampbrush chromosome stage, Dumont (1972, stage III), but not in larger oocytes. Thus, apparently, the dense fibrogranular material observed in primordial germ cells, oogonia, and oocytes is not present continuously throughout oogenesis.

The reappearance of typical germinal granules appears to occur during the hormonally-induced period of maturation. In *Rana pipiens*, small electron dense masses, reminiscent of nuage, can be observed within clusters of mitochondria located just beneath the vegetal hemisphere cortex (Fig. 4A). After the induction of maturation with progesterone (see Smith and Ecker, 1970; Smith, 1974), and coincident with breakdown of the oocyte nucleus (germinal vesicle), larger electron dense masses with associated ribosomes become a prominent feature within the mitochondrial aggregates (Fig. 4B). At progressively later times, the electron dense masses assume the more rounded compact appearance characteristic of germinal granules observed in unfertilized and fertilized eggs (Fig. 4D; also Mahowald and Hennen, 1971). In *Xenopus*

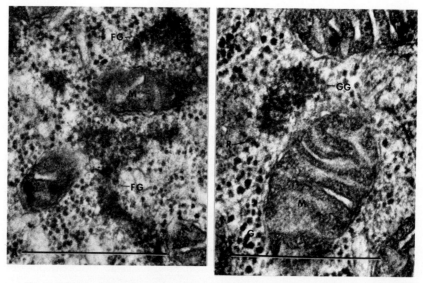

Fig. 3. Mitochondria (M) aggregates containing electron dense bodies in a vitellogenic stage III oocyte (3A) and a full grown maturing oocyte (3B) of *Xenopus laevis*. The fibrogranular (FG) material in 3A closely resembles the germinal granule (GG) observed during maturation (6 hours after exposure to progesterone). Both ribosomes (R) and glycogen (G) are observed in the immediate area. Fixation according to Williams and Smith (1971). Scale represents 0.5 μ.

laevis we have not observed nuage-like material even in full grown (stage VI) oocytes. However, dense bodies essentially identical to characteristic germinal granules do appear during maturation, again coincident with breakdown of the germinal vesicle (Fig. 3B).

The above observations, which correlate the appearance of definitive germinal granules with breakdown of the germinal vesicle might be interpreted to mean that contents of the germinal vesicle are involved in the formation of these granules. However, the observation (Fig. 4C) that germinal granules appear in maturing enucleated oocytes at about the same time as in nucleated controls rules out a nuclear contribution to their formation during maturation. Accordingly, the results suggested either that (1) the cytoplasm of full grown oocytes contains "precursors" which coalesce during maturation or (2) components of the germinal granules are synthesized *de novo* during maturation (Williams and Smith, 1971). Since subsequent autoradiographic experiments failed to show incorporation either of nucleosides or amino acids into germinal granules during maturation (Williams and Smith, unpublished data), the former suggestion seems more reasonable at this time. The identity of such "precursors" remains unknown.

Both Fawcett (1972) and Eddy (1974) have pointed out the similarities in form and distribution between nuage, observed during mammalian gameto-

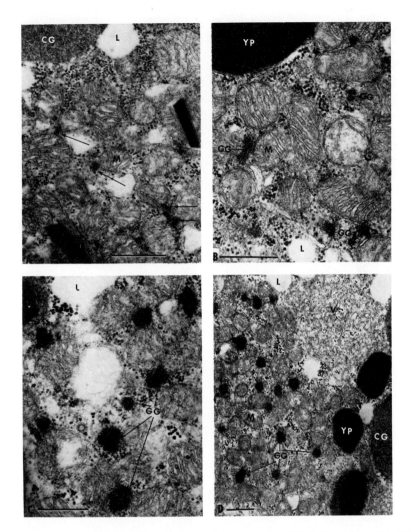

Fig. 4. Vegetal polar regions in *Rana pipiens* oocytes. (A) Fullgrown oocyte. Small electron dense areas (arrows) are present within the aggregations of mitochondria which lie below the cortical granules. (B) Oocyte undergoing maturation, 11 hours after progesterone treatment. Germinal granules are present within the subcortical aggregations of mitochondria. (C) Enucleated oocyte 23 hours after progesterone treatment. Germinal granules are present within an aggregation of mitochondria. Yolk platelets and lipid droplets surround the mitochondria. Ribosomes and glycogen are interspersed between the mitochondria. (D) Unfertilized egg. Numerous germinal granules are present within a mitochondrial cluster lying below the cortical granules. Yolk platelets and vesicles surround the mitochondrial cluster. GG, germinal granule; CG, cortical granule; YP, yolk platelet; R, ribosomes; G, glycogen; L, lipid; V, vesicles. Scale represents 0.5 μ. From Williams and Smith (1971).

genesis, and the germinal granules or polar granules of amphibians and insects. Kalt (1973) argues strongly for the concept that nuage and germinal granules in *Xenopus* are a single entity, largely because both are interpreted as having the same basic fine structure and association with mitochondria. In this sense, the fibrogranular material associated with mitochondria in stage III *Xenopus* oocytes and the germinal granules observed in maturing oocytes (Fig. 3A, B) appear to be virtually identical. On the other hand, nuage material, particularly in smaller oocytes and primordial germ cells, does not resemble the germinal granules seen in unfertilized or fertilized eggs in that the dense masses appear to be composed of finer fibrillar components (Williams and Smith, unpublished data; Mahowald and Hennen, 1971). Likewise, both in *Drosophila* and *Miastor,* fibrous material equivalent to nuage coexists in the mature egg with typical polar granules, but the former apparently disappears at the time of fertilization or the maturation division (Mahowald, 1971a, 1974).

Variability in fixation and staining, possible morphological transformations, and degree of aggregation or dispersal of presumed precursors easily could produce discontinuities in the appearance of equivalent structures at different stages. Thus, whether the so-called nuage (or other fibrogranular material) is equivalent to germinal granules remains conjectural, and one is left with individual interpretations of static electron micrographs. Therefore, as in *Drosophila,* doubt concerning continuity centers around two periods, the time when primordial germ cells migrate to the genital ridges and oogenesis. It is unlikely that further electron microscopy will unequivocally resolve these doubts. On the other hand, one may reasonably question whether it is even necessary that components of the germ plasm maintain continuity between generations, provided that the germ line itself remains intact.

In most cases in which cytoplasmic organelles contain their own genetic material (e.g. mitochondria), it is not difficult to discuss continuity from generation to generation. In fact, Dawid and Blackler (1972) have reported that *Xenopus* mitochondrial DNA is inherited maternally and cytoplasmically through the primordial germ cells. Incidentally, this may explain partially the apparent segregation of large mitochondrial aggregates to germinal plasm and hence to developing primordial germ cells. On the other hand, germ cell determinants conceivably need be present only until certain cells are determined to become germ cells; subsequently they could be synthesized *de novo* within the germ line for the next generation. In this context, Blackler (1966, 1970) considers germ cell determination to have occurred prior to neurulation, and this corresponds rather nicely to the time period in which typical germinal granules, with associated ribosomes, are observed. Obviously, the problems of the origin and continuity of germinal granules may fall between these two extremes. There may be continuity of a portion of the germinal granule throughout the life cycle with acquisition of the other part at a specific developmental stage. This view is supported by Mahowald (1971a, b) as applied to insect polar granules.

In summary, it seems clear that characteristic electron dense bodies are a consistent and unique feature (see Hay, 1968; Morita et al., 1969; Coward, 1974 for exception) of the germ plasm, both in *Drosophila* and *Amphibia*. The more extensive work with *Drosophila* supports the idea of origin during oogenesis and continuity throughout the life cycle. Likewise, in those instances in which studies have been made, the changes in amphibian germinal granules mimic those in *Drosophila* very closely. Are, then, the granular inclusions equivalent to germ cell determinants as first postulated by Hegner (1914)?

III. EXPERIMENTAL STUDIES ON THE ROLE OF GERM PLASM IN THE FORMATION OF PRIMORDIAL GERM CELLS

A. *Destruction or Deletion of Germ Plasm*

Classically, two kinds of experimental approaches have been used to verify the role of germ plasm in the formation of amphibian primordial germ cells: removal of germ plasm surgically and destruction by irradiation (reviews by Blackler, 1966; 1970). In the former case, the most complete experiments have been performed by Buehr and Blackler (1970) using eggs of *Xenopus laevis*. Incisions were made into the vegetal pole of 2-cell and 4-cell eggs, resulting in the formation of exudates shown to contain germ plasm in some cases. Further cytological analysis revealed the absence of stainable germ plasm in about one-third of the embryos (blastulas) which developed from operated eggs. Correspondingly, about one-third of the tadpoles which developed from operated eggs were totally sterile. Additional numbers of animals exhibited reduced numbers of primordial germ cells, presumably resulting from only partial removal of germ plasm. Thus, not only do these studies support a determinative role for germ plasm, they suggest that once removed, the egg can not produce more plasm.

Bounoure (1934) first reported that ultraviolet irradiation of the vegetal hemisphere of frog eggs resulted in germ cell elimination. These studies were extended by Bounoure et al. (1954) and by Padoa (1963). However, in no case was complete sterility obtained (no germ cells) in *all* larvae derived from irradiated eggs. In the later work on *Rana pipiens* (Smith, 1966), complete sterility was always obtained when UV above a certain dose was applied just before or during the first cleavage division. In addition, since it was possible to visualize rapidly and count primordial germ cells *in situ* (Smith, 1965), quantitative studies on germ cell "survival" after irradiation at different wavelengths became practical (Smith, 1966). This "action spectrum" supported the hypothesis of a specific UV sensitive component containing nucleic acid localized within the germ plasm. Finally, the observation that unirradiated vegetal pole cytoplasm could restore germ cell formation when microinjected into the vegetal hemisphere of UV irradiated eggs supported the belief that the UV sensitive component was a germ cell determinant.

Recently, Okada et al. (1974) have essentially repeated, on *Drosophila,* the microinjection experiments reported by Smith (1966). The posterior pole plasm of *Drosophila* eggs was irradiated with UV light and normal pole plasm then was injected back into the same region of the irradiated eggs. A high percentage of the irradiated injected eggs failed to survive, but among the survivors sterility (usually resulting from UV) was prevented in 42% of the cases. Preliminary results reported by Okada et al. (1974) further indicated that the "germ cell determinants" were not species specific. Taken together with data from amphibians, it would appear justifiable to conclude that UV sensitive germ cell determinants exist within the germ plasm (pole plasm) of both vertebrate and invertebrate organisms. Unfortunately, identification of the UV sensitive component remains unresolved in both cases.

B. The Effect of UV Irradiation on Germinal Granules

Recently, we have begun ultrastructural studies on the morphology and distribution of germinal granules after UV irradiation, principally in *Xenopus laevis* eggs (Williams and Smith, in preparation). No consistent effect of UV irradiation on germinal granule morphology has been seen. Both control and irradiated blastulas (Fig. 5) and gastrulas (Fig. 6) appear to contain granules in comparable numbers and of comparable size, composed in both cases of electron dense foci about 17 nm in diameter. Again, perhaps this is not surprising as UV induced changes at the macromolecular level (i.e. pyrimidine dimers) would not necessarily be expected to produce gross morphological changes in germinal granules. Likewise, UV irradiation does not appear to prevent mitochondrial aggregates, containing the germinal granules, from undergoing the characteristic intracellular shift to a juxtanuclear position, (Fig. 7) as already indicated for stainable germ plasm observed in the light microscope (Nieuwkoop and Faber, 1956; Blackler, 1958). On the other hand, Ikenishi et al. (1974) have reported direct UV effects on *Xenopus* germ plasm, resulting in swelling and vacuolation of mitochondria and fragmentation of germinal granules. The latter interpretation of their results, however, appears to rest on arguments concerning fine structural organization of the electron dense bodies.

In *Xenopus,* the electron dense germinal granules frequently contain light areas which occupy a central or excentric position (Fig. 5; also Czolowska, 1972; Kalt, 1973). Ikenishi et al. (1974) interpret germinal granule structure as representing a cylinder, folded back or branched irregularly, composed of

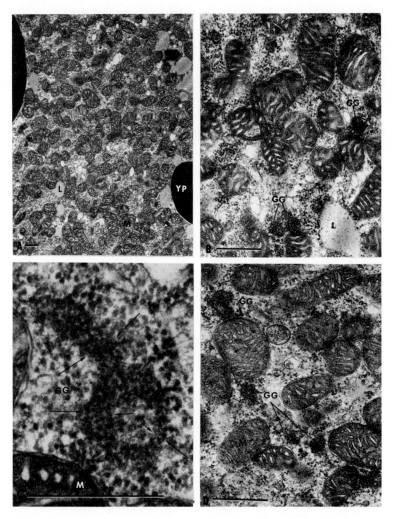

Fig. 5. Germ plasm regions of *Xenopus laevis* blastulas derived from control and UV irradiated eggs. A and B. Portion of control germ plasm regions showing the large aggregates of mitochondria (M) and distinct germinal granules (GG). C. Higher magnification of a single germinal granule indicating the loosely woven structure with internal spaces. D. Portion of the germ plasm after UV irradiation (7200 ergs/mm^2), showing germinal granules of the same size and shape as in controls. Arrows in (C) indicate possible points at which the diameter of the so-called cylinder could be measured (see text). YP, yolk platelet; L, lipid. Scale represents 0.5 μ. Fixation as per Kalt and Tandler (see Kalt, 1973).

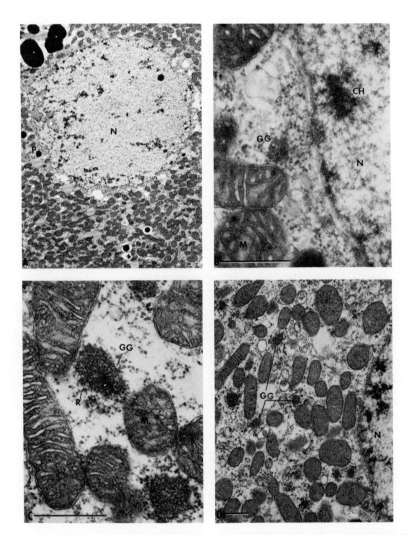

Fig. 6. Germ plasm regions of *Xenopus laevis* gastrulas derived from control and UV irradiated eggs. A and B. Portions of the germ plasm from controls at stage 11 (Nieuwkoop and Faber, 1956) showing the juxtanuclear position of mitochondria aggregates (A) and a germinal granule (B). Nuclear condensed chromatin (CH) can be seen close to the nuclear envelope. C and D. Portions of the germ plasm from stage 12 control (C) and UV irradiated (7200 crgs/mm^2) embryos. Note the similarity in appearance and structure of the germinal granules (GG) in both cases. N, nucleus; P, pigment; R, ribosomes. Scale represents 0.5 μ. Fixation as per Kalt and Tandler (see Kalt, 1973).

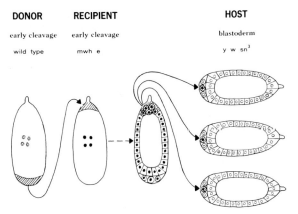

Fig. 7. Schematic illustration of the procedure for transplanting posterior pole plasm and anterior cells in *Drosophila*. From Illmensee and Mahowald (1974).

multiple numbers of electron dense foci, 7 nm or 14 nm in diameter. This is analogous in some respects to a bundle of rods aligned parallel to each other. According to this view, the light areas appear when the cylinder folds back on itself. Thus, the diameter of "germinal granules" is taken to be the diameter of the so-called cylinder (the distance from the edge of an internal space to the outer edge of the electron dense body). On this basis, UV irradiation reportedly reduces *average* cylinder diameter by about 7 nm, with considerable overlap in the range of measured diameters between irradiated and control embryos. These differences appear to be marginal and depend largely on the degree of compactness of the entire electron dense mass. This of course could be a function of fixation procudures, as suggested by others (see Czolowska, 1972; Kalt, 1973), as well as the plane of section through a dense body. Furthermore, the so-called cylinder can have different diameters even within a single electron dense body, dependent on which points are selected for measurement (see Fig. 5D). Thus, while the idea that UV irradiation does in fact have a direct effect on electron dense germinal granules is very appealing, the data supporting this contention are not compelling.

C. Commentary on UV Studies.

In spite of the fact that only about 12% of irradiated *Drosophila* eggs survived to become adults, Okada et al. (1974) report that gross morphogenesis was essentially normal in those eggs which did survive. However, histological studies revealed that, in irradiated eggs, migration of nuclei into the posterior plasm was delayed and constriction of pole cells was prevented. This is similar to observations of Bantock (1970) who reported that UV irradiation of the pole plasm of gall midge eggs caused a delay in pole cell separation. Since both delayed nuclear migration and pole cell separation could account for UV induced sterility, Okada et al. (1974) question whether UV irradiation specifically destroys germ cell determinants or whether UV acts indirectly to block pole cell formation. In the latter case, it becomes more difficult to explain the nature and mode of action of corrective agents reintroduced into irradiated *Drosophila* (and amphibian) eggs by microinjection.

Several years ago, Nieuwkoop (1974) advanced the argument that UV irradiation could prevent germ cell formation in amphibians by creating progressively unfavorable conditions for germ cell differentiation. Blackler (1966, 1970) has indicated that UV irradiation has no effect on the physical appearance of germ *plasm* observed with the light microscope (perhaps not surprisingly) but does lead to later abnormalities, largely by affecting the intracellular shift of germinal plasm to a juxtanuclear position. Since either this restriction or, in theory, delayed migration of germ cells to the genital ridges could result in sterility, the specificity of UV action on "germ cell determinants" again is questioned.

It should be pointed out that in amphibians, unlike *Drosophila,* UV irradiation of the vegetal hemisphere at doses sufficient to cause sterility results in no lethality (Smith, 1966), although irradiation shortly after fertilization does cause abnormalities associated with gastrulation (Grant and Wacaster, 1972; Malacinski, 1972). Likewise, irradiation with a wavelength of 254 nm, at doses resulting in sterility, causes no obvious retardation in developmental events, while irradiation with a wavelength of 230 nm resulted in severe developmental abnormalities with little effect on germ cell formation (Smith, 1966).

A comparison of the effects of UV at different stages of development (*Rana pipiens*) also showed that, after the first cleavage division, germ cell formation was progressively less affected by irradiation (Smith, 1966); by the 8 cell stage, no sterile larvae resulted from irradiation. Since the germ plasm observed with light microscopy moves away from the oocyte surface during cleavage (see Blackler, 1966) and UV has low penetratibility, the suggestion was made that UV sensitive "targets" were restricted within the germ plasm (Smith, 1966). Recently, this hypothesis has been approached in greater detail by Tanabe and Kotani (1974), using fertilized *Xenopus laevis* eggs.

In this work, the effect of UV irradiation at different stages of development on subsequent germ cell numbers was correlated directly with the subcortical depth of stainable germ plasm. As distance of the germ plasm from the vegetal hemisphere surface increased, the sterilizing effect of UV irradiation decreased. Likewise, displacement of germ plasm into the interior by centrifugation, followed by UV irradiation, revealed a substantially decreased effect on germ cell formation not attributable to centrifugation alone (Tanabe and Kotani, 1974). Taken together, these studies provide stronger support for the hypothesis that a UV-sensitive component, present within the germinal plasm, is necessary for germ cell formation.

D. *Definitive Demonstration of Germ cell Determinants*

Viewed against the background discussed above, the elegant experiments recently reported by Illmense and Mahowald (1974) provide unequivocal evidence for the existence of germ cell determinants within the pole plasm of *Drosophila* eggs. These experiments avoided the use of any deletion or irradiation procedures and their attendant uncertainties. Rather, the basic experimental design consisted of transferring pole plasm from wild type eggs into the anterior pole of mutant eggs, followed later (after blastoderm formation) by reintroduction of anterior blastoderm cells into the posterior pole cell region of another genetically marked host (Fig. 7). Ultrastructural examination of the first host series revealed, in several cases, the presence of polar granules in anterior cells which now resembled typical pole cells normally seen at the posterior tip. More important, among the fertile flies which developed from hosts receiving anterior "pole cells", 4 cases (4%) were found to be germ line mosaics when tested by appropriate crosses. Other controls consisting of anterior "somatic cells" from noninjected (no pole plasm) donors implanted into the posterior pole of mutant hosts failed to give rise to germ line mosaics. Thus, these experiments show directly that the posterior pole plasm contains a component(s) which not only induces the formation of morphologically recognizable pole cells in a region normally destined to produce only somatic cells, but that these cells can develop into functional gametes when placed back into the appropriate region for migration into the gonads. Whether the cytoplasm *in toto* or a specific portion, such as polar granules, functions in germ cell determination remains unknown. However, whatever its nature this component(s) fulfills all of the criteria expected of a germ cell determinant.

IV. NATURE AND ACTION OF GERM CELL DETERMINANTS

Cytochemical studies have revealed that the stainable germ plasm contains RNA, both in amphibian eggs (Blackler, 1958; Czolowska, 1969) and insect eggs (see Mahowald, 1971c). Likewise, in both cases the RNA moiety appears to have a transient association with the germ plasm since the affinity for basophilic stains disappears during early development (gastrula stage in amphibians, Blackler, 1958; blastoderm stage in *Drosophila*, Mahowald, 1971c). However, such procedures most likely stain only ribosomal RNA, known to be associated with the electron dense germinal (polar) granules during early embryogenesis. Thus, cytochemistry at the light microscope level is not necessarily informative.

Mahowald (1971b) has reported that polar granules in *Drosophila* stain positively with the indium trichloride procedure and that the stainability was removed by cold perchloric acid extraction. Thus, at the ultrastructural level, the presence of RNA within polar granules is indicated. Later, during the formation of pole cells, when polar granules fragment, they lose the ability to bind indium and by the blastoderm stage no longer stain positively at either the light or electron microscope level. Curiously, the loss of RNA from polar granules parallels closely the loss of ribosomes from the polar granule periphery. However, because the polar granules lack the particulate staining characteristic of ribosomes, Mahowald (1971b) argues against the RNA positive staining as being due to ribosomes.

Ultrastructural studies on germinal granules in frogs have led to conflicting views. Mahowald and Hennen (1971) report "moderate" staining of germinal granules in *Rana pipiens* with indium. Likewise, Kalt (1973) obtained "faint" staining of the dense bodies in *Xenopus,* but attributes this to background density not due to nucleic acids. He further states that, while RNase digestion partially disrupts ribosomes, it has no effect on the structure of germinal granules. Our own studies on glycolmethacrylate embedded tissues of *Rana pipiens* (Williams and Smith, unpublished data) generally support this view; RNase had no effect on germinal granule structure while pepsin digestion resulted in extensive disruption. Obviously, negative results do not constitute proof for the absence of a specific component, particularly at the cytochemical level, when a component such as RNA might comprise a very small percentage of the total mass. Thus, perhaps the strongest indication for involvement of RNA in germ cell formation is the UV "action spectrum" reported by Smith (1966), which suggests that the UV sensitive material contains nucleic acid. However, as already pointed out, the identity of this UV target could be other than the electron dense bodies.

The presence of ribosomes adjacent to the electron dense bodies, both in *Drosophila* and amphibians, led to the suggestion that the germinal (polar) granules are engaged in protein synthesis during early embryogenesis (Mahowald, 1968, 1971a, b; Mahowald and Hennen, 1971). Based on the cytochemical detection of RNA within the germinal (polar) granules, they are hypothesized to be a site for the localization of maternal messenger RNA which codes for the synthesis of certain proteins necessary to the formation of primordial germ cells. Furthermore, in this hypothesis, only the proteinaceous moiety of polar granules would be present continuously throughout the life cycle, with RNA added in oogenesis and disappearing once it had been translated. As already indicated, there is no direct evidence and very little indirect evidence to support such a suggestion. On the other hand, the cytological studies of Blackler (1958) suggest an alternative to the active role outlined above of germ plasm components in directing the formation of germ cells. In this case, when the germ plasm viewed with light microscopy becomes juxtanuclear during the gastrula stage, the cells that contain it are reported to become mitotically inhibited and retain their embryonic character (Blackler, 1958, 1966). Thus the germ plasm can be viewed as "protecting" the unspecialized nature of this group of cells from the stimuli which mediate differentiation of the surrounding somatic cells (Blackler, 1958; Fischberg and Blackler, 1963).

In summary, the evidence discussed in the preceding pages leaves little doubt that the fertilized eggs of both a vertebrate and an invertebrate organism contain a substance directly responsible for the differentiation of a discrete cell type, the primordial germ cell. This germ plasm continues to provide the clearest example of the cytoplasmic localizations long suspected of playing a determinative role in early organogenesis. However, the nature, origin, and mode of action of the germ cell determinants remains problematical. Progress along these lines has been mainly in the form of identifying, within the germinal plasm region, discrete electron dense bodies, the existence of which can be related to germ cell determination. It is, of course, conceivable that the dense bodies play no role in germ cell determination. Thus, what remains to be accomplished is a direct demonstration that these structures alone can or cannot direct germ cell formation. By isolating these structures and testing their activity directly, it should be possible to identify their biological function as well as their chemical indentity. It can readily be appreciated that this is no small task.

ACKNOWLEDGMENTS

The original work described here was supported by a research grant from the National Institutes of Health (HD04229). We thank Drs. A. P. Mahowald, Karl Illmensee and A. C. Webb for many helpful comments during preparation of this paper. We also thank Mrs. Kirsten Keem for technical assistance.

REFERENCES

Al-Mukhtar, K. and Webb, A. C. (1971). An ultrastructural study of primordial germ cells, oogonia, and early oocytes in Xenopus laevis. *J. Embryol. Exp. Morph.* **26,** 195–217.
Balinsky, B. I. (1966). Changes in the ultrastructure of amphibian eggs following fertilization. *Acta Embryol. Morphol. Exp.* **9,** 132–154.
Bantock, C. R. (1970). Experiments on chromosome elimination in the gall midge, *Mayetiola destructor. J. Embryol. Exp. Morph.* **24,** 257–286.
Blackler, A. W. (1958). Contribution to the study of germ cells in the anura. *J. Embryol. Exp. Morph.* **6,** 491–503.
Blackler, A. W. (1962). Transfer of primordial germ cells between two subspecies of *Xenopus laevis. J. Embryol. Exp. Morph.* **10,** 641–651.
Blackler, A. W. (1965). The continuity of the germ line in amphibians and mammals. *Ann. Biol. IV,* 627-635.

Blackler, A. W. (1966). Embryonic sex cells of amphibia. *Adv. Reprod. Physiol.* **1**, 9–28.

Blackler, A. W. (1970). The integrity of the reproductive cell line in the amphibia. *Current Topics Dev. Biol.* **5**, 71–87.

Blackler, A. W. and Gecking, C. A. (1972a). Transmission of sex cells of one species through the body of a second species in the genus *Xenopus*. I. Intraspecific matings. *Develop. Biol.* **27**, 376–384.

Blackler, A. W. and Gecking, C. A. (1972b). Transmission of sex cells of one species through the body of a second species in the genus *Xenopus*. II. Interspecific matings. *Develop. Biol.* **27**, 385–394.

Bounoure, L. (1934). Recherches sur la lignée germinale chez la Grenouille rousse aux premiers stades du développement. *Ann. Sci. Nat., 10e Ser.* **17**, 67–248.

Bounoure, L., Aubry, R. and Huck, M. L. (1954). Nouvelles recherches experimentales sur les origines de la lignée reproductrice chez la Grenouille rousse. *J. Embryol. Exp. Morph.* **2**, 245–263.

Buehr, M. L. and Blackler, A. W. (1970). Sterility and partial sterility in the South African clawed toad following the pricking of the egg. *J. Embryol. Exp. Morph.* **23**, 375–484.

Coggins, L. W. (1973). An ultrastructural and radioautographic study of early oogenesis in the toad *Xenopus laevis*. *J. Cell Sci.* **12**, 71–93.

Counce, S. J. (1963). Developmental morphology of polar granules in *Drosophila* including observations on pole cell behavior and distribution during embryogenesis. *J. Morph.* **112**, 129–145.

Coward, S. J. (1974). Chromatoid Bodies in Somatic Cells of the Planarian: observations on their behavior during mitosis. *Anat. Rec.*, in press

Czolowska, R. (1969). Observations on the origin of the germinal cytoplasm in *Xenopus laevis*. *J. Embryol. Exp. Morph.* **22**, 229–251.

Czolowska, R. (1972). The fine structure of the "germinal cytoplasm" in the egg of Xenopus laevis. *Wilhelm Roux'Arch.* **169**, 335–344.

Dawid, I. B. and Blackler, A. W. (1972). Maternal and cytoplasmic inheritance of mitochondrial DNA in *Xenopus*. *Develop. Biol.* **29**, 152–161.

DuMont, J. N. (1972). Oogenesis in *Xenopus laevis* (Daudin) I. Stages of oocyte development in laboratory maintained animals. *J. Morphol.* **136**, 153–180.

Eddy, E. M. (1974). Fine structural observations on the form and distribution of nuage in germ cells of the rat. *Anat. Rec.* **178**, 731–758.

Eddy, E. M. and Ito, S. (1971). Fine structural and radioautographic observations on dense perinuclear cytoplasmic material in tadpole oocytes. *J. Cell Biol.* **49**, 90–108.

Fawcett, D. W. (1972). Observations on cell differentiation and organelle continuity in spermatogenesis. *In,* "Edinburg Symposium of Genetics of the Spermatozoan" (R. A. Beatty and S. Gluecksohn-Waelsch, eds.), pp. 37–68. Edinburg, New York.

Fischberg, M. and Blackler, A. W. (1963). Loss of nuclear potentiality in the soma versus preservation of nuclear potentiality in the germ line. *In,* "Biological Organization at the Cellular and Supercellular Level" (R. J. C. Harris, ed.), pp. 111–127. Academic Press, N. Y.

Grant, P. and Wacaster, J. F. (1972). The amphibian gray crescent region - a site of developmental information? *Develop. Biol.* **28**, 454–471.

Gurdon, J. R. and Woodland, H. R. (1968). The cytoplasmic control of nuclear activity in animal development. *Biol. Rev.* **43**, 233–267.

Hay, E. D. (1968). Dedifferentiation and metaplasia in vertebrate and invertebrate regeneration. *In,* "The Stability of the Differentiated State" (H. Ursprung, ed.), pp. 85–108. Springer-Verlag, N. Y.

Hegner, R. W. (1914). Studies on germ cells. *J. Morphol.* **25**, 375–510.

Ikenishi, K., Kotani, M. and Tanabe, K. (1974). Ultrastructural changes associated with UV irradiation in the "germinal plasm" of Xenopus laevis. *Develop. Biol.* **36**, 155–168.

Illmensee, K. and Mahowald, A. P. (1974). Transplantation of posterior polar plasm in *Drosophila.* Induction of germ cells at the anterior pole of the egg. *Proc. Nat. Acad. Sci. U.S.* **71**, 1016–1020.

Kalt, M. R. (1973). Ultrastructural observations on the germ line of *Xenopus laevis. Zeit. Zellforsch.* **138**, 41–62.

Kessel, R. G. (1969). Cytodifferentiation in the *Rana pipiens* oocyte. I. Association between mitochondria and nucleolus like bodies in young oocytes. *J. Ultrastruct. Res.* **28**, 61–77.

Kessel, R. G. (1971). Cytodifferentiation in the *Rana pipiens* oocyte II. Intramito-chondrial yolk. *Zeit. Zellforsch.* **112**, 313–332.

Mahowald, A. P. (1962). Fine structure of pole cells and polar granules in *Drosophila melanogaster. J. Exp. Zool.* **151**, 201–215.

Mahowald, A. P. (1968). Polar granules of *Drosophila* II. Ultrastructural changes during early embryogenesis. *J. Exp. Zool.* **167**, 237–262.

Mahowald, A. P. (1971a). Polar granules of *Drosophila* III. The continuity of polar granules during the life cycle of *Drosophila. J. Exp. Zool.* **176**, 329–344.

Mahowald, A. P. (1971b). Origin and continuity of polar granules. *In,* "Origin and Continuity of Cell Organelles" (J. Reinert and H. Ursprung, eds.), pp. 159–169. Springer-Verlag, N. Y.

Mahowald, A. P. (1971c). Polar granules of Drosophila IV. Cytochemical studies showing loss of RNA from polar granules during early stages of embryogenesis. *J. Exp. Zool.* **176**, 345–352.

Mahowald, A. P. and Hennen, S. (1971). Ultrastructure of the "germ plasm" in eggs and embryos of *Rana pipiens. Develop. Biol.* **24**, 37–53.

Mahowald, A. P. (1974). Ultrastructural changes in the germ plasm during the life cycle of *Miastor. Wilhelm Roux'Arch.* (in press).

Massover, W. H. (1968). Cytoplasmic cylinders in bullfrog oocytes. *J. Ultrastract. Res.* **22**, 159–167.

Malacinski, G. M. (1972). Identification of a presumptive morphogenetic determinant from the amphibian oocyte germinal vesicle nucleus. *Cell Diff.* **1**, 253.

Mintz, B. (1961). Formation and early development of germ cells. *In,* "Symposium on the Germ Cells and Earliest Stages of Development". Instit. Intern. d'Embryol. pp. 1–24, Fondazione A. Baselli, Instituto Lombardi, Milano.

Morita, M., Best, J. B., and Noel, J. (1969). Electron Microscopic studies of planarian regeneration. I. Fine structure of neoblasts in *Dugesia dorotocephala. J. Ultrastruct. Res.* **27**, 7–23.

Nieuwkoop, P. D. (1947). Experimental investigations on the origin and determination of the germ cells and on the development of the lateral plate and germ ridges in Urodeles. *Arch. neerl. Zool.* **8**, 1–205.

Nieuwkoop, P. D. and Faber, J. (1956). *In,* "Normal Table of *Xenopus laevis* (Daudin) (P. D. Nieuwkoop and J. Faber, eds.) North Holland Publ. Co., Amsterdam.

Okada, M., Kleinman, A. and Schneiderman, H. A. (1974). Restoration of fertility in steri-lized *Drosophila* eggs by transplantation of polar cytoplasm. *Develop. Biol.* **37**, 43–54.

Padoa, E. (1963). Le gonadi di girini de *Rana esculenta* da nova irradiate con ultraviolette. *Monit. Zool. Ital.* **71**, 238–249.

Reed, S. C. and Stanley, H. P. (1972). Fine structure of spermatogenesis in the South African clawed toad *Xenopus laevis* Daudin. *J. Ultrastruct. Res.* **41**, 277–295.

Reynaud, G. (1969). Transfert de cellules germinales primordiales de dindon a' l'embryon de poult par injection intravasculaire. *J. Embryol. Exp. Morphol.* **21**, 485–507.

Simon, D. (1957). La localization primaire des cellules germinales de l'embryon de poulet: prevues expe'rimentales. *C. R. Soc. Biol.* **151**, 1010–1012.

Smith, L. D. (1964). A test of the capacity of presumptive somatic cells to transform into primordial germ cells in the Mexican axolotl. *J. Exp. Zool.* **156**, 229–242.

Smith, L. D. (1965). Transplantation of the nuclei of primordial germ cells into enucleated eggs of *Rana pipiens. Proc. Nat. Acad. Sci. U.S.* **54**, 101–107.

Smith, L. D. (1966). The role of a "germinal plasm" in the formation of primordial germ cells in *Rana pipiens. Develop. Biol.* **14**, 330–347.

Smith, L. D. (1974). Molecular events during oocyte maturation. *In,* "The Biochemistry of Animal Development (R. Weber, ed.) Vol. 3, Chapter 1. Academic Press, N. Y.

Smith, L. D. and Ecker, R. E. (1970). Regulatory processes in the maturation and early cleavage of amphibian eggs. *Current Topics. in Develop. Biol.* **5**, 1–38.

Spiegelman, M. and Bennett, D. (1973). A light and electron-microscopic study of primordial germ cells in the early mouse embryo. *J. Embryol. Exp. Morphol.* **30**, 97–118.

Tanabe, K., and Kotani, M. (1974). Relationship between the amount of the "germinal plasm" and the number of primordial germ cells in Xenopus laevis. *J. Embryol. Exp. Morphol.* **31**, 89–98.

Williams, M. A. and Smith, L. D. (1971). Ultrastructure of the "germinal plasm" during maturation and early cleavage in *Rana pipiens. Develop. Biol.* **25**, 568–580.

Willier, B. H. (1937). Experimentally produced sterile gonads and the problem of the origin of germ cells in the chick embryo. *Anat. Rec.* **70**, 89–112.

Wilson, E. B. (1925). The Cell in Development and Heredity. pp. 310–328. Macmillan Co., N. Y.

Gametogenesis in the Male: Prospects for its Control

Don W. Fawcett, M.D.

Department of Anatomy
and
Laboratory of Human Reproduction
and Reproductive Biology
Harvard Medical School
Boston, Massachusetts

I. Introduction 25
II. Endocrine Control of Spermatogenesis 26
III. Organization of the Seminiferous Epithelium 28
    A. Provisions for upward mobility of the germ cells 30
    B. Compartmentation of the epithelium and its implications ... 32
IV. Syncytial Nature of the Germ Cells 36
V. Vulnerable Events in Spermatid Differentiation 40
VI. Post-testicular Maturation of Spermatozoa 46
VII. Regional Specialization of the Sperm Membrane 47
VIII. Concluding Comment 48
    References 50

I. INTRODUCTION

In a world already facing a serious food crisis we continue to add each year 80 million new mouths to feed. Surely one of the most urgent problems confronting mankind is the control of its own numbers. While there has been significant progress in the past twenty years in the development of contraceptive methods of women, the male has been largely neglected in the search for safe and effective means of fertility control. The state of our knowledge of male reproductive physiology is so fragmentary that there remain fascinating fundamental problems for the cell biologist.

In their efforts to understand gametogenesis in the male, reproductive biologists have long been preoccupied with the role of hormones. Therefore, the endocrine requirements for spermatogenesis have been known in broad outline for many years but the cyclic nature of the process remains unexplained. Approaches to the biochemistry of spermatogenesis have been frustrated by the inability to dissociate the testis and to isolate pure cell

25

fractions from its highly heterogeneous population of cell types. We still know very little about the control of germ cell differentiation. Relatively little investigative attention has been devoted to the complex cellular relationships involved or to the possible dependence of gamete differentiation upon specific patterns of histological organization within the gonads. There are indications that these are not trivial considerations and that the key to further progress in understanding and controlling spermatogenesis may reside in a penetrating analysis of the organization of the seminiferous epithelium, and in study of the short range interactions among its several cell types. Developmental biologists have much to contribute to this effort because of their experience in cell population dynamics, growth regulation, and the interaction of cells with each other and with their immediate environment.

I will review here very briefly current progress in our understanding of the endocrinological and morphological aspects of spermatogenesis and I shall try to identify, in the complex process of differentiation, release, and post-testicular maturation of spermatozoa, some vulnerable steps that might lend themselves to pharmacological interruption. If this essay engages the interest of more cell and developmental biologists in this challenging and socially relevant area of research it will have fulfilled its purpose.

II. ENDOCRINE CONTROL OF SPERMATOGENESIS

The understanding of the endocrine regulation of spermatogenesis that has been gained over the past four decades can be very briefly stated. Two hypophyseal gonadotropins, luteinizing hormone (LH) and follicle stimulating hormone (FSH) are required, as in the female, but LH appears to be the more important. It stimulates the interstitial cells of Leydig to synthesize androgens, mainly testosterone. Androgen is required for spermatogenesis, for maintenance of the accessory glands of the male reproductive tract and to sustain libido. The location of the Leydig cells in the interstices among the seminiferous tubules favors the maintenance of high local concentration of androgen that is essential for spermatogenesis while the lower concentration in the systemic circulation is sufficient to maintain the accessory glands. The release of LH is under negative feed-back control by blood-borne testosterone acting upon the hypothalamus (Fig. 1). It has not been practical in the male to take advantage of this for fertility control because administration of progestational compounds such as are used in the female, would inevitably result in loss of libido. The role of FSH in the male is less clear than that of LH. It seems to be necessary for initiation of fertility at puberty but since spermatogenesis can be maintained in hypophysectomized laboratory rodents with testosterone alone (Nelson and Merckel, 1973; Ahmad, et al., 1973) the significance of FSH in the adult male has been controversial. The situation has been somewhat clarified by the recent demonstration of stimulation of protein synthesis in the tubules by FSH in the prepuberal rat (Means and Hall, 1968)

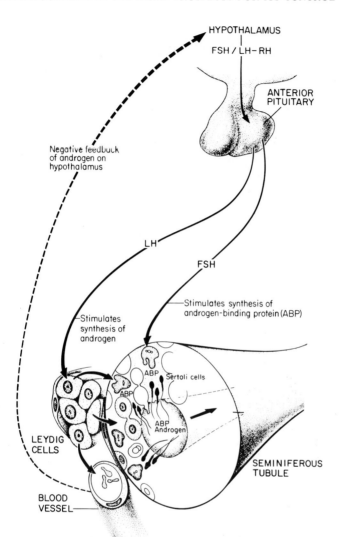

HYPOTHALAMUS
|
FSH / LH-RH

ANTERIOR
PITUITARY

Negative feedback
of androgen on
hypothalamus

LH

FSH

Stimulates synthesis of
androgen-binding protein (ABP)

Stimulates
synthesis of
androgen

ABP

ABP   Sertoli cells

ABP
Androgen

LEYDIG
CELLS

SEMINIFEROUS
TUBULE

BLOOD
VESSEL

Fig. 1. Schematic representation of the endocrine regulation of spermatogenesis. The high local concentrations of androgen required are maintained by the Leydig cells in response to release of LH from the pituitary. In addition to the well established feed-back of circulating androgen on LH levels, it has been proposed that FSH levels are similarly controlled by a hypothetical substance "inhibin" originating in the seminiferous tubules.

and the finding that in post-puberal animals, FSH stimulates Sertoli cells to synthesize an androgen-binding protein (French and Ritzen, 1973; Hansson, *et al.*, 1974). The binding protein is believed to contribute to the maintenance of high androgen concentrations in the seminiferous epithelium and to serve as a vehicle for transport of androgen through the excurrent ducts of the testis to the epididymis where it may be essential for maintenance of the normal function of that organ. Thus spermatogenesis takes place in an androgen rich environment and the spermatozoa are transported through the rete testis, ductuli efferentes, and initial segment of the epididymis in a fluid containing androgen in a concentration as much as 15 times that in the peripheral blood (Harris, 1973). Such an environment appears to be necessary for the continued maturation of spermatozoa and for their acquisition of fertilizing capacity (Lubicz-Nawrocki and Glover, 1970; Blaquier, *et al.*, 1973).

A negative feed-back control on levels of FSH in the male has been postulated and attributed to a hypothetical substance "inhibin" produced in the seminiferous tubules (McCullagh and Schneider, 1940). Extracts obtained from the rete testis fluid (Setchell, 1974) from bovine testis (DeKretser, *et al.*, 1974) and from semen (Franchimont, 1974), have been reported to lower FSH levels in experimental animals. Vigorous efforts are now underway to verify the existence of inhibin and to isolate and characterize it. If FSH is necessary to support spermatogenesis in the adult, a substance capable of selectively inhibiting FSH secretion would be a promising approach to suppression of spermatogenesis without affecting LH secretion and decreasing libido.

If FSH and a high local concentration of androgen were the only humoral requirements for differentiation of the germ cells, it should not be difficult to create *in vitro* an appropriate environment for maintenance of spermatogenesis. Yet all efforts to get mammalian spermatogenesis to progress in organ culture beyond the pachytene spermatocyte stage have thus far failed (Steinberger, *et al.*, 1970). This strongly suggests that there are other important requirements. Among these may be special topographical relationships of cells within the epithelium-relationships which are not maintained *in vitro*. Further evaluation of this possibility requires that we first consider certain unique features in the organization of the seminiferous epithelium.

III. ORGANIZATION OF THE SEMINIFEROUS EPITHELIUM

The lining of the seminiferous tubules has a number of characteristics that make it unique among epithelia. It is made up of two distinct categories of cells – a non-dividing population of supporting Sertoli cells, and a proliferating population of germ cells in various stages of their differentiation into spermatozoa. The Sertoli cells are columnar and extend from base to free surface and are provided with elaborate lateral processes that extend into the interstices between the associated spermatogonia, spermatocytes and spermatids. The Sertoli cells are long lived and cease to proliferate before puberty; the germ cells are a constantly renewing population with the stem cells located at the base and more advanced stages at successively higher levels in the epithelium. The germ cells slowly move upward along the sides of the supporting cells as they differentiate. The spermatids approach the surface they elongate and establish a new relationship to the supporting cells – now occupying deep recesses in their apical cytoplasm (Fig. 2). The occurrence of two distinct populations, one fixed and the other constantly moving upward, is a feature peculiar to the seminiferous epithelium. The continually changing

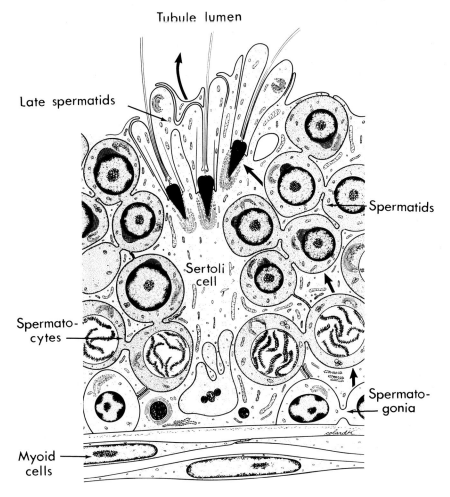

Fig. 2. Schematic representation of the organization of the mammalian seminiferous epithelium. The non-proliferating supporting cells extend the full thickness of the epithelium, while the germ cells, occupying interstices between the supporting cells, slowly move upward as they differentiate. (From D. W. Fawcett, In, *Male Fertility and Sterility*," Serono Symposium, 1975.)

relationships between the germ cells and the supporting cells create problems of cell coherence and cell communication which may have important implications for the maintenance of spermatogenesis (Fawcett, 1974).

Provisions for upward mobility of the germ cells

In most columnar epithelia, other than the seminiferous epithelium, the cells have on their lateral surfaces, local membrane specializations that serve to maintain cohesion and communication between the cells. The mechanical function of attachment is subserved by *zonulae occludentes*, bands of membrane fusion that encircle the apical ends of the cells, and by *maculae adherentes* or desmosomes scattered at random over their lateral surfaces. The communication between cells which is essential for coordination of function throughout the epithelium is maintained through *nexuses* or *gap-junctions*. These are specialized sites of close membrane apposition that permit passage of ions and small molecules from cell to cell. In addition to providing for low-resistance electrical coupling of cells, the gap-junctions are also sites of very firm attachment of the opposing membranes. If one examines electron micrographs of the seminiferous epithelium, one finds no zonulae occludentes, nor typical desmosomes, nor nexuses on the interface between the supporting cells and germ cells (Fig. 3). Careful search of extensive areas of these membranes in freeze-fracture preparations confirms the absence of these junctional specializations (Fawcett, 1974). The absence of gap-junctions is especially surprising, for if one considers the highly integrated nature of the spermatogenic cycle in mammals, and the precise timing of events in germ cell differentiation, one might expect to find abundant and highly developed specializations for cell-to-cell communication. It is clear, however, that the presence of such firm attachments on the interface between germ cells and supporting cells would prevent their movement relative to one another. Thus it seems that in the seminiferous epithelium, upward mobility of the germ cells is maintained at the sacrifice of one of the commonest and most efficient mechanisms for communication among epithelial cells (Fawcett, 1974). In the absence of such membrane specializations, it seems reasonable to conclude that whatever influence the Sertoli cells have no differentiation of the germ cells, it is probably mediated indirectly by modification of the extracellular fluid environment in which the germ cells develop and by release of molecules to which the germ cell membranes are freely permeable.

The ability of the Sertoli cells to maintain the appropriate microenvironment for germ cell development may well depend upon the fact that the differentiating germ cells are isolated from the general extracellular fluid compartment of the testis by a permeability barrier located near the base of the epithelium (Dym and Fawcett, 1970).

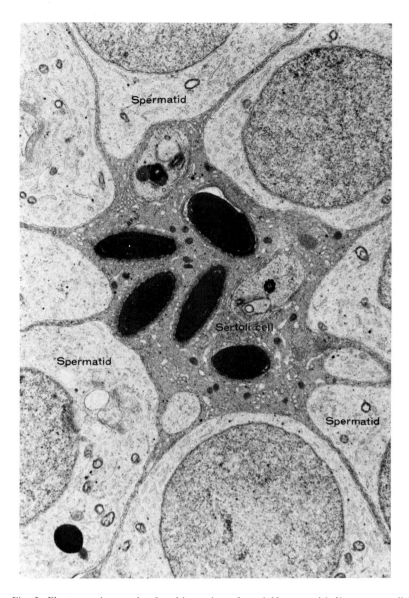

Fig. 3. Electron micrograph of a thin-section of seminiferous epithelium cut parallel to the basal lamina. Cross sections of several late spermatids are seen in the columnar portion of the Sertoli cell. Numerous slender cell processes extend between the surrounding early spermatids. No desmosomes, gap-junctions or other junctional complexes are seen along the interface between the Sertoli cell and the germ cells.

Compartmentation of the epithelium by occluding junctional complexes

In the past decade a number of physiological studies have shown that vital dyes and a broad spectrum of other compounds administered intravenously, escape from the interstitial capillaries and venules of the testis and rapidly appear in the testicular lymph. The majority of these substances cannot be detected in the tubules or in fluid collected from the rete testis. It was concluded from such observations that there is a *blood-testis permeability barrier* located somewhere immediately around or within the seminiferous tubules (Kormano, 1968; Setchell, *et al.*, 1969). A more precise localization of this barrier has been achieved by morphological investigations using horseradish peroxidase and other electron opaque probes of the extracellular space (Fawcett, *et al.*, 1970; Dym and Fawcett, 1970). When such a probe is injected intravenously and the testes are then fixed after various time intervals for electron microscopy, the electron opaque material is found in low power electron micrographs at early time intervals in the interstitial spaces but it is largely excluded from the seminiferous tubules. In micrographs at higher magnification, one finds that the tracer has, in fact, entered the intercellular clefts at the base of the epithelium clearly outlining the spermatogonia. It may extend a short distance further into the cleft between the over-arching Sertoli cells, but there it stops abruptly. At the site where penetration of the tracer is stopped, one always finds a special junctional complex on the opposing membranes of adjacent Sertoli cells (Fawcett, *et al.*, 1970; Dym and Fawcett, 1970).

These unusual junctions were described a number of years ago from electron micrographs of thin sections (Flickinger and Fawcett, 1967; Nicander, 1967) but an appreciation of their significance has been a more recent development. They bear some resemblance to the zonula occludens found in the juxtaluminal region of other epithelia. The features that set them apart from other epithelial cell junctions are: 1) their location near the base of the epithelium, 2) the presence of parallel bundles of filaments in the subjacent cytoplasm, 3) the constant association of cisternae of the endoplasmic reticulum with the junctional areas of membrane (Fig. 4), and 4) the occurrence of focal sites of fusion of the opposing membranes, spaced at more-or-less regular intervals along the junction (Figs. 5 and 6). Whether these sites of membrane fusion were punctate or linear could not be clearly resolved by the study of thin sections, but the question has now been settled by the technique of freeze-cleaving which exposes to view extensive areas of half-membrane (Fawcett, 1974; Gilula and Fawcett, 1975). In such preparations, many long parallel rows of particles are found within the Sertoli cell membranes (Figs. 4 and 7). These are generally associated with the inwardly directed outer half-membrane (the B-face) but where there are discontinuities in the rows on this face, the missing particles are found on the A-face (Fig.

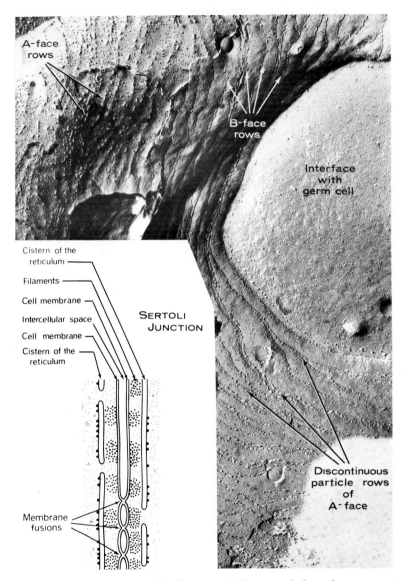

A-face rows

B-face rows

Interface with germ cell

Cistern of the reticulum

Filaments

Cell membrane

Intercellular space

Cell membrane

Cistern of the reticulum

SERTOLI JUNCTION

Discontinuous particle rows of A-face

Membrane fusions

Fig. 4. The inset at the lower left illustrates in diagrammatic form the components of the unique occluding junctions found on the boundary between adjacent Sertoli cells near the base of the epithelium. The electron micrograph presents a freeze-fracture preparation showing the long parallel rows of membrane-intercalated particles found principally on the B-face of the Sertoli membrane. These correspond in their distribution to the lines of membrane fusion which close the intercellular clefts. The interface of the Sertoli cell with a germ cell, at the right, shows no such particle aggregates.

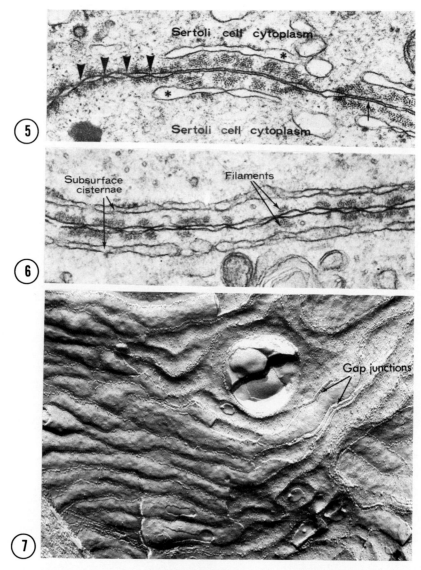

Fig. 5. A thin section of the occluding junction between two Sertoli cells in rat testis, showing multiple sites of membrane fusion (at arrows) and the bundles of filaments interposed between the cell surface and associated cisternae of the endoplasmic reticulum (marked by asterisks). (From Dym and Fawcett, 1970.)

Fig. 6. A similar Sertoli cell junction from ram testis.

Fig. 7. A freeze-cleaving preparation from guinea pig testis showing multiple rows of B-face particles corresponding to lines of membrane fusion. In addition, on the A-face at the upper right, are some closely packed arrays of particles constituting an atypical gap-junction.

34

4). The rows of intramembranous particles correspond to the sites of membrane fusion observed in thin sections of the Sertoli junctions. The extensive views presented in freeze-fracture preparations make it clear that circumferential lines of membrane fusion extend for long distances around the base of the Sertoli cell parallel to the basal lamina. These specializations are exceptional among occluding epithelial junctions: 1) in the large number of parallel rows of membrane fusion, 2) the absence of anastomosis between rows, and 3) the preferential association of the membrane intercalated particles with the B-face. These unique junctional complexes between adjacent Sertoli cells are now generally accepted as the principal structural basis of the blood-testis permeability barrier.

Situated, as many of them are, above the spermatogonia these junctional specializations partition the epithelium into a *basal compartment* occupied by the spermatogonia and an *adluminal compartment* containing the more advanced germ cells (Fig. 8). Thus the stem cells in the basal compartment are exposed to the same fluid environment as the interstitial tissue. The differentiating germ cells, on the other hand, are isolated in the adluminal compartment by the intraepithelial barrier. Therefore, the meiotic and post-meiotic stages of development probably reside in an entirely different microenvironment. The Sertoli cells that form the walls of the adluminal compartment are strategically situated to create, by their metabolic and secretory activities, a special environment that may favor germ cell differentiation. The composition of the fluid in that compartment is not known in detail but some of its properties can be inferred from micropuncture studies of the rete testis and seminiferous tubules. It appears to be somewhat hyperosmotic and unusually rich in potassium (Tuck, *et al.*, 1973). The speculation that the special environment of the adluminal compartment is essential for completion of germ cell development, derives some support from the observation that in organ culture, where germ cell development does not progress, the Sertoli cell junctions are not maintained (Baccetti, *et al.*, 1975).

Still lacking is an explanation for the periodic local modification or dissolution of the occluding junctions that permits clusters of preleptotene spermatocytes to move up from the basal compartment into the adluminal compartment. It is conceivable that dissolution of the junctions on the adluminal side of the ascending germ cells is accompanied by simultaneous formation of new occluding junctions on their abluminal side so that a barrier exists at all times during the translocation of the next generation of germ cells from one compartment to the other. This process needs more study but it is a difficult problem for technical reasons.

In addition to the possible importance of the barrier in maintenance of a suitable environment for germ cell differentiation, it also has immunological significance in preventing antigenic products of post-meiotic germ cells from reaching the blood and inducing an autoimmune reaction (Johnson, 1970).

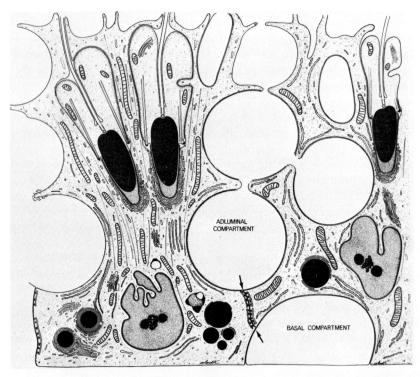

Fig. 8. Drawing showing how the Sertoli cell junctional complexes divide the epithelium into a basal compartment containing stem cells and an adluminal compartment containing the meiotic and post-meiotic stages of the germ cells. (From D. W. Fawcett, The Sertoli Cell, In, *"Handbook of Physiology*," 1975.)

IV. THE SYNCYTIAL NATURE OF THE GERM CELLS

In the early days of biological electron microscopy, in the 1950's, it was not uncommon to find that we were describing structural details that were observed or anticipated by discerning cytologists of the 19th century. When I participated in the 16th Growth Society Symposium in 1958, we had recently rediscovered the intercellular bridges between developing male germ cells (Burgos and Fawcett, 1955; Fawcett, 1959). It now seems likely that these were first seen in teased preparations of fresh testis by LaValette, in 1865, but his report appears to have been overlooked or discounted by his contempories and the observation never found its way into textbooks. These bridges are formed by an unusual process of incomplete cytokinesis which leaves patent channels between daughter cells after the microtubules in the spindle bridges have dispersed. We tentatively suggested 15 years ago that the syncytial nature of the germ cells might be the basis for the synchrony of their differentiation (Fawcett, *et al.*, 1959). Because only a few bridges were included in any single ultra-thin section, it was difficult to estimate the size of the syncytia. Influenced by the diagrams in contemporary textbooks, and by an inborn conservatism, we were thinking in terms of 2 conjoined spermatogonia, 4 primary spermatocytes, 8 secondaries and 16 spermatids. Later studies of stem cell renewal in spermatogenesis (Clermont and Bustos-Obregon, 1968; Huckins, 1971) established that there are at least 6 and possibly 7 generations of spermatogonia prior to the onset of meiosis and further electron microscopic study showed that the progeny of all but the earliest of these spermatogonial divisions are also connected by bridges (Dym and Fawcett, 1971; Huckins, 1973). A large upward revision in our estimates of the numbers of interconnected cells was therefore necessary (Fig. 9). It is now evident that if the daughter cells of all divisions subsequent to commitment of the stem cells to differentiation were to remain connected, there would ultimately be as many as 512 conjoined spermatids. Thus mammalian spermatogenesis is basically similar to that of insects and lower vertebrates in which comparable numbers of spermatids develop in cysts from a single encapsulated primordial germ cell. In the mammal, however, the theoretical numbers may never be achieved because the normal degeneration of up to 30% of spermatogonia and spermatocytes probably creates gaps in the long chains of cells, thus reducing somewhat the actual upper size limits of the syncytia. Nevertheless, their actual size remains very large. A recent study of serial sections oriented parallel to the base of the epithelium has found 80 conjoined spermatocytes where 128 might have been expected (Moens and Hugenholtz, 1974). The demonstration of such large numbers of conjoined cells has strengthened the original interpretation of these syncytia as a device for synchronization of differentiation. It clearly does not explain the synchronization of development over larger areas than that occupied by a single syncytium. It has been estimated that some 40 syncytia enter meiosis and spermatid development simultaneously in each square millimeter of seminiferous epithelium (Moens and Go, 1971). Coordination of differentiation over these larger areas may depend upon communication among Sertoli cells mediated via the atypical nexuses within their unique junctional complexes (Gilula and Fawcett, 1975).

Persistence of spindle bridges and synchronization of differentiation has been reported for the cnidoblasts of hydroids (Fawcett, *et al.*, 1959) but in very few other biological materials. The clusters of male germ cells are unique in certain respects. A *clone* is defined as a group of cells of *identical genetic constitution* derived by proliferation from a single cell of origin. The germ cell syncytia conform to this definition only until the spermatocyte stage. Then, as a result of crossing-over and disjunction, the nuclei within the spermatid syncytia differ genetically, not only with respect to their sex chromosomes, but in many autosomal traits as well. Thus, after meiosis, the syncytia might be more accurately described as a naturally occurring *heterokaryon* inasmuch as nuclei of differing genotype are contained in the same cytoplasm. Although

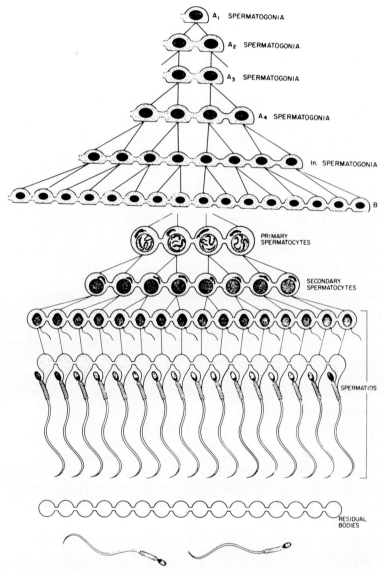

Fig. 9. Schematic representation of the syncytial nature of the mammalian germ cells. Cytokinesis is incomplete in all but the earliest spermatogonial divisions, resulting in expanding clones of germ cells which remain joined by intercellular bridges. (Modified after Dym and Fawcett, 1971.)

there is, as yet, no clear cut evidence of haploid gene expression in spermatogenesis, it has not been entirely excluded. The full biological significance of the syncytial nature of the developing male germ cells probably is yet to be discovered.

Whatever other consequences the syncytial nature of the germ cells may have, it has important implications for the dynamics of cell movement within the epithelium and for the mechanism of sperm release. We can no longer envision the upward movement of *individual* germ cells along the sides of the Sertoli cells. Our interpretations must provide for the upward translocation of very large branching chains of interconnected cells. It seems unlikely that vis-a-tergo resulting from proliferation at the base of the epithelium can entirely account for the movements observed. The elaborate interdigitation of Sertoli cell processes with the germ cells suggests that the supporting cells have an active role in this process. The complex mechanism by which individual spermatozoa are separated from the syncytium also appears to depend upon active movements and precisely controlled changes in shape of the Sertoli cells (Fawcett and Phillips, 1969). The eccentric lobules of excess spermatid cytoplasm are retained in the epithelium by Sertoli cell processes while the condensed nucleus, connecting piece and flagellum are gradually extruded into the lumen (Fig. 10). The slender strand of cytoplasm connecting the residual cytoplasm to the neck region of the nascent spermatozoon is increasingly attenuated and ultimately breaks (Fig. 10, C and D). Its distal portion retracts to form the cytoplasmic droplet of the free sperma-

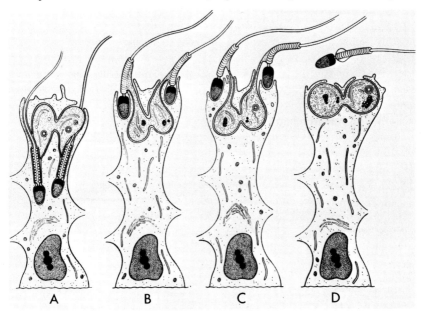

A   B   C   D

Fig. 10. Diagram of the stages of sperm release. The conjoined cell bodies of the late spermatids are retained in the epithelium, while the nucleus, neck region and tail are gradually extruded into the lumen. The narrow stalk connecting the neck region with the cell body becomes increasingly attenuated and finally gives way. Individual spermatozoa are thus separated from the syncytial cell bodies. (From D. W. Fawcett, In, *"Regulation of Reproduction"* S. J. Segal, *et al.*, ed., Charles C. Thomas, Springfield, 1973.)

tozoon, and the distal portion of the stalk retracts into the residual body in the epithelium. The spermatid is relatively passive in this process which seems to depend largely upon the motility mechanisms of the Sertoli cells. The large size of the clusters of late spermatids makes it certain that they extend beyond the limits of any one supporting cell and probably span the apices of several. If spermiation is mainly an activity of the Sertoli cells, the synchronized release of spermatozoa would seem to require a rather precise coordination of many such cells.

This complex process by which individual sperm are separated off from syncytial chains of residual bodies would seem to be a vulnerable step in spermatogenesis which might lend itself to pharmacological interruption. Anything that would prevent casting off of individual sperm would certainly seriously impair male fertility. To date no chemical agent with a selective action upon this process has been found.

V. VULNERABLE EVENTS IN SPERMATID DIFFERENTIATION

We do not as yet have sufficient understanding of the biochemistry of germ cell differentiation to attempt rational development of drugs for selective interruption of vulnerable steps in spermatogenesis. Moreover, the existence of the blood-testis permeability barrier complicates efforts to develop a male antifertility agent. To be effective such a compound would either have to act primarily upon the supporting cells or have solubility properties that would enable it to diffuse through these cells to the adluminal compartment. Nevertheless, in the routine toxicity testing of various drugs that were being developed as amoebocides, antibiotics or cancer chemotherapeutic agents a number of compounds have been found which do suppress spermatogenesis (Jackson, 1961; Nelson and Patanelli, 1965).

Some of these drugs have been found to destroy spermatogonia. These have not been pursued further, since the objective of developing a reversible male antifertility agent would not be served by a compound lethal to the stem cells. Other compounds were found to interrupt the process by acting upon primary spermatocytes. These cells spend many days in meiotic prophase and at this stage are believed to be especially susceptible to genetic damage. The fact that the majority of the compounds acting upon the spermatocytes are alkylating agents, has raised the spectre of possible mutagenic effects upon spermatozoa, if suppression of spermatogenesis were to be incomplete. This is certainly a reasonable concern, and it is now virtually impossible to obtain support for investigation of the mode of action of any alkylating agent on the seminiferous epithelium.

The exclusion of the premeiotic and meiotic stages of spermatogenesis as acceptable targets for drug action, leaves the period of spermatid differentiation as the only segment of the process where pharmacological intervention would seem to be relatively safe. The morphogenetic events of this period have been thoroughly studied by transmission electron microscopy (Burgos and Fawcett, 1955; Fawcett and Hollenberg, 1963; Ploen, 1971) and more recently, the method of freeze-fracturing has provided new insights into the process by exposing to view surfaces inaccessible in thin sections (Fawcett, 1974).

The early phases of spermatid development involve the synthesis of the enzyme rich contents of the future acrosome and segregation of the product in a number of proacrosomal granules in the Golgi complex (Fig. 11, A and B). These granules enlarge by coalescence and are ultimately reduced to a single, membrane-limited, acrosomal granule which becomes fixed to the nuclear envelope (Fig. 11C). This organelle continues to grow by accretion, as many small vesicles are formed in the Golgi complex and coalesce with it, adding their contents to the expanding acrosomal vesicle (Figs. 11, D and C, and 12). The number and the distribution of the fusing Golgi vesicles during the growth of the acrosome is seen most dramatically in freeze-fracture preparations that provide a surface view of the acrosomal vesicle (Figs. 14 and 15). The vesicle extends its area of contact with the nuclear envelope as it enlarges, ultimately forming a cap closely applied to the entire anterior hemisphere of the spermatid nucleus (Fig. 13). When the acrosomal cap has attained its definitive volume, the Golgi complex leaves its surface and, during spermatid elongation, migrates caudally into the residual cytoplasm. This terminates the phase of active protein and carbohydrate synthesis associated with formation of the acrosomal cap. The subsequent phase of spermatid development includes condensation of the chromatin, flattening of the nucleus, and shaping of the apical portion of the acrosome (Fig. 11, F to L). Nuclear condensation involves cytoplasmic synthesis of arginine-rich histones which replace the lysine-rich histones in the nucleus and bring about a remarkable reordering and compaction of the DNA in a pattern that results in gradual acquisition of the nuclear shape characteristic of the species (Fawcett, et al., 1971). Concurrent with these changes in the nucleus, there are progressive changes in the configuration of the associated acrosomal cap.

The complex interplay of morphogenetic forces within the nucleus, in the acrosome, and in the surrounding Sertoli cell seems to make this phase of development especially vulnerable to damage by a variety of chemical and physical agents. The antifertility compound bis (dichoroacetyl) diamine (WIN 18,446), for example, when given to guinea pigs in relatively small dosage over a two week period permits the normal development of spermatids up to the cap phase (Fig. 12) but results in remarkable distortion of the acrosome and nucleus of later stages (Flores and Fawcett, 1972). The acrosomes take on extremely bizarre shapes with many finger-like processes of the Sertoli cell invading and distorting them (Fig. 17). The nuclei undergo condensation but

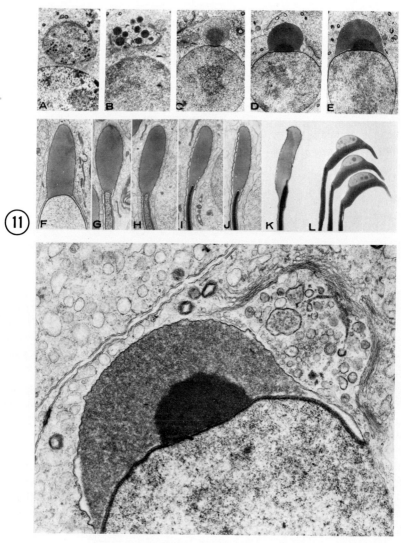

Fig. 11. Successive stages in the formation of the acrosome, and condensation of the nucleus in guinea pig spermiogenesis. Steps A to J take place in the testis; K and L represent continued shape change occurring during passage through the epididymis. (From D. W. Fawcett and D. Phillips, 1969.)

Fig. 12. Electron micrograph of a cap-phase spermatid from a guinea pig treated for 14 days with bis (dichloroacetyl) diamine in a relatively low dosage. Development of the acrosome to this stage has proceeded quite normally.

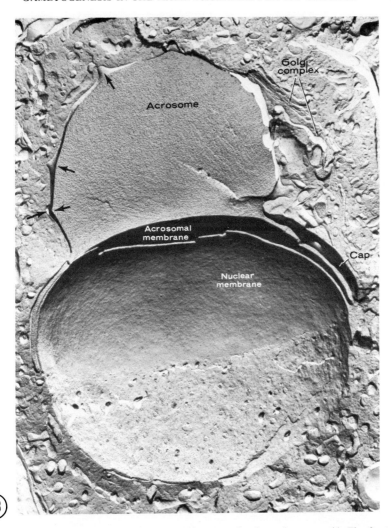

Fig. 13. A freeze-cleaving preparation of a normal guinea pig spermatid. The fracture plane has broken sagitally through the acrosomal vesicle and then split the inner acrosomal membrane and the outer membrane of the nuclear envelope. The Golgi complex can be identified and some nipple-like protrusions of the acrosomal membrane (at arrows) which represent vesicles in process of fusing with the membrane to add their contents to the growing acrosome.

tend to roll up around their long axis so that in longitudinal section they may present a double profile (Fig. 17). In transverse section, they are often curved into C-shapes and in other instances Y-shapes are observed. This highly characteristic pathology of the spermatids does not appear to be specific for this drug, for almost identical distortions of sperm nuclei can be produced by

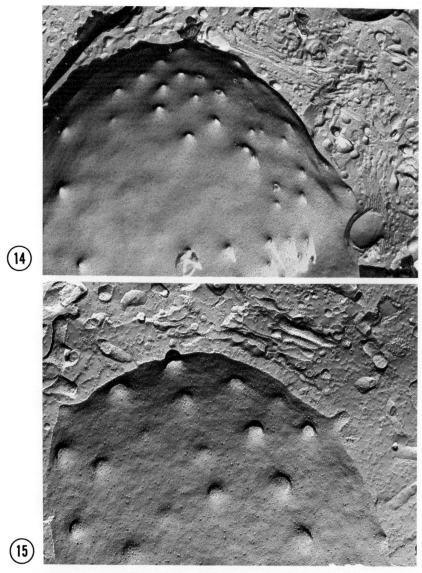

Figs. 14 and 15. Freeze-cleaving preparations of guinea pig spermatids in which the fracture plane has cleaved the membrane of the acrosomal vesicle. A large number of Golgi vesicles are revealed in the process of fusing with the acrosome, to contribute their contents to its growth. Thin sections are much less satisfactory for observing the number and distribution of these transport vesicles.

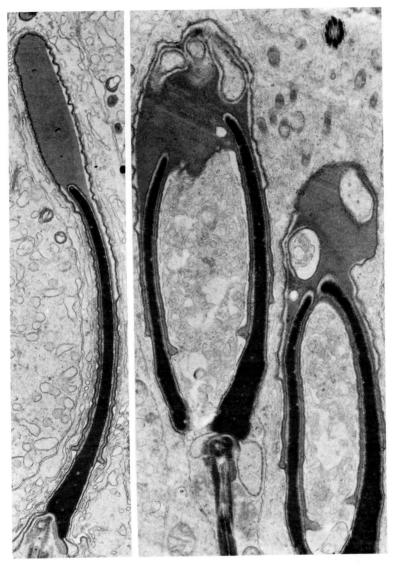

Fig. 16. A sagittal section of a large spermatid from normal guinea pig testis for comparison with Fig. 17.

Fig. 17. Two spermatids from a guinea pig treated fourteen days with bis (dichloroacetyl) diamine (WIN 18,446). Spermatids in the stages of nuclear condensation and acrosome shaping are especially vulnerable. The nuclei often roll up on their long axis so that they present a double profile in longitudinal section. The acrosomes are grossly distorted.

45

brief cryptorchidism (Ploen, 1973) and by either active immunization of guinea pigs with testicular tissue, or passive transfer of antitestis antiserum (Nagano and Okumura, 1973). While no common mechanism of action of heat, antibodies, and bis (dichloroacetyl) diamines can be suggested, their similar effects emphasize that nuclear condensation and acrosomal shaping are processes especially vulnerable to interference. This raises the hope that other compounds may be found that will be free of systemic toxicity but capable of selectively interfering with these late events in spermatid differentiation.

VI. POST-TESTICULAR MATURATION OF SPERMATOZOA

Gametogenesis does not end with the release of spermatozoa from the seminiferous epithelium. In some species there is continued morphological differentiation of the sperm as they pass through the epididymis (Fawcett and Hollenberg, 1963). In many species there are also progressive changes in the pattern of motility (Gaddum, 1968; Fray, et al., 1972) and, most importantly, there is a gradual acquisition of fertilizing capacity by spermatozoa in the body and tail of the epididymis (Orgebin-Crist, 1969). These changes are believed to depend, at least in part, upon the specific environment maintained in the lumen of the ductus epididymis by its lining epithelium. This portion of the tract should be an ideal site of action for a male antifertility agent. By the time spermatozoa reach the epididymis their nuclei are condensed, their DNA is metabolically inert and probably immune to chemical mutagenesis. Much investigative interest is now concentrating on the physiology of the epididymis in the belief that a better understanding of its biochemistry may make it possible to alter its function so as to prevent maturation of spermatozoa.

Recent studies indicate that the normal function of the initial segment of the rodent epididymis is dependent upon a high intraluminal concentration of androgen carried downstream from the testis in the testicular fluid. The more distal segments of the duct are maintained by the lower levels of androgen in the peripheral circulation. This has suggested the possibility that an anti-androgen, such as cyproterone acetate, in relatively small doses might permit spermatogenesis and normal libido while suppressing epididymal function sufficiently to produce infertility (Prasad, et al., 1970). Encouraging results have been obtained with this approach in animals and clinical trials are now in progress.

Enthusiasm for the epididymis as the preferred site of action for a male antifertility compound has been greatly stimulated by the discovery that α-chlorohydrin produces reversible infertility in rats, monkeys, rams, and boars (Coppola, 1969; Ericsson and Baker, 1970). This relatively simple compound, or its metabolites, is concentrated in the lumen of the cauda epididymis and there appears to act upon the spermatozoa (Crabo and Appelgren, 1972). There is no effect upon spermatogenesis, and there is complete recovery of fertility after withdrawal of the drug. A closely related compound 1-amino, 3-chloro, 2-propanol hydrochloride (Lederle) has a similar action (Coppola and Saldarini, 1974) and a striking separation of toxicity from the antifertility effects is achieved when the l- and -d isomers are separated (Paul, *et al.*, 1974). It is hoped that the l-isomer may prove sufficiently free of toxicity for trial in humans.

It is still not clear whether these compounds block a specific enzymatic step in the energy generating glycolytic pathway of the spermatozoa or whether they have a deleterious effect upon the membrane of the spermatozoon resulting in impaired motility and vitality. In this context there is now heightened interest in the membranes of spermatozoa.

VII. REGIONAL SPECIALIZATION OF THE SPERM MEMBRANE

The plasmalemma of the spermatozoon has usually been described as a simple trilaminar unit membrane like that of other cells. However, recent studies of its surface charge, involving binding of colloidal iron hydroxide (Cooper and Bedford, 1969; Yanigamachi, *et al.*, 1972) or identification of its saccharide residues by specific binding of lectins (Edelman and Millette, 1971; Nicolson and Yanigamachi, 1974) have demonstrated striking regional differences. There is reason to believe that these local membrane specializations are related to the specific functions of the different regions of the spermatozoon in fertilization. The membrane over the acrosome is involved in the acrosome reaction which releases hydrolytic enzymes and facilitates penetration of the extrinsic coats of the egg. The membrane of the postacrosomal region recognizes and fuses first with the oolemma in syngamy. Sperm motility requires the integrity of the membrane of the midpiece and the principal piece of the tail. The local properties of the flagellar membrane may regulate access of substrates to the energy generating mechanism and may be involved in propagation of waves of bending along the tail.

The method of freeze-fracturing permits the electron microscopist to split membranes in half and to examine their internal organization. The most dramatic examples, to date, of regional specializations within the plane of the plasma membrane have been provided by this new approach. The membrane over the guinea pig sperm acrosome has a carbohydrate-rich coat that exhibits a high degree of order. This seems to reflect a regular periodic structure within the membrane that is revealed in freeze-fracture preparations (Fig. 18). Instead of the usual random dispersion of intramembranous protein particles there are large particle-free areas that exhibit a very regular pattern suggesting a crystal-lattice. Similar crystalline domains have been seen in the corresponding region of rat sperm heads (Friend and Fawcett, 1974). The outer acrosomal membrane displays similar highly ordered regions but these have a different pattern. It seems likely that these unusual membrane specializations of the anterior portion of the sperm head are related, in some way, to the unique behavior of these two adjoining membranes in the acrosome reaction.

The membrane overlying the mitochondria of the midpiece in the guinea pig spermatozoon (but not in the rat) contains circumferentially oriented, beaded strands of 60 to 80 Å particles. These are concentrated over the gyres of the underlying mitochondrial sheath with particle-poor areas over the grooves between adjacent mitochondria (Fig. 19). This is an interesting example of a local membrane specialization in precise topographical relation to a subjacent cytoplasmic organelle. It is difficult to escape the conclusion that this is a functionally meaningful association.

In the membrane of the principal piece, the usual random pattern of membrane intercalated particles is observed (Fig. 20), but in addition, a double row of larger (90 Å) particles runs longitudinally over outer dense fiber number one (Fig. 21). This intriguing localization of an intramembranous differentiation in precise and constant relationship to an internal component of the sperm tail still awaits an explanation.

These studies of local differentiations of the sperm membrane by freeze-fracturing have provided a basis for further investigation of their functional correlates and their alterations in different experimental conditions. It has recently been reported that the strands of intramembranous particles in the midpiece of the guinea pig spermatozoon dissociate during prolonged incubation in physiological media which are known to promote "capacitation" of spermatozoa. Temporarily correlated with these structural changes within the membrane there is a modification of the pattern of flagellar beat, recorded cinematographically (Koehler and Gaddum-Rosse, 1975).

These intriguing observations focus renewed investigative attention upon the sperm membrane as a possible site of action for antifertility agents. There is a considerable degree of specificity in fertilization. The specificity in the recognition and fusion of the gametes in fertilization no doubt depends upon specific properties of the egg membrane and of the postacrosomal region of the spermatozoon. It has been shown that exposure of mammalian eggs to wheat germ agglutinin so alters the properties of the zona pellucida that *in vitro* fertilization is blocked (Oikawa, *et al.*, 1973). With a better understanding of membrane biology it may be possible to prevent normal maturation of the sperm membrane or to alter its internal organization during development in such a way as to suppress sperm motility, to alter its capacity to penetrate the zone, or to impair its ability to recognize and fuse with the egg.

VIII. CONCLUDING COMMENT

Studies of spermatozoa and of methods for their long term preservation have been of inestimable value in animal breeding to produce food and fiber for an expanding world population. There is now an urgent need to mobilize our scientific resources to develop safe and effective means of controlling

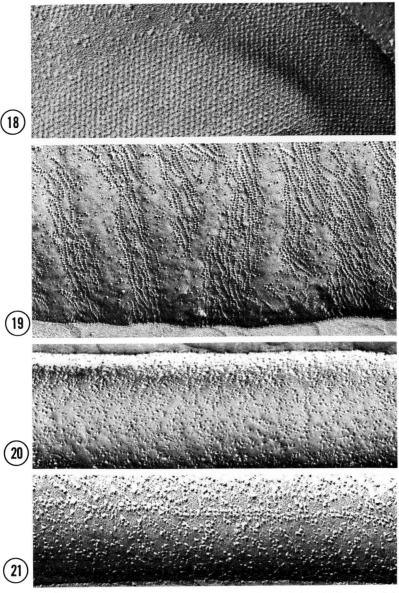

Fig. 18. Freeze-cleaving preparation of the cell membrane (A-face) overlying the acrosome of a rat spermatozoon. A very regular pattern is seen, suggesting crystalline areas within the membrane. (Micrograph courtesy of Dr. D. Friend.)

Fig. 19. Membrane of the mid-piece of guinea pig spermatozoon (A-face). Particles are associated in long rows oriented circumferentially and concentrated over the gyres of the underlying mitochondrial sheath. The membrane over the interstices between mitochondria contains randomly distributed single particles.

Fig. 20. The membrane of the principal piece of guinea pig sperm has the usual random pattern of 70 to 90 Å particles.

Fig. 21. Along one side of the principal piece, a double row of intramenbranous particles runs longitudinally. This membrane specialization is always over outer dense fiber number one.

human fertility. The investment in endocrinological research in the past twenty-five years brought rich dividends in relatively safe oral contraceptives for women. Compared to the time, effort and fiscal resources that have gone into research on the female only a very small fraction has been expended in seeking reversible means of fertility control in the male. Yet development of a greater variety of methods for both sexes is needed to meet the demands of greater safety and to conform to the diverse social, economic, and religious restraints in different parts of the world.

It is probably unrealistic to expect that any pharmacological agent taken by either sex for months or years will be entirely free of undesirable side-effects. It seems evident however, that if a male pill as well as a female pill were available, partners could alternate responsibility for contraception every few weeks or months, and in this way the risk of cumulative effects of long-term continuous medication would be reduced for both.

An oral contraceptive for the male is still many years in the future, but encouraging progress is being made in acquiring the basic understanding of male reproductive physiology that will be necessary to develop one. In this presentation, I have briefly indicated some of the current approaches to the problem and have attempted to identify a few of the vulnerable steps in spermatogenesis, sperm release, and post-testicular maturation that might lend themselves to pharmacological interruption. I hope I have left the impression that this is not merely applied science but that there remain intriguing fundamental problems to challenge the imagination and ingenuity of cell and developmental biologists as well as those already committed to reproductive biology.

REFERENCES

Ahmad, N., Holtmeyer, G. C., and Eik-Nes, K. B. (1973). Maintenance of spermatogenesis in rats with intratesticular implants containing testosterone or dihydrotestosterone. *Biol. Reprod.* **8,** 411–419.

Baccetti, B., Bigliardi, E. and Vegni Talluri, M. (1975). Ultrastructural observations on rat male germ cells *"in vitro"* and on the effects of TEPA. (in press).

Blaquier, J. A., Cameo, M. S., and Burgos, M. H. (1972). The role of androgens in the maturation of epididymal spermatozoa in the guinea pig. *Endocrinology* **90,** 839.

Burgos, M. H. and Fawcett, D. W. (1955). Studies on the fine structure of the mammalian testis. I. Differentiation of the spermatids in the cat. *J. Biophys. Biochem. Cytol.* **1,** 287–299.

Clermont, Y. and Bustos-Obregon, E. (1968). Re-examination of spermatogonial renewal in the rat by means of seminiferous tubules mounted *"in toto."* *Am. J. Anat.* **122,** 237–247.

Cooper, C. W. and Bedford, J. M. (1971). Acquisition of surface charge by the plasma membrane of mammalian spermatozoa during epididymal maturation. *Anat. Rec.* **169,** 300–303.

Coppola, J. A. (1969). An extragonadal male antifertility agent. *Life Sciences* **8,** 43.

Coppola, J. A. and Saldarini, R. J. (1974). A new orally active male antifertility agent. *Contraception* **9**, 459–470.

Crabo, B. and Appelgren L. E. (1972). Distribution of (^{14}C) α-chlorohydrin in mice and rats. *J. Reprod. Fertil.* **30**, 161.

DeKretser, D., Hudson, B., Lee, R. and Keogh J. (1974). Selective suppression of FSH by testis extracts. *Internat. Res. Comm. System Physiol. of Reprod. – Obs. and Gyn.,* **Vol. 2,** p. 1406.

Dym, M. (1973). The fine structure of the monkey Sertoli cell and its role in maintaining the blood-testis barrier. *Anat. Rec.* **175**, 639–655.

Dym, M. and Fawcett, D. W. (1970). Observations on the blood-testis barrier of the rat and on the physiological compartmentation of the seminiferous epithelium. *Biol. Reprod.* **3**, 308–326.

Dym, M. and Fawcett, D. W. (1971). Further observations on the numbers of spermatogonia, spermatocytes and spermatids joined by intercellular bridges in mammalian spermatogenesis. *Biol. Reprod.* **4**, 195–215.

Edelman, G. M. and Millette, C. F. (1971). Molecular probes of spermatozoon structure. *Proc. Nat. Acad. Sci.* U.S. **68**, 2436.

Ericsson, R. J. and Baker, V. F. (1970). Male infertility compounds: biological properties of U-5897 and U-15, 646. *J. Reprod. Fertil.* **21**, 267.

Fawcett, D. W. (1959). Changes in the fine structure of the cytoplasmic organelles during differentiation. *In*, "Developmental Cytology," (D. Rudnick, ed.), pp. 161–190. Ronald Press, N.Y.

Fawcett, D. W. (1974). Observations on the organization of the interstitial tissue of the testis and on the occluding cell junctions in the seminiferous epithelium. Schering Symp. on Contraception-Male. Advances in Bioscience 10, Pergamon Press, Oxford.

Fawcett, D. W. (1975). Interactions between Sertoli cells and germ cells. Serono Symp. on Male Fertility and Sterility. (in press).

Fawcett, D. W. (1975). Morphogenesis of the mammalian sperm acrosome in new perspective. Wenner Gren Symposium, *Functional Anatomy of the Spermatozoon,* Pergamon Press, Oxford. (in press).

Fawcett, D. W., Ito, S. and Slauterback, D. L. (1959). The occurrence of intercellular bridges in groups of cells exhibiting synchronous differentiation. *J. Biophys. Biochem. Cytol.* **5**, 453–460.

Fawcett, D. W. and Hollenberg, R. (1963). Changes in the acrosome of guinea pig spermatozoa during passage through the epididymis. *Zeit. Zellforsch.* **60**, 276.

Fawcett, D. W. and Phillips, D. M. (1969). Observations on the release of spermatozoa and on changes in the head during passage through the epididymis. *J. Reprod. Fertil.* **(Suppl. 6),** 405–418.

Fawcett, D. W., Leak, L. V. and Heidger, P. N. (1970). Electron microscopic observations on the structural components of the blood-testis barrier. *J. Reprod. Fertil.* **(Suppl. 10),** 105–122.

Fawcett, D. W., Anderson, W. A. and Phillips, D. M. (1971). Morphogenetic factors influencing the shape of the sperm head. *Develop. Biol.* **26**, 220–251.

Flickinger, C. and Fawcett, D. W. (1967). The junctional specializations of Sertoli cells in the seminiferous epithelium. *Anat. Rec.* **158**, 207–222.

Flores, N. M. and Fawcett, D. W. (1972). Ultrastructural effects of the anti-spermatogenic compound WIN-18446 (bis dichloroacetyl diamine). *Anat. Rec.* **172**, 310.

Franchimont, P. (personal communication)

Fray, C. S., Hoffer, A. P. and Fawcett, D. W. (1972). A re-examination of motility patterns of rat epididymal spermatozoa. *Anat. Rec.* **173**, 301–308.

French, F. S. and Ritzen, E. M. (1973). Androgen binding protein in efferent duct fluid of rat testis. *J. Reprod. Fertil.* **32**, 479–483.

Friend, D. S. and Fawcett, D. W. (1974). Membrane differentiations in freeze-fractured mammalian spermatozoa. J. Cell Biol. (in press).

Gaddu, P. (1968). Sperm maturation in the male reproductive tract: development of motility. *Anat. Rec.* **161**, 471.

Gilula, N. B. and Fawcett, D. W. (1975). Junctional membrane specializations in the mature and immature rat testis. (in manuscript).

Hansson, V., French, F. S., Weddington, S. C., McLean, W. S., Tindall, D. J., Smith, A. A., Nahfeh, S. N. and Ritzen, E. M. (1973). Androgen transport mechanisms in the testis and epididymis. *Internat. Res. Comm., Sept., Nov.*

Harris, M. E. (1973). Concentration of testosterone in testis fluid of the rat. *Proc. Soc. Study of Reprod.,* p. 61, Abstr. 55.

Howard, P. J. and James, L. P. (1973). Immunological implications of vasectomy. *J. Urol.* **109**, 76–94.

Huckins, C. (1971). Spermatogonial stem cell population in adult rats. I. Their morphology, proliferation, and maturation. *Anat. Rec.* **169**, 533–558.

Jackson, H., Fox, B. W. and Craig, A. W. (1961). Antifertility substances and their assessment in the male rodent. *J. Reprod. Fertil.* **2**, 447–465.

Johnson, M. H. (1970). Changes in the blood-testis barrier in the guinea pig in relation to histological damage following isoimmunization with testis. *J. Reprod. Fertil.* **22**, 119–127.

Koehler, J. K. and Gaddum-Rosse, P. (1975). Media induced alterations of the membrane associated particles of the guinea pig sperm tail. *J. Ultrastr. Res.* (in press).

Kormano, M. (1968). Penetration of intravenous trypan blue into the rat testis and epididymis. *Acta Histochem.* **30**, 133–136.

Lubicz-Nawrocki, C. M. and Glover, T. D. (1970). Effects of gonadectomy and testosterone replacement on the viability of epididymal spermatozoa in golden hamster. *J. Endocr.* **48**, xxii–xxiii.

McCullagh, D. R. and Schneider, I. (1940). The effect of a non-androgenic testis extract on the estrous cycle in rats. *Endocrinology* **27**, 899–902.

Means, A. R. and Hall, P. (1968). Protein biosynthesis in the testis. I. Comparison between stimulation by FSH and glucose. *Endocrinology* **82**, 597–602.

Moens, P. B. and Go, V. L. W. (1971). Intercellular bridges and division patterns of rat spermatogonia. *Zeit. Zellforsch.* **127**, 201–208.

Moens, P. B. and Hugenholtz, A. D. (1975). Rat spermatogenesis, a numerical analysis based on quantitative electron microscopy. *Biol. Reprod.* (in press).

Nagano, T. and Okumura, K. (1973). Fine structural changes of allergic spermatogenesis in the guinea pig. I. Similarity in the initial changes induced by passive transfer of anti-testis serum and by immunization with testicular tissue. *Virchow's Arch. Abt. B. Zellpath.* **14**, 223–235.

Nelson, W. O. and Merckel, C. (1973). Maintenance of spermatogenesis in testis of the hypophysectomized rat with sterol derivatives. *Proc. Soc. Exp. Biol. Med.* **36**, 825–828.

Nelson, W. O. and Patanelli, D. J. (1965). Chemical control of spermatogenesis. *In*, "Agents Affecting Fertility" C. R. Austin and J. S. Perry, eds.), Little Brown and Co., Boston.

Nicander, L. (1967). An electron microscopical study of cell contacts in the seminiferous epithelium of some mammals. *Zeit, Zellforsch.* **83**, 375–397.

Nicolson, G. L. and Yanigamachi, R. (1974). Terminal saccharides on sperm plasma membranes. *Science* **184**, 1295–1297.

Oikawa, T., Yanigamachi, R. and Nicolson, G. L. (1973). Wheatgerm agglutinin blocks mammalian fertilization. *Nature* **241**, 256–259.

Orgebin-Crist, M. C. (1969). Studies on the function of the epididymis. *Biol. Reprod.* **(Suppl. 1)**, 155–175.

Paul, R., Williams, R. P. and Cohen, E. (1974). Structure activity studies with chlorohydrins as orally active male antifertility agents. *Contraception* **9**, 451–457.

Ploen, L. (1971). A scheme of rabbit spermateleosis based upon electron microscopical observations. *Zeit. Zellforsch.* **115**, 553–564.

Ploen, L. (1973). An electron microscope study of the delayed effects on rabbit spermateleosis following experimental cryptorchidism for twenty-four hours. *Verchow's Arch. B. Abt. Zellpath.* **14**, 159–184.

Prasad, M. R. N., Singh, S. P. and Rajalakshmi, M. (1970). Fertility control in male rats by continuous release of microquantities of cyproterone acetate from silastic capsules. *Contraception* **2**, 165–177.

Setchell, B. and Jacks, F. (1974). Inhibin-like activity in rete testis fluid. *J. Endocr.* **62** (in press).

Steinberger, E., Steinberger, A. and Ficher, M. (1970). Study of spermatogenesis and steroid metabolism in cultures of mammalian testes. *Rec. Prog. Hormone Res.* U.S.A. **26**, 547–588.

Yanigamachi, R., Noda, Y. D., Fujimota, M. and Nicolson, G. I. (1972). The distribution of negative surface charges on mammalian spermatozoa. *Am. J. Anat.* **135**, 497–570.

Developmental Mechanisms in *Volvox* Reproduction

Gary Kochert

Department of Botany
University of Georgia
Athens, Georgia 30602

I. Introduction 55
II. Asexual Reproduction 58
   A. General Comments 58
   B. Differentiation of Reproductive Cells 60
   C. Nuclear Activity During the Asexual Cycle 67
   D. Biochemical Events During Asexual Reproduction 70
III. Sexual Reproduction 74
   A. General Features 74
   B. The Fertilization Reaction 75
   C. Zygospore Formation and Germination 78
IV. Sexual Inducing Substances 78
   A. Purification and Characterization 78
   B. Production of Sexual Inducers 82
   C. Mode of Action of Inducers 83
V. Summary 86
   References 88

I. INTRODUCTION

The genus *Volvox* is a large one with more than 20 species and a world-wide distribution. The various species present a fascinating array of developmental patterns in their relatively simple life cycles. Work on developmental mechanisms in *Volvox* under controlled laboratory conditions has only just begun, but increasingly scientists are realizing that this organism presents unique opportunities for the study of developmental mechanisms in a eukaryotic system.

Despite the variability among species of *Volvox* there are some common features. Mature spheroids[1] range from nearly spherical to ovoid in shape and

[1] The term spheroid is used here to denote individuals belonging to the genus *Volvox*. These are often termed "colonies" in algological or protozoological literature. It is clear, however, that each "colony" does not represent a group of organisms of the same kind living in close association, but is in fact an individual with differentiated cell types.

consist of hundreds or thousands of cells embedded in a transparent sheath (Figs. 1 and 2). The sheath itself is composed of protein and carbohydrate. Treatment of most species with pronase will solubilize the sheath material and yield a suspension of isolated cells. Ethanol precipitation of the supernate from such a preparation yields a fibrous white precipitate consisting of about one-third protein and two-thirds carbohydrate. The principle sugars present after acid hydrolysis are arabinose and galactose (Darley and Kochert, unpublished observations). Amino acid analysis of the material reveals that a large amount (21%) of the residues are hydroxproline (Lamport, personal communication). This fraction of the sheath material thus appears to represent a glycoprotein similar in overall characteristics to "extensin" described from higher plant cell walls (Lamport, 1969). It is not known at present how many components are present in the sheath, but based on histochemical techniques there appears also to be a sulfated polysaccharide component, possibly localized in a surface layer over the spheroid (Soyer, 1973; Burr and McCracken, 1973).

In all *Volvox* species mature spheroids contain cells of two basic types: a large number of somatic cells and a much smaller number of reproductive cells (Figs. 1 and 2). Somatic cells are biflagellate and contain a single nucleus, a single chloroplast with basal pyrenoid, and are similar in general morphology and ultrastructure to cells of the unicellular alga *Chlamydomonas* (Kochert and Olson, 1970). In some species of *Volvox* the somatic cells in mature

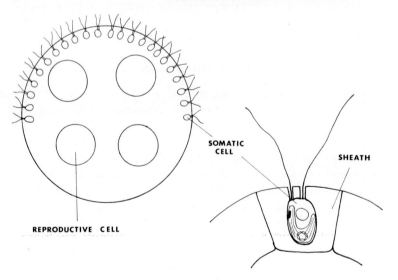

Fig. 1. Diagram of *Volvox* structure.

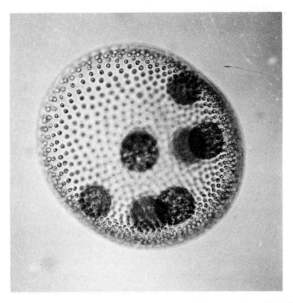

Fig. 2. Asexual spheroid with 8 uncleaved gonidia in two tiers of 4. The anterior end of the colony is toward the upper left.

spheroids appear connected to one another by cytoplasmic strands while in others mature somatic cells display no such connections (Smith, 1944). Reproductive cells are always larger at maturity than are somatic cells, and in mature spheroids are located inside the spheroid just beneath the surface layer of somatic cells. Reproductive cells have traditionally been classified into three types: asexual reproductive cells called gonidia, which undergo a series of divisions to produce new spheroids; androgonidia, also called male initial cells which cleave to form packets of sperm; and eggs, which are fertilized by the sperm to form zygotes.

Volvox spheroid types are named on the basis of the type of reproductive cell they contain. All species produce asexual spheroids which contain somatic cells and gonidia. In some species, sexual spheroids may contain both androgonidia and eggs, while in others separate male and female spheroids are formed (see below).

In addition to the obvious differentiation between somatic and reproductive cells seen in Volvox spheroids, there are other indications of differentiation within the spheroid. There is a definite anterior-posterior axis which is reflected both in the cellular construction of the spheroid and in the pattern of motility. Spheroids swim with this anterior end forward while rotating about their longitudinal axis. Reproductive cells are usually confined to the posterior two-thirds of the spheroid (Fig. 2). Somatic cells of an individual spheroid also show signs of differentiation among themselves and among these is a gradient in eyespot size. Eyespots in the anterior-most somatic cells are large and conspicuous. Cells at positions progressively further from the anterior end of the spheroid contain smaller and smaller eyespots; cells in the posterior portion of the spheroid contain very small eyespots or no eyespots at all (Pocock, 1933).

II. ASEXUAL REPRODUCTION

A. General Comments

The process of asexual reproduction in *Volvox* is accomplished by cleavage of gonidia in asexual spheroids to form young spheroids. Two general types of gonidial growth and cleavage are found among *Volvox* species. In one type, gonidia begin to divide when they are quite small. A period of growth separates each mitosis, and thus the developing spheroid grows in size as the division process takes place. In another group of species, gonidia undergo a long period of growth before divisions begin (Fig. 3). In the latter case a true cleavage appears to occur; the cytoplasm of the gonidium is cleaved into progressively smaller units in a rapid series of divisions with no apparent growth occurring during this process. In both types of spheroid formation the patterns of cell division are basically similar. The pattern of cell division has been extensively studied in several *Volvox* species and will not be reviewed in depth here (Janet, 1923; Pocock, 1933; Kochert, 1968; Starr 1969, 1970). In general, however, all the cells of a young spheroid are formed in a series of

Fig. 3. Diagram of the two basic patterns of juvenile colony formation in *Volvox* species. See text for details.

synchronous cell divisions which lead to the formation of a hollow ball of cells which have their anterior ends directed toward the inside of the sphere; if reproductive cells have been differentiated by this time they will be on the outer surface of the embryo (Figs. 4 and 5). All the cells which will be present in the mature spheroid are formed by these divisions; no new cells are added during the subsequent growth and maturation of the young spheroid. After the division period, the embryo undergoes a complex series of morpho-genetic movements which result in the spheroid turning itself inside out in the process called inversion (Figs. 6 and 7). The anterior end of the embryo now becomes the posterior end of the developing spheroid. Inversion is a relatively rapid process usually requiring only 30-60 minutes. The motive force is not understood, but striking changes in cell shape occur along with the appearance of an extensive system of cytoskeletal microtubules (Kelland, 1964; Pickett-Heaps, 1970).

Several important changes occur in the post-inversion maturation period. Following inversion the cells are closely packed and angular from mutual compression (Fig. 8). After a short lag period, synthesis of the sheath material by the somatic cells begins. This material is apparently formed inside the somatic cells in vesicles which later discharge their contents outside the cell (Bisalputra and Stein, 1966; Kochert and Olson, 1970). Mutants have been isolated which lack the ability to synthesize the sheath material; these dissociate to form a suspension of somatic cells and gonidia at this stage

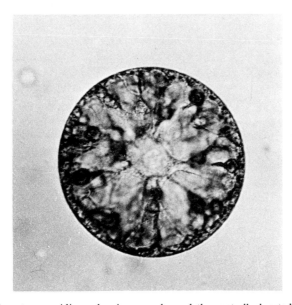

Fig. 4. A mature gonidium, showing vacuoles and the centrally located nucleus.

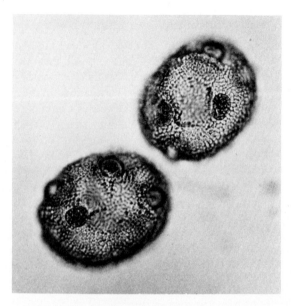

Fig. 5. Embryos just before the beginning of inversion. The larger cells are gonidial initials.

(Starr, 1971; Sessoms and Huskey, 1973). In the normal case, however, the somatic cells move apart from one another as the sheath material is produced and the young spheroid consequently grows in size. During this period eyespots appear in the anterior somatic cells, and the 2 flagella (which have been present on the somatic cells since before inversion) elongate, first one and then the other.

All the stages in spheroid formation described above have taken place while the embryo is still enclosed within the parent spheroid in a vesicle apparently derived from the parent gonidial cell wall. When the young spheroids are mature they can often be seen rotating slowly within this vesicle. Release of the young spheroids occurs by a dissolution of the parent matrix (Fig. 9). In most species, spheroids escape from the parent through individual pores dissolved in the adjacent area of the spheroid matrix, but in *V. aureus*, *V. obversus*, and others, all the young spheroids escape through a common pore which is formed in the rear of the parent.

Pocock (1933) was one of the first to speculate that young spheroids form their escape pores by secreting an enzyme. The enzyme could be proteolytic in nature since the sheath material contains protein and in some species treatment with pronase will cause premature release even of gonidia and young embryos (Fig. 10). Direct proof of the existance of such a "release enzyme" however remains to be demonstrated. In addition it is clear that

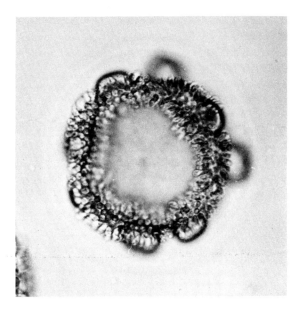

Fig. 6. Anterior polar view of an inverting embryo. The anterior lobes are folding back over the surface of the embryo.

release of young spheroids from the parent does not take place at a definite stage in the development of young spheroids; the time of release varies with environmental conditions. Often release of young spheroids can be induced by shutting off the aeration in a culture or by placing the culture in the dark.

After release of a young spheroid, the parent (which now consists of only somatic cells) remains viable for a time as a "ghost", but eventually the parental spheroid dissociates. The individual somatic cells, which are terminally differentiated, undergo senesence and eventually die.

Somatic cells of the newly released progeny spheroids continue to produce sheath material and the spheroids gradually enlarge. Meanwhile their reproductive cells enlarge and begin to divide and another cycle of asexual reproduction begins.

B. Differentiation of Reproductive Cells

The relationship between somatic cells and germ cells has long been a favorite subject of research for developmental biologists. The first organisms studied in this respect were animals in which embryogenesis takes place externally. Very early, researchers noted and distinguished two general types of cells: germ cells, representing the only physical link between successive

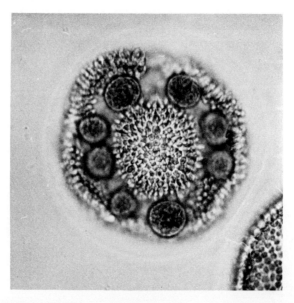

Fig. 7. Posterior polar view of an inverting embryo. The lobes which are folding back around the margins will meet to form the posterior end of the colony.

generations; and somatic cells, which perish when the organism dies. Various theories have been advanced to explain this fascinating dichotomy, including selective loss of genetic material. Most early concepts in this area stem from the writings of August Weismann. He postulated the setting aside of a "germ plasm" which provided continuity between successive generations (Weismann, 1889). In the original concept only the germ plasm contained all the information for the formation of a complete organism. Somatic cells lacked some of the "determinants" (genes) and hence could not form complete new organisms. Weismann's theories were postulated at a time when there was a virtual absence of biochemical knowledge of cellular components and little understanding of genetic concepts. It is not surprising then that most of his theories survive today only in a modified form. However, some organisms do indeed appear to maintain a full complement of genetic material only in germ cells, and not in somatic cells. In some nematodes and insects, loss of chromosomes or parts of chromosomes occurs in cells destined to form somatic tissue (Boveri, 1899; Geyer – Duszynska, 1966). In no case has a qualitative loss of genetic information been rigorously proven, but several experiments have shown that if all cells of such an organism are induced to loose chromosomes or parts of chromosomes the resultant animals lack viable gonads. This carries the implication that the chromosome parts which are lost contain genes necessary for gonad differentiation. The only biochemical

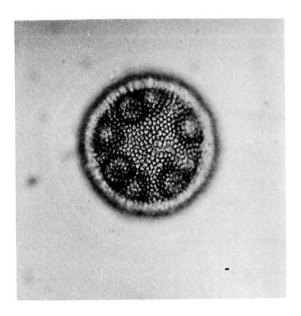

Fig. 8. Embryo which has just finished inversion. Note angular shape of the somatic cells and interior position of the gonidia.

evidence comes from studies on *Parascaris*, where 27% of the total DNA is lost in the formation of somatic cells and both repetitive and unique DNA are apparently lost (Tobler *et al.*, 1972).

In most organisms, however, the differentiation between somatic and germ cells clearly does not involve selective loss of genetic material. Nuclear transplant experiments in animals (Gurdon, 1963) and single-cell regeneration experiments in plants (Steward *et al.*, 1964) have demonstrated that nuclei in somatic or potentially somatic tissues contain the full genetic potential for the formation of a complete new organism.

Weismann's germ plasm concept has, however, survived in a modified form. During early embryogenesis of certain amphibians, insects, and birds a differentially staining cytoplasmic region can be discerned. This material is incorporated into the germ cells of the adult. Experiments have demonstrated that if one removes the germ plasm or inactivates it by ultraviolet irradiation, no functional germ cells are formed (Smith, 1966 and this volume). The clear implication is that the germ plasm causes nuclei which come in contact with it to initiate the developmental pattern which leads to gamete formation.

The algae of the class Volvocaceae, and particularly the genus *Volvox*, have long been recognized as especially favorable organisms for the study of differentiation into "somatic" and "germ" lines. Weismann himself recognized this (Weismann, 1889) as have numerous later researchers. A species of *Volvox*

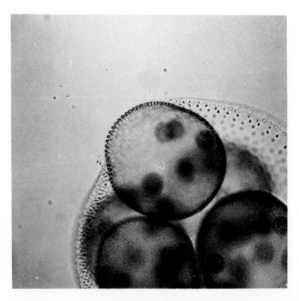

Fig. 9. Young spheroid being released from the parent.

(V. weismannia) was named after Weismann by Powers (1908) because it represented particularly favorable material for experimental study of germ and somatic cell differentiation (now called *V. carteri* f. *weismannia*). This organism has only two cell types and these are very different in both morphology and function. Reproductive cells, which are large and easily distinguishable, retain the power to undergo cell division to produce another generation of organisms. The small somatic cells are terminally differentiated and gradually undergo senescence, eventually dying without dividing.

In the genus *Volvox* there are species in which the reproductive cells do not become morphologically distinguishable from the somatic cells until after the juvenile spheroid has been released from the parent; in other species the reproductive cells differentiate very early during cleavage to form juvenile spheroids (Smith, 1944). The problem of reproductive cell differentiation is most easily approached in the latter group of species, one of which is *V. carteri* f. *weismannia*.

In *V. carteri* unequal divisions differentiate the gonidial initials from the prospective somatic cells at the division of the 16 celled stage or at the division of the 32 celled stage during juvenile spheroid formation (Kochert, 1968). In *V. carteri* f. *weismannia* gonidia are located in the spheroid in a very regular pattern consisting of tiers of 4 gonidia (Fig. 11). Gonidia in each tier are arranged symmetrically in a plane perpendicular to the longitudinal axis of the spheroid. Eight gonidia are usually present (two tiers of 4). One

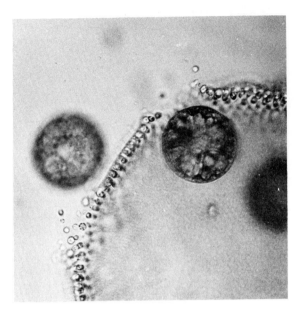

Fig. 10. Gonidia being prematurely released from the parent by Pronase (50 µg/ml) treatment.

tier of 4 gonidia will be located just forward of the equator of the spheroid; the other tier is located farther back toward the posterior portion. Gonidia in the posterior tier are located behind the spaces between individual gonidia in the front tier so that all 8 gonidia are visible in an anterior view of the spheroid (Fig. 12).

This very regular distribution of gonidia in asexual spheroids is traceable to the pattern of early cleavage during juvenile spheroid formation. The first 4 cleavages of the gonidium produce 16 cells in 4 tiers of 4. At the 16-celled stage all the cells are morphologically identical, but when the cleavage to form 32 cells takes place, 8 of the cells divide unequally. The larger of the two cells in each pair will form a gonidial initial, the smaller will continue to cleave and form somatic cells. In this way the symmetrical pattern of gonidial localization is established very early in juvenile spheroid formation. In fact it may be that the pattern is already established in mature gonidia prior to their cleavage. We have shown that if mature, uninucleate gonidia are irradiated with unilateral ultraviolet (UV) light just prior to the beginning of cleavage and then allowed to cleave, spheroids are produced which lack one or more gonidia (Kochert and Yates, 1970). With short UV doses, the most frequently encountered alteration is a lack of one gonidium in either the anterior or posterior tier (1T1) in Fig. 11. Longer doses give progressively more drastic effects. With the shorter doses all the remaining gonidia in a spheroid are in

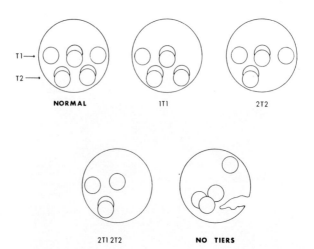

NORMAL   1T1   2T2

2T1 2T2   NO TIERS

Fig. 11. Diagram of the structure of *V. carteri* asexual spheroids. The "normal" spheroid has 8 gonidia in two tiers of 4. The other spheroids illustrate the results of UV-irradiation during the precleavage period (1T1 = 1 gonidium missing in Tier 1; 2T2 = 2 gonidia missing in Tier 2 etc.).

the proper locations and in many cases all the somatic cells appear normal (Fig. 13). If the remaining gonidia are allowed to cleave to form a second generation, all these have the normal number and distribution of reproductive cells. Thus the change brought about by the UV-irradiation is clearly not a permanent genetic change. One explanation which is consistent with these results is that the pattern of location of gonidia in the mature spheroid is predetermined in the mature gonidium. This predetermination could be in the form of a cytoplasmic pattern of localization of a "germ plasm". If the hypothetical germ plasm were localized in tiers of four in the mature gonidium, nuclei coming into this material might be induced to form a gonidium (or protected from forming a somatic cell (Fig. 14). If one of the active components of the germ plasm areas in the undivided gonidium were to be destroyed by UV irradiation, the resultant spheroid would lack a gonidium on the side corresponding to that irradiated. Germ plasm has been shown to be UV-labile in insect and amphibian eggs and irradiation of the germ plasm results in the formation of organisms which lack functional gametes (for review see Gurdon and Woodland, 1968).

We have also carried out experiments in which mature gonidia were centrifuged to displace the cytoplasm. All the centrifuged gonidia cleaved to produce spheroids which were viable, but which had altered distribution of gonidia. This again implicated the existence in mature gonidia of a cyto-plasmic "prepatterning" determining the location of gonidia in the ensuing

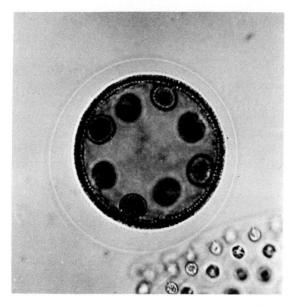

Fig. 12. Anterior polar view of a young spheroid showing tiered arrangement of gonidia. The spheroid is still enclosed in the cleavage vesicle.

spheroid. If the cytoplasmic localization theory is correct, one would predict that the cytoplasmic pattern is programmed during the long period of gonidial growth and enlargement which precedes cleavage.

In any case the differentiation into reproductive and somatic cells probably does not involve loss of genetic material by the somatic cells. Starr (1970) and Sessoms and Huskey (1973) have reported the isolation of mutant *V. carteri* strains in which the somatic cells behave as reproductive cells and undergo cleavage to form miniature colonies. Two non-linked mutants which carry this phenotype have been isolated by Griffin and Huskey (1974). In crosses the trait segregates as a single allele and the most straight-forward interpretation would be a change in an epigenetic regulatory system.

C. Nuclear Activity During the Asexual Cycle

In the classical cases of cytoplasmic programming in animals (sea urchins, amphibians) experimental evidence indicates that nuclear activity is relatively unimportant during early development of the embryo. This is demonstrated both by the fact that fertilized eggs placed in actinomycin-D continue relatively normal development for many hours through many cell divisions and by the observation that in interspecific hybrids, male parental characters do not appear until late in development (for review see Davidson, 1968).

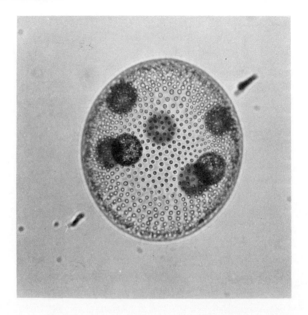

Fig. 13. A 1T2 spheroid (missing 1 gonidium in tier 2) produced by UV treatment of a precleavage gonidium. Note that the somatic cells appear normal.

Since asexual development of *V. carteri* embryos superficially resembles cleavage of an amphibian or sea urchin egg, we undertook inhibitor experiments to try to ascertain the importance of nuclear transcription in the asexual developmental process (Weinheimer, 1973). All stages of spheroid formation in *V. carteri* f. *weismannia* were found to be relatively sensitive to actinomycin-D treatment. With low actinomycin-D doses (0.5 μg/ml) differen-

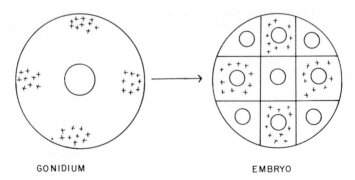

GONIDIUM   EMBRYO

Fig. 14. Diagramatic representation of proposed localization pattern of "germ plasm" in *V. carteri* early embryogenesis. Nuclei enclosed in cells containing the substance indicated by "+" would be "induced" by it to form a gonidium.

tial sensitivity was seen. There appeared to be two very sensitive times in the life cycle: all stages of gonidial enlargement prior to the first gonidial cleavage, and the stage just prior to inversion of the embryo. When placed in low concentrations of actinomycin-D, small gonidia did not enlarge and mature gonidia usually failed to cleave even once. Successive stages became more and more actinomycin resistant; when 16-celled embryos were placed in the actinomycin-D they underwent an average of five further cleavages (to form approximately 512-celled embryos) (Table 1). Cleavage stages after the 16-celled stage became progressively more sensitive to actinomycin-D. No embryos were seen to undergo inversion in the presence of the drug when placed in actinomycin-D prior to the beginning of inversion but after the last cleavage had occurred. The same general pattern of inhibition was seen with 5-fluorouracil. All embryo stages placed in a high concentration (325 μg/ml) of 5-fluorouracil underwent normal cleavage and development to the late preinversion stage; then they stopped development just before or during inversion. Undivided gonidia failed to enlarge or divide in the presence of the drug. These experiments may be interpreted to mean that there is maximum immediate need for genetic transcription during the growth period before the first gonidial cleavage and just prior to inversion. However, since nothing is at present known about permeability changes during development, trivial explanations are also possible for some of the observed phenomena.

Another very interesting effect was observed in the actinomycin-D but not the 5-fluorouracil experiments. These embryos, it will be recalled, differentiate somatic cells from reproductive cells by unequal divisions at either the 16 or the 32-celled stage. When these cleavages took place in the

TABLE 1

*The Mean Number of Cell Divisions Completed after
Addition of 0.5 μg/ml Actinomycin D*

Stage of development	Mean number of divisions completed
Undivided gonidia	0.40
2−celled embryo	1.33
4−celled embryo	3.33
8−celled embryo	4.00
16−celled embryo	5.00
32−celled embryo	3.00
64−celled embryo	2.33
128−celled embryo	2.00
256−celled embryo	1.00
Greater than 256−celled embryo	0.00

presence of actinomycin-D, 32 or 64-celled embryos, were sometimes produced, all the cells of which were equal in size. No differentiation into somatic and reproductive cells had taken place. Thus it would appear that just prior to the divisions which differentiate reproductive from somatic cells there is a requirement for synthesis of RNA. If this is so the divisions that separate gonidia from somatic cells may be under immediate genetic control and RNA's needed for differentiation of the two cell types would then be synthesized just before the actual differentiation takes place.

D. Biochemical Events During Asexual Reproduction

The only *Volvox* strain from which DNA has thus far been isolated and characterized is *V. carteri* f. *weismannia*. Two main components are present (Fig. 15): a major component of nuclear DNA (ρ = 1.715 gm/cm^3) and a minor component, a satellite DNA assumed to be chloroplast DNA (ρ = 1.707 gm/cm^3) (Kochert and Jaworski, 1972). Synthesis of DNA through the asexual cycle has been studied in synchronized cultures of whole colonies and also in isolated somatic and reproductive cells (Fig. 16) prepared by pronase dissociation of colonies followed by differential filtration (Yates, 1974).

Somatic cells contain 1.14 x 10^{-7}μg of DNA per cell. About 6.5% of the total DNA is found in the chloroplast DNA peak after extraction and analysis by analytical ultracentrifugation (Fig. 17). Both these values are in close agreement with the values reported for *Chlamydomonas*, a unicellular green alga similar in morphology to *Volvox* somatic cells (Chiang and Sueoka, 1967).

Nearly mature gonidia (6-8 hours before the first cleavage) contain about 25 times as much DNA as do somatic cells. Also the relative proportion of

Fig. 15. Analytical ultracentrifugation of isolated *V. carteri* DNA. Marker DNA = 1.731 gm/cm^3; nuclear DNA = 1.715 gm/cm^3; chloroplast DNA = 1.707 gm/cm^3.

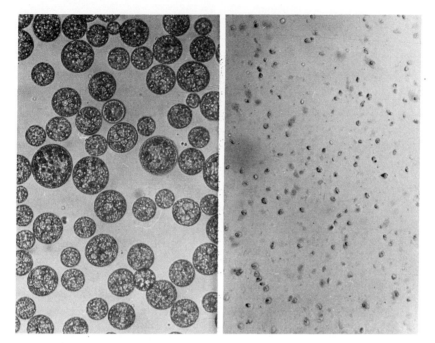

Fig. 16. Preparations of isolated somatic cells and gonidia. Spheroids were dissociated by protease treatment, then separated by filtration through nylon filters.

nuclear and chloroplast DNA is different from somatic cells. Gonidia contain about 30% chloroplast DNA (Fig. 17). Thus there appears to be preferential synthesis of chloroplast DNA during the period of gonidial enlargement prior to cleavage. However, the amount accumulated is apparently not enough to furnish all the cells of the resultant spheroid with chloroplast DNA. If one considers the amount of chloroplast DNA present in a single somatic cell to represent a basic chloroplast genome then each mature gonidium would contain 100 to 150 chloroplast genomes, apparently all contained in one large chloroplast.

DNA synthesis (as measured by $^{32}PO_4$ incorporation and total DNA/spheroid) in synchronized cultures of whole spheroids is maximum during the period of gonidial cleavage. A small but significant amount of DNA synthesis also occurs in the early stages of the enlargement of young spheroids after inversion. Experiments with isolated somatic and reproductive cells show that synthesis during this period is largely confined to the gonidia and may represent the differential synthesis of the excess chloroplast DNA noted above.

Ribosomal RNA's have also been isolated and characterized in *V. carteri* (Kochert and Sansing, 1970, Kochert, 1971). Total nucleic acid extractions

Fig. 17. Analytical centrifugation of *V. carteri* DNA from isolated somatic cells (upper) and gonidia (lower). Vertical line denotes marker DNA (ρ = 1.731 gm/cm^3).

show, in addition to DNA, 4 high-molecular-weight rRNA components: two from the cytoplasmic ribosomes and two from chloroplast ribosomes. Proportions of the various nucleic acid components are different in gonidia and somatic cells (Fig. 18). On a per spheroid basis the 8-10 gonidia contain as much or more ribosomal RNA than do the 35000 somatic cells, but less than 1/10th of the amount of DNA (Yates, 1974). Therefore, it seems likely that the mature gonidium has enough rRNA to supply all the cells of the developing embryo with ribosomes during the cleavage period. Also the ratio of chloroplast rRNA to cytoplasmic rRNA is higher in gonidia than in somatic cells, again indicating some differential synthesis of chloroplast material during pre-cleavage gonidial enlargement.

Ribosomal RNA synthesis was measured by $^{32}PO_4$ incorporation followed by isolation of the RNA and fractionation on polyacrylamide gels (Yates, 1974). The most active period of rRNA synthesis in both somatic cells and gonidia appeared to be during the period of early spheroid enlargement in newly released spheroids. Synthesis slows in the gonidia as they become more mature and drops to a relatively low level during the later stages of gonidial

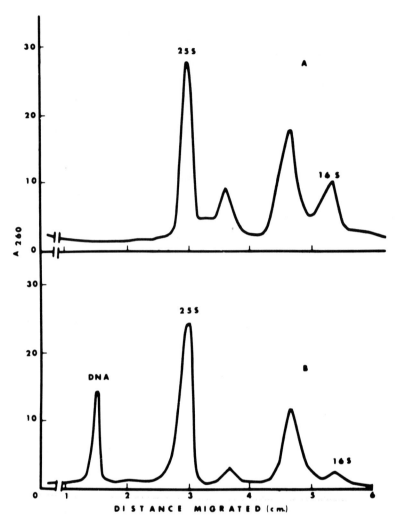

Fig. 18. Polyacrylamide gel electrophoresis of total nucleic acids isolated from somatic cells (lower) and gonidia (upper). The four peaks present in addition to the DNA peak are rRNA's (25S = a cytoplasmic rRNA; 16S - a chloroplast rRNA). Note high proportion of 16S/25S rRNA in the gonidium and virtual absence of DNA.

cleavage. Somatic cells show a fairly steady decrease in rRNA synthesis after the high point during early spheroid enlargement. Some rRNA synthesis apparently continues in parent somatic cells until at least the time of young spheroid release.

The initial stages of spheroid expansion before gonidial cleavage also represent the period most active in protein synthesis. All stages of the life cycle apparently require protein synthesis since cycloheximide (1 μg/ml) blocks further development at all stages of the asexual cycle.

III. SEXUAL REPRODUCTION

A. General Features

All *Volvox* species are assumed to be basically haploid organisms. The only diploid stage is the zygote, with meiosis occurring during zygote maturation or germination. Only one product of meiosis survives; in this aspect *Volvox* is not a particularly favorable organism for genetic studies.

In *Volvox*, sperm packets and eggs are contained in spheroids produced by gonidial divisions in the same manner as are asexual spheroids. In most species formation of sexual spheroids is under the control of specific sexual inducing substances which are liberated into the surrounding medium (see below).

Several general patterns of sexual reproduction are found among the species of *Volvox*. All the species for which information is now available fit into one of the following categories:

Type I Sexuality (Monoclonic[2], dioecious) Male spheroids containing sperm packets are produced. Sperm fertilize young gonidia in asexual spheroids to form zygotes. In this type no morphologically differentiated female colonies are formed and the full cycle of sexual reproduction can occur in a single clone. Examples of this sexual pattern include *V. aureus, V. spermatosphaera* and *V. tertius.*

In *V. aureus* a sexual inducing substance is produced which induces the formation of male colonies. No sexual inducers have yet been identified in other species of this type:

Type II Sexuality (Monoclonic, dioecious) Separate male spheroids with sperm packets and female spheroids with eggs are formed in the same clonal culture. Sexual reproduction occurs within a single clone. This type is represented by one strain of *V. africanus* and by *V. powersii*. No sexual hormones have yet been demonstrated.

Type III Sexuality (Diclonic, dioecious) This type always involves two clones; male spheroids are formed in one clone and female spheroids in the other. Thus two clones must be mixed for the full cycle of sexual reproduction to occur. When crosses are made with species of this type, 50% of the zygotes produce male clones on germination and 50% produce female clones.

---

[2] The terms monoecious and dioecious are here used in their traditional sense, i.e. monoecious = both male and female gametes produced on one individual; dioecious = male and female gametes produced on separate individuals. "Monoclonic" means that sexual reproduction can occur within a clone; diclonic means that two clones must be mixed for fertilization to occur.

In *V. carteri, V. rouselettii, V. gigas,* and *V. obversus* male spheroids produce a sexual inducer. In one strain of *V. dissipatrix* both male and female spheroids produce inducer. In all the species the inducer causes the formation of both male and female spheroids.

Type IV Sexuality (Monoclonic, monoecious) Spheroids are produced which contain both sperm packets and eggs. Fertilization and zygote formation occur in a single clone. No sexual inducing substances have been demonstrated in this type.

Type V Sexuality (Monoclonic, monoecious-dioecious) In this type both spheroids with sperm and eggs and spheroids with either sperm or eggs are formed in the same clone. This type is at present represented only by strains of *V. africanus.* These produce pure male spheroids and monoecious spheroids. No sexual inducing substances have been demonstrated in this strain.

B. The Fertilization Reaction

Whatever the type of sexual reproduction exhibited, the series of steps that take place in the sexual process is similar in general features in all species. Androgonidia in male or monoecious spheroids undergo a series of cleavages to form a packet of sperm (Fig. 19). Sperm packets may contain 16-512 sperm in the various species of *Volvox.* The initial stages in cleavage to form sperm packets are similar to those which occur in the formation of juvenile spheroids from gonidia, and an inversion occurs at the end of the cleavage

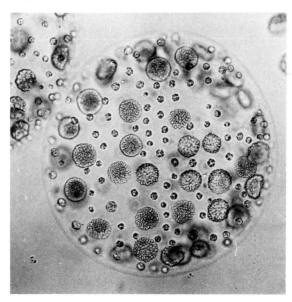

Fig. 19. Male colony containing mature or nearly mature sperm packets.

process. For this reason several workers have speculated that sperm packets represent spheroids modified during evolution so that all cells are sperm (Pocock, 1933).

The sequence of steps in the fertilization process has been described in only a few species but appears to be basically similar in all. The following description is based on *V. carteri* f. *weismannia* which has been analyzed in the greatest detail (Kochert, 1968; Hutt and Kochert, 1972; Hutt, 1972).

Sperm packets are released by the dissolution of a pore through the somatic cell layer of the parent spheroid. This probably occurs as a response to secretion of a proteolytic enzyme by the mature sperm bundle. After release the sperm packets are actively motile in the culture medium and swim in a coordinated manner. Although attempts have been made, we have not been able to demonstrate any chemical attraction of sperm packets to receptive female spheroids. In cultures the sperm appear to swim about randomly colliding with all types of spheroids.

When sperm packets contact a female spheroid, however, an extremely interesting series of steps ensues (Figs. 20). The first interaction is apparently between the flagella of the sperm packet and the flagella of the somatic cells of the female spheroid and results in the binding of the sperm packet to the female spheroid. Experimentally deflagellated females will not bind sperm

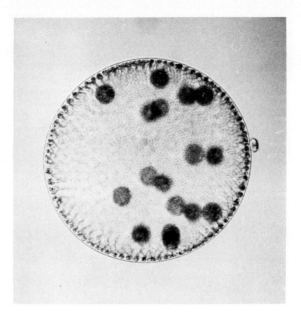

Fig. 20. Female colony containing eggs, showing a sperm packet attached to the posterior portion of the colony.

packets; the flagella isolated from the female spheroids will bind to the sperm packets. The ability of female spheroids to bind sperm is also destroyed by heating the females for two minutes at 40°C or by brief pre-treatment with pronase or trypsin. It would appear, then, that protein (or glycoprotein) receptors are present on the flagella of the female spheroid.

Two-three minutes after sperm bundles bind to female colonies, they begin to undergo dissolution to individual sperm (Fig. 21). As this occurs a fertilization pore is formed through the somatic cell layer (Fig. 22). In this process the sheath material between somatic cells is solubilized and several somatic cells are liberated into the culture medium. As the pore is formed the sperm packet becomes completely disorganized to individual sperm. These enter the spheroid through the fertilization pore (about 6 minutes after the initial attachment) and burrow through the spheroid to the eggs. Actual fertilization has been seen only in *V. carteri* f. *nagariensis*, but there is genetic evidence that a true fertilization and meiosis occurs in other species (Kochert, 1968; VandeBerg and Starr, 1971).

Formation of the fertilization pore by sperm packets is almost certainly effected by a proteolytic enzyme secreted by the sperm packet in response to interaction with the female spheroid. In a series of experiments we have shown that soybean trypsin inhibitor (50 μg/ml) will completely block fertilization pore formation. Also cycloheximide (10 μg/ml) and certain other inhibitors of protein synthesis will block fertilization pore formation and hence penetration of the female spheroid. Attachment of sperm bundles and

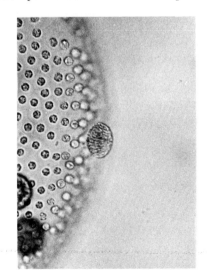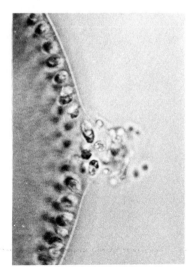

Fig. 21. Sperm packet attachment and dissolution. *Left.* Sperm packet newly attached; *Right.* Sperm packet undergoing dissolution.

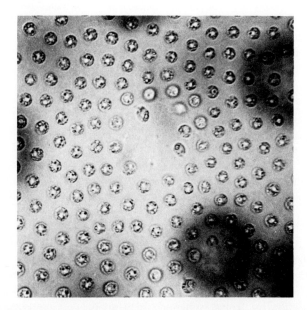

Fig. 22. Fertilization pore formed in a female colony by a sperm packet.

dissolution of the bundle to individual sperm almost always proceeds in the presence of the inhibitor, but protein synthesis is apparently required either for the actual synthesis of the proposed proteolytic enzyme or for its release from the sperm packet.

C. Zygospore Formation and Germination

In all species of *Volvox* the fertilized zygote develops into a thick-walled, orange-colored zygospore (Fig. 23). This structure is resistant to desiccation and allows the organism to survive unfavorable environmental conditions. Germination of zygospores has been accomplished in some species by transfer to fresh culture medium after a maturation period. A single cell emerges from the germinating zygote spore; in some species it is flagellated and may even be motile. Soon after its release from the zygospore it undergoes a series of divisions to form a miniature spheroid (Fig. 24). Full adult size is reached after one or two cycles of asexual reproduction. The complete life cycle of *V. carteri* is diagrammed in Fig. 25.

IV. SEXUAL INDUCERS

A. Purification and Characterization

In many species of *Volvox* the formation of sexual spheroids or the differentiation of gametes in existing spheroids is controlled by specific sexual inducers which are released into the surrounding medium. These substances

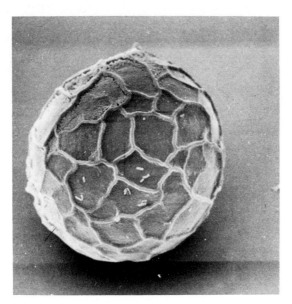

Fig. 23. Scanning electron micrograph of a zygospore, showing the heavy, ornamented wall.

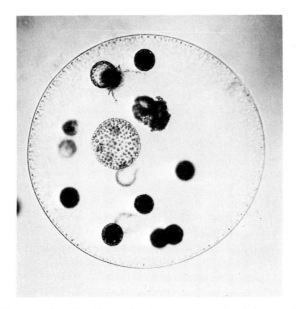

Fig. 24. Female spheroid with germinating zygospores. A miniature colony has been formed from a zygote near the center of the spheroid.

FEMALE CLONE   MALE CLONE

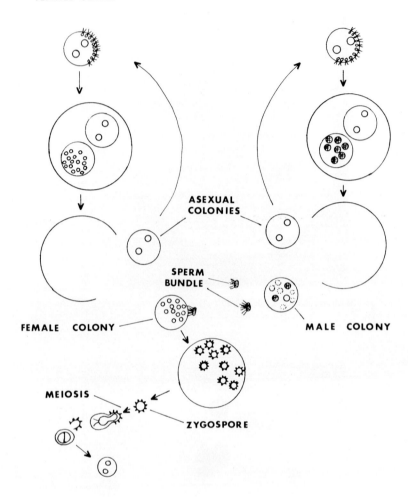

Fig. 25. A diagramatic presentation of the life cycle of *V. carteri*.

are species-specific; in no case has a sexual inducer produced by one species been shown to be effective in inducing sexuality in another. Inducers from the closely related forms *V. carteri* f. *weismannia* and *V. carteri* f. *nagariensis* do not cross-induce (Starr, 1970). Strains of *V. carteri* f. *weismannia* from various locations in the United States will, however, cross-induce.

All the sexual inducers described so far are relatively large molecules, with reported molecular weights of the inducers from various species ranging from about 10,000 to >200,000. In general the inducers are relatively heat resistant (the *V. gigas* inducer can be autoclaved without loss of biological activity);

biological activity is not destroyed by treatment with DNase, RNase, trypsin, or chymotrypsin, but is reduced or destroyed by pronase treatment (Darden, 1966, 1970; Kochert, 1968; Starr, 1969; McCracken and Starr, 1970; Vande-Berg and Starr, 1971).

The inducers which have been analyzed in greatest detail are those from *V. aureus* (Ely and Darden, 1972), *V. carteri* f. *nagariensis* (Starr and Jaenicke, 1974), and *V. carteri* f. *weismannia* (Kochert and Yates, 1974). In each of these cases the inducer is demonstrated to be a glycoprotein, with a relatively high carbohydrate content (about 40% in the *V. carteri* strains). The inducers from the *V. carteri* strains have been purified to near homogeneity (Fig. 26). Both monomer molecular weights of $28\text{-}32\text{x}10^3$ readily form higher molecular weight aggregrates. Aggregate formation is the probable explanation of the multiple forms of inducer previously reported in *V. carteri* and other species (Ely and Darden, 1972; McCracken and Starr, 1970).

The purified inducers from the *V. carteri* strains are active in extremely low concentrations (10^{-14} M for f. *weismannia,* 10^{-16} M for f. *nagariensis*). On the basis of their bioassay conditions, Starr and Jaenicke (1974) calculate that 1800 molecules of inducer/gonidium resulted in 14.4% of the gonidia being induced to form females. Pall (1973) has shown that at low inducer

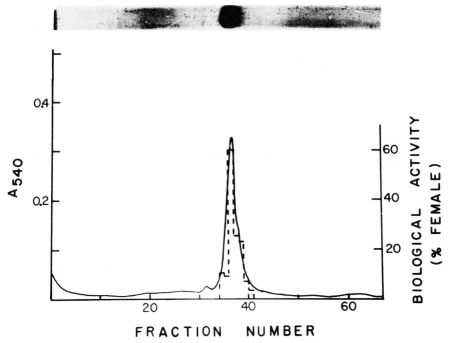

Fig. 26. SDS – polyacrylamide gel electrophoresis of purified sexual inducer from *V. carteri* f. *weismannia*. The inducer material forms a single band which coincides with the biological activity.

concentrations the frequency of induction in *V. carteri* f. *nagariensis* is proportional to the square of the inducer concentration. These results are consistent with the idea that only two molecules of inducer are required per gonidium.

Amino acid analysis of purified *V. carteri* inducers reveal little or no hydroxyproline, so the inducer is probably not a form of sheath glycoprotein. Carbohydrate components were analyzed only in the preparations of the *V. carteri* f. *nagariensis* inducer where arabinose (6.6%), xylose (25.5%), mannose (15.6%), galactose (4.6%), glucose (32.5%), N-acetyl-glucosamine (3.9%), and an unknown peak (possibly N-acetyl-galactosamine) were present. The number of carbohydrate chains per inducer molecule has not yet been reported.

B. Production of Sexual Inducers

In every species so far characterized male spheroids produce large amounts of the sexual inducer. It is not known whether the male somatic cells or the sperm from the male spheroids are responsible for production of the inducer, but the inducer becomes most concentrated in the medium only after release and disintegration of sperm packets.

It is probable that asexual spheroids also synthesize small amounts of the inducer, since this seems the most reasonable way to explain the "spontaneous" appearance of the first male in those species where male formation is under the control of an inducer. In such cases if males are induced by a substance produced only by males, how does the first male arise? The male clone can typically be kept asexual by frequent transfer to fresh medium; if the culture is allowed to grow for a longer period of time, males will appear and induce the formation of more males (Darden, 1966; McCracken and Starr, 1970). Starr (1970, 1972) has postulated that the first male arises in *V. carteri* f. *nagariensis* as a consequence of genetic mutation. This is difficult to prove directly since male spheroids cannot reproduce asexually; it is therefore impossible to isolate a male spheroid and to see if only males are produced in subsequent generations.

The problem can be approached indirectly, however. Spontaneous females are produced in *V. carteri* f. *nagariensis* in the apparent absence of inducer at a rate of 1 in 20,000 (Starr, 1972). Some of these will, when isolated, continue to produce only females in subsequent generations and thus appear to represent mutants. Starr reported that when spontaneous females were crossed with the male strain some progeny appeared to represent spontaneous males (but, again, they could not be maintained asexually).

Some spontaneous female clones, however, apparently do not continue to produce female spheroids in subsequent generations. These clones may represent a transient escape from repression. This sort of epigenetic regulatory mechanism seems a more probable explanation for the appearance of most spontaneous males than mutation. If the assumption is made that synthesis of the inducer or some regulatory protein involved in expressing the male potential is repressed in asexual spheroids, there would still be presumably some low level of synthesis occurring. Transient bursts of synthesis could also be expected to occur, perhaps in response to environmental conditions. This is the case in bacterial and viral systems at any rate (Lewin, 1970). Certainly, spontaneous males could and do arise by spontaneous mutations; the basic question is whether *most* spontaneous males arise by mutation. It seems more reasonable to assume that such a fundamental developmental process as sexuality would depend on some control more precise than random mutation. The final answer will depend on the discovery of a method or the isolation of a new strain in which male spheroids can be propagated asexually.

C. Mode of Action of Inducers

Although very little is known about the molecular mode of action of any *Volvox* sexual inducer, some useful information about the general mode of action is available. This includes data on the period during the life cycle which is susceptible to the inducer and the time of exposure necessary to induce sexuality.

There seem to be two basic extremes when one considers the time in the life cycle at which the sexual inducers exert their effect. One extreme is represented by *V. carteri*, one of the species in which gonidia undergo a long period of enlargement prior to cleavage. Induction of sexuality in *V. carteri* involves a modification in the cleavage pattern. Gonidia, it will be recalled, are normally differentiated from somatic cells by unequal cleavages early in embryogenesis of *V. carteri*. Induced embryos delay the unequal cleavages one or more stages and when these occur the larger of the two products develops into an egg or androgonidium rather than a gonidium (Kochert, 1968; Starr, 1969). As an example, *V. carteri* f. *weismannia* gonidia normally produce eight gonidia by unequal division when the 16-celled stage cleaves to form the 32-celled stage. In the presence of the sexual inducer, gonidia (from the female clone) do not undergo unequal divisions until the 32-64 celled stage, at which time 16-20 egg initials are formed. To achieve this effect gonidia must be in contact with inducer during the early stages of gonidial enlargement prior to cleavage. If one waits until the later stages of gonidial maturation, or the early stages of cleavage, an asexual spheroid will be formed. Thus the *V. carteri* inducers cause a gonidium to cleave to form a female spheroid (or a male in the male clone) rather than an asexual colony.

The other extreme is represented by *V. rousseletii* (McCracken and Starr, 1970). In this organism reproductive cells are not morphologically differentiable from somatic cells during cleavage but become apparent soon after inversion. Shortly before release from the parent, spheroids contain three types of cells: a large number of small somatic cells, 8-12 or more large gonidial initial cells, and a somewhat greater number of intermediate-sized cells. In asexual reproduction the gonidial initials enlarge and later undergo a series of divisions to form juvenile spheroids. The intermediate-sized cells revert to somatic cells. In the presence of the sexual inducer the gonidial initials and the intermediate-sized cells divide to form sperm packets (or function as eggs in the female clone). Gonidial initials can be induced to form sperm packets if exposed to the inducer during their development prior to cleavage or as late as the 2 or 4 celled stage of cleavage. Since the induction systems in *V. carteri* and *V. rousseletii* appear superficially to be quite different, two basic types of induction systems have been proposed (McCracken and Starr, 1970). In one of these (*V. carteri*) inducers act on gonidia and cause them to cleave to form sexual rather than asexual spheroids; in the other type (*V. rousseletii*) the inducer acts on reproductive cells in existing spheroids, either causing them to form sperm packets or to function as eggs.

The two types of induction system may not be as different as it might initially appear, however. Some evidence consistent with this view can be derived from some of our recent experiments on female induction in *V. carteri* f. *weismannia* (Kochert and Yates, unpublished). In this strain, exposure to the inducer early in the precleavage enlargement period causes the gonidium to cleave to form a female rather than an asexual spheroid. If one waits until the gonidium is mature before placing it in inducer it is too late to induce that gonidium to form a female. Experiments were done in which young spheroids were placed in a solution of the inducer when their gonidia were just beginning enlargement and were allowed to develop in the inducer solution until their gonidia were mature. Then some of the spheroids were removed, placed on an agar surface and irradiated with ultraviolet light from a germicidal light. Following irradiation the spheroids were placed back in inducer solution to continue development. Control colonies were placed on the agar surface but not irradiated, then returned to the inducer solution. A five second dose of UV irradiation reduced by 50% the number of females formed when the gonidia cleaved, and a ten second dose resulted in only asexual spheroids being formed. Controls always formed nearly 100% females.

This result can be interpreted in a number of ways. One possibility is that the inducer must, in fact, be present during the early stages of cleavage to exert its effect, but penetration of the inducer is so slow that one must put gonidia in inducer solution long before the actual susceptible stage (to give some time to build up an effective concentration of inducer by the time the susceptible period arises). In this interpretation the action of UV irradiation would be to inactivate inducer molecules in or on the surface of the mature gonidium and "allow" the gonidium to develop into an asexual rather than a female spheroid. The effect of the inducer in *V. carteri* would then be to prevent differentiation of gonidia and to cause certain other cells (part of which would have become somatic cells) to differentiate into eggs. The effect of the *V. rousseletii* inducer can be described in the same terms: it prevents the differentiation of gonidial initials (and turns them into eggs) and induces the formation of eggs by cells that would otherwise become somatic.

Darden (1970) has previously suggested just such a role for the *V. aureus* male inducer. Darden was the first to note that the first observable event in induction of sexual spheroids in some species was the inhibition of gonidial differentiation. He proposed that inducers act by preventing gonidial differentiation and that this results directly or indirectly in the formation of sperm or eggs by cells which would otherwise have been somatic, and that *Volvox* "inducers" might better be termed repressors. Starr (1972) has elaborated on this basic idea and proposed a model in which the inducer acts as a co-repressor of an operon controlling gonidial differentiation.

The UV irradiation experiments mentioned above can also be used to support a repressor hypothesis. It is also possible that the ultraviolet irradiation inactivates a repressor of gonidial development which is formed in the gonidium in response to the inducer. In this case the action of ultraviolet irradiation on induced *Volvox* gonidia would be like the "induction" of bacteriophage lambda from lysogenic bacterial strains by short UV doses. In the lambda case it has been shown that the ultraviolet irradiation inactivates a repressor and allows transcription of the portions of the lambda genome necessary for viral replication (Martin, 1970).

The precise molecular role of *Volvox* sexual inducers will of course await more detailed knowledge at the molecular level, both from biochemical and from genetic studies. The recent purification and characterization of sexual inducers from two strains of *V. carteri* has made the biochemical studies feasible. We have succeeded in labeling the purified *V. carteri* f. *weismannia* inducer with ^{125}I *in vitro* without apparent loss of biological activity. When the labeled inducer preparation is incubated with target asexual spheroids from the female clone, uptake of the ^{125}I label can be demonstrated. This uptake also appears to be species-specific; non-target spheroids of *V. aureus* take up much less label than do the *V. carteri* spheroids (Fig. 27). It is not yet known whether inducer molecules must penetrate the susceptible cells to exert their effect or whether they act at the cell surface.

The next step will be to attempt to determine whether or not inducer molecules actually enter the target cells. Two main mechanisms of hormone action have been demonstrated as a result of pioneering studies with animal hormones. In one of these, hormone molecules act at the cell surface and change the concentration inside the target cell of a "second-messenger". A classical case of such a mechanism would be the effect of glucagon on cyclic

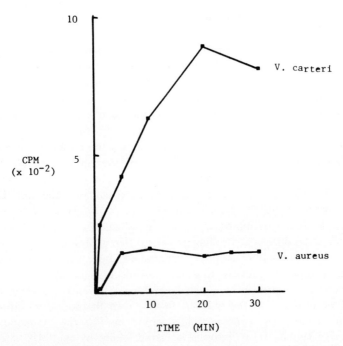

Fig. 27. Binding of purified, ^{125}I-labelled inducer to colonies of *V. carteri* f. *weismannia* and to *V. aureus.*

AMP. concentrations in mammalian liver cells (Jost and Rickenberg, 1971). In this case externally added cyclic AMP or related substances can mimic the hormone effect. We find that cyclic AMP, cyclic GMP, dibuturyl-cyclic AMP, or theophylline will not induce females in *V. carteri* f. *weismannia* (nor will any of these prevent the induction of females by the inducer), but we do not know whether these compounds are taken up by the target cells. The second general mechanism of hormone action is exemplified by the action of steroid hormones on chick oviduct tissue (O'Malley *et al*. 1972). In this system hormone molecules enter the cytoplasm and complex with a cytoplasmic receptor protein. The hormone-receptor complex then enters the target cell and presumably exerts a direct effect at the transcriptional level. Whether *Volvox* sexual inducers act by either of the general mechanisms discussed above will be revealed only by further research.

V. SUMMARY

The various species of *Volvox* provide a fascinating array of systems for study by developmental biologists. The basic differentiation into somatic and reproductive cells which occurs during the formation of all types of *Volvox* spheroids is a baseline system for studies of cell differentiation. In most species the cleavage of a gonidium leads to the formation of only two cell types in any one spheroid. There are no complications involving nervous systems, circulatory systems, or complex growth requirements. However, the two cell types which are formed are very different in both morphology and function. As an additional advantage, one can isolate the two cell types clearly by dissociation of the colony followed by differential filtration. If one uses one of the mutant strains which fail to produce sheath material, even the preliminary dissociation step can be dispensed with. Lastly, and perhaps of greatest significance, one can control the pathway of differentiation in most species by using the sexual induction system.

Considerable progress has been made toward an eventual understanding of basic developmental mechanisms in *Volvox*. The basic biology and lifecycle of most species has been documented under controlled conditions. Detailed studies have been made of the morphological aspects of development such as gonidial cleavage, the fertilization reaction, and zygote germination. Methods for the genetic analysis of *Volvox* development have been developed.

In *V. carteri* f. *weismannia* the process of asexual reproduction has been studied in some detail. In this organism a large gonidium undergoes a series of changes to form a juvenile spheroid. During this process the nucleus and the chromosomes are seen to become progressively smaller as cleavages proceed. It would be an attractive hypothesis to propose that the organism synthesizes enough DNA during the long growth period before cleavage to supply all the cells of the developing embryo with DNA. Virtually all the DNA is, however, synthesized during the actual cleavage period. There does appear to be some "prepackaging" of chloroplast genomes, however. The precleavage gonidium has one (or more) large chloroplasts with many pyrenoids, containing a total of more than 100 times the amount of DNA contained in a somatic cell chloroplast.

There is indirect evidence that ribosomes may be prepackaged in gonidia also. The rRNA/DNA ratio is very much higher in gonidia than in somatic cells. Also, the total amount of rRNA contained in the 8-10 gonidia of a late-precleavage asexual spheroid is more than the total of the 35,000 somatic cells which make up the remainder of the spheroid. The rate of rRNA synthesis is high in growing gonidia, but falls sharply during cleavage. It would appear that most of the ribosomes needed to "stock" the juvenile colony are made before cleavage begins. If so, this pattern of rRNA synthesis would be similar to that observed in sea urchins and in amphibians (Davidson, 1968). The amphibian *Xenopus* has enough ribosomes in its fertilized egg to carry it through to the tailbud stage of development (for review see Macgregor, 1972).

Another intriguing aspect of asexual reproduction in *V. carteri* is the precise localization of gonidia in the mature spheroid. We have shown that if mature gonidia are exposed to unilateral UV irradiation, some of the gonidia cleave normally except that they are missing one or more gonidia. There is apparently no genetic damage since the remaining gonidia in the spheroid cleave to produce normal spheroids in the next generation. These experiments parallel those of Smith (1966) on frog eggs, where it was demonstrated that UV-irradiation of the egg vegetal pole resulted in a failure of reproductive cell differentiation at a later stage of development. Smith presented convincing evidence from cytoplasmic transfers that the effect of the UV was to inactivate a "germ plasm" in the vegetal pole cytoplasm of the egg. We have no such evidence, but our results may mean that such a cytoplasmic morphogenetic substance may be involved in the differentiation of *V. carteri* gonidia during cleavage.

Sexual differentiation in many *Volvox* species is controlled by species – specific sexual inducers. These have been purified and partially characterized in *V. carteri* f. *nagariensis* and *V. carteri* f. *weismannia*. Although these do not cross-induce, they appear to be similar in general properties. The active principle in both cases is a glycoprotein with a molecular weight of about 30,000. The extreme stability of these molecules and the relative ease of producing them in large quantities should aid efforts to determine their mode of action at the cellular and molecular level. The purified inducer can be radioactively labelled without loss of biological activity and species – specific uptake can be demonstrated. We have great hope that further work along this line will eventually unravel the fascinating complexities of the developmental mechanisms involved in *Volvox* reproduction.

REFERENCES

Bisalputra, T., and Stein, J. R. (1966). The development of cytoplasmic bridges in *Volvox aureus. Canad. Jour. Bot.* **44**, 1697–1702.
Boveri, T. (1899). "Die Entwicklung von *Ascaris megalocephala* mit besonder Rucksicht auf die Kernverhaltnisse". Festchr, F. C. Von Kupffer, Jena.
Burr, F. A., and McCracken, M. D. (1973). Existence of a surface layer on the sheath of *Volvox. Jour. Phycol.* **9**, 245–346.
Chiang, K. S. and Sueoka, N. (1967). Replication of chloroplast DNA in *Chlamydomonas reinhardi* during vegetative cell cycle: its mode and regulation. *Proc. Nat. Acad. Sci. U.S.* **57**, 1506–1513.
Darden, W. H. (1966). Sexual differentiation in *Volvox aureus. J. Protozool.* **13**, 239–255.
Darden, W. H. (1970). Hormonal control of sexuality in the genus *Volvox. Ann. N.Y. Acad. Sci.* **175**, 757–763.
Davidson, E. (1968) "Gene Activity in Early Development" Academic Press, New York.
Ely, T. H. and Darden, W. H. (1972). Concentration and purification of the male-inducing substance from *Volvox aureus* M5. *Microbios* **5**, 51–56.
Geyer – Duszynska, I. (1966). Genetic factors in oogenesis and spermatogenesis in Cecidiomyideae. *Chromosomes Today* **1**, 174–190.

Griffin, B. E., and Huskey R. J. (1974). Genetic control of differentiation in *Volvox*. *Genetics* **77**, Suppl. s27. (abstract).

Gurdon, J. B. (1963). Nuclear transplantation in amphibia and the importance of stable nuclear changes in promoting cellular differentiation. *Quart. Rev. Biol.* **38**, 54–78.

Gurdon, J. B., and Woodland, H. (1968). The cytoplasmic control of nuclear activity in animal development. *Biol. Rev.* **43**, 233–267.

Hutt, W. (1972). Ultrastructural and experimental studies of fertilization in *Volvox carteri*. Ph.D. thesis, University of Georgia, Athens, Georgia.

Hutt, W., and Kochert, G. (1971). Effects of some protein and nucleic acid synthesis inhibitors on fertilization in *Volvox carteri*. *J. Phycol.* **7**, 316–320.

Janet, C. (1923). "Le *Volvox*. Troisieme memorie." Protat Freres. Macon.

Jost, J. P., and Rickenberg, H. V. (1971). Cyclic AMP. *Ann. Rev. Biochem.* **771**.

Kelland, J. (1964). Inversion in Volvox. Ph.D. thesis, Princeton University, Princeton, New Jersey.

Kochert, G. (1968). Differentiation of reproductive cells in *Volvox carteri*. *J. Protozool.* **15**, 438–452.

Kochert, G. (1971). Ribosomal RNA synthesis in *Volvox*. *Arch. Biochem. Biophys.* **147**, 318–322.

Kochert, G. and Jaworski, A. J. (1972). Isolation and characterization of *Volvox* DNA. *Jour. Phycol.* **8** suppl. 16. (abstract).

Kochert, G. and Olson, L. W. (1970). Ultrastructure of *Volvox carteri*. I. The asexual colony. *Arch. Mikrobiol.* **24**, 19–30.

Kochert, G. and Sansing, N. G. (1971). Isolation and characterization of nucleic acids from *Volvox carteri*. *Biochem. Biophys. Acta* **238**, 397–405.

Kochert, G., and Yates, I. (1970). A UV-labile morphogenetic substance in *Volvox carteri*. *Devel. Biol.* **23**, 128–135.

Kochert, G. and Yates, I. (1974). Purification and partial characterization of a glycoprotein sexual inducer from *Volvox carteri*. *Proc. Nat. Acad. Sci.* U.S. **71**, 1211–1214.

Lamport, D. T. A. (1969). The isolation and partial characterization of hydroxyproline-rich glycopeptides obtained by enzymic degradation of primary cell walls. *Biochemistry* **8**, 1155–1163.

Lewin, B. (1970). Molecular basis of gene expression. Wiley Interscience. New York.

Macgregor, H. (1972). The nucleolus and its genes in amphibian oogenesis. *Biol. Rev.* **47**, 177–206.

McCracken, M. D., and Starr, R. C. (1970). Induction and development of reproductive cells in the K-32 strains of *Volvox rousseletii*. *Arch. Protistenk.* **112**, 262–282.

O'Malley, B. W., Spelsberg, T., Schrader, W., Chytel, F. and Steggles, A. (1972). Mechanisms of interaction of a hormone-receptor complex with the genome of a eukaryotic target cell. *Nature (London)* **235**, 141–146.

Pall, M. (1973). Sexual induction in *Volvox carteri*: a quantitative study. *J. Cell Biol.* **59**, 238–241.

Pickett-Heaps, J. D. (1969). Some ultrastructural features of *Volvox* with particular reference to the phenomenon of inversion. *Planta* (Berl) **90**, 174–190.

Pocock, M. A. (1933). *Volvox* in South Africa. *Ann. South Afr. Mus.* **16**, 523–646.

Powers, J. H. (1908). Further studies in *Volvox*, with descriptions of three new species. *Trans. Amer. Microsc. Soc.* **27**, 123–149.

Sessoms, A. and Huskey, R. J. (1973). Isolation and characterization of morphogenetic mutants of *Volvox carteri* f. *nagariensis*. *Proc. Nat. Acad. Sci.* U.S. **70**, 1640–1644.

Smith, G. M. (1944). A comparative study of the species of *Volvox*. *Trans. Amer. Microsc. Soc.* **63**, 265–310.

Smith, L. D. (1966). The role of a "germinal plasm" in the formation of primordial germ cells in *Rana pipiens. Develop. Biol.* **14**, 330–347.

Soyer, M. (1973). Complement à l'etude ultrastructurale des Volvocales. Etude des colonies femelles de *Volvox aureus* E. *Ann. Sci. Nat. Zool.* Paris 15, 231–258.

Starr, R. C. (1969). Structure, reproduction, and differentiation in *Volvox carteri* f. *nagariensis Iyengar*, strains HK9 & HK10. *Arch. Protistenk.* **111**, 204–222.

Starr, R. C. (1970). Control of differentiation in *Volvox. Develop. Biol. Suppl.* **4**, 59–100.

Starr, R. C. (1972). A working model for the control of differentiation during development of the embryo of *Volvox carteri* f. *nagariensis. Soc. Bot. Fr.*, memoires 1972, 175–182.

Starr, R. C. and Jaenicke, L. (1974). Purification and characterization of the hormone initiating sexual morphogenesis in *Volvox carteri* f. *nagariensis. Proc. Nat. Acad. Sci.* U.S. **71**, 1050–1054.

Tobler, H., Smith, K. D. and Ursprung, H. (1972). Molecular aspects of chromatin elemination in *Ascaris lumbricoides. Develop. Biol.* **27**, 190–203.

VandeBerg, W. J., and Starr, R. C. (1971). Structure, reproduction and differentiation in *Volvox gigas* and *Volvox powersii. Arch. Protistenk.* **113**, 195–219.

Weinheimer, E. W. (1973). Nucleic acid and protein metabolism in the asexual life cycle of Volvox carteri. Ph.D. thesis, University of Georgia, Athens, Georgia.

Weismann, A. (1889). The continuity of the germ-plasm as the foundation of heredity. *In* "Essays upon Heredity" (Poulton, E. B., Schonland, S., and Shipley, A. E., eds.), Clarendon Press, Oxford.

Yates, I. (1974). Nucleic acid and protein metabolism in synchronized cultures of *Volvox carteri*. Ph.D. thesis, University of Georgia, Athens, Georgia.

II. Parthenogenesis

Teratocarcinogenesis and Spontaneous Parthenogenesis in Mice

Leroy C. Stevens

The Jackson Laboratory,
Bar Harbor, Maine 04609

I.   Introduction 93
II.   Teratocarcinogenesis in Male Primordial Germ Cells 94
      A.   Genetic Influences on Teratocarcinogenesis 95
      B.   Development of Teratomas from Primordial Germ Cells 95
      C.   Experimental Production of Teratomas from Primordial
           Germ Cells 96
III. Embryoid Bodies Derived from Transplantable Teratomas 97
IV.   Spontaneous Ovarian Teratocarcinogenesis 99
V.   Embryo-derived Teratomas 99
VI.   Spontaneous Parthenogenesis 101
      A.   Implantation of Spontaneous LT Parthenotes 102
VII. Summary and Conclusions 104
      References 105

I. INTRODUCTION

Parthenogenesis may be regarded as the development of an embryo from an egg without genetic contribution from a male gamete (Beatty, 1967). It may occur either before or after ovulation. In the latter case, the embryo (parthenote; Webster, 1961) may implant in the uterus where it may survive as long as about 8 days. The best documented example of spontaneous parthenogenesis occurring within the ovary is described for LT mice (Stevens and Varnum, 1974) in which parthenotes may develop to the blastocyst, egg cylinder, or primitive streak stage. Subsequent growth of ovarian parthenotes is disorganized and they give rise to teratomas (or teratocarcinoma or embryonal carcinoma, depending upon their histological composition and growth potential). Thus, in LT mice, parthenogenesis has been shown to be the initial stage of teratocarcinogenesis.

Spontaneous teratomas in male strain 129 mice also originate from germ

93

cells, but at a much earlier stage of development than in females (Stevens, 1967).

Parthenogenesis and teratocarcinogenesis are similar processes in that they both result in the activation of germ cells to develop without participation of cells of the opposite sex. Gaillard (1974) has recently reviewed the literature on parthenogenesis, embryogenesis, and teratocarcinogenesis in the ovaries and testes of human beings and other mammals.

In this article I will try to summarize current knowledge on teratocarcinogenesis and embryogenesis in the gonads of male and female mice, and spontaneous parthenogenesis in females.

II. TERATOCARCINOGENESIS IN MALE PRIMODIAL GERM CELLS

Teratocarcinogenesis in the testes of mice is extremely rare. Testicular teratomas in mice were not described until 1954 (Stevens and Little). It was the fortuitous finding that they were a characteristic of an inbred strain that made it possible to study their development.

Spontaneous testicular teratomas were found in a small percentage of strain 129 males. In adults they were composed of many kinds of tissues derived from all three germ layers (Fig. 1) (Stevens and Hummel, 1957). Usually the tumors were benign, but many have been maintained as progressively growing transplantable teratomas (Stevens, 1958). The undifferentiated embryonic stem cells of these tumors (Fig. 2) are pluripotent and can give rise

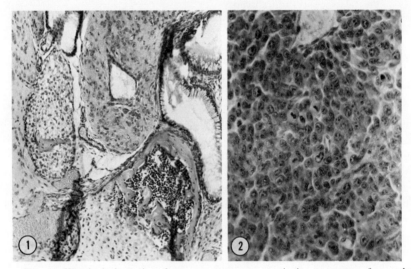

Fig. 1. Histological section from a spontaneous testicular teratoma of an adult showing notochord with a spicule of bone (left); neural tissue, cartilage, and bone with marrow (middle); and an epithelial cyst (right).

Fig. 2. Undifferentiated embryonic cells in a teratoma of an adult.

to many kinds of differentiated tissues, as well as to more embryonic cells like themselves (Kleinsmith and Pierce, 1964).

A. Genetic Influences on Teratocarcinogenesis

Attempts were made to increase the incidence of teratomas in strain 129 mice so that a developmental study could be made. The gene yellow (A^y) increases the incidence of several kinds of tumors in mice (Heston, 1965), and for this reason it was introduced onto the strain 129 genetic background. Rather than increasing the incidence of teratomas, the A^y/+ mice had only one-tenth as many as their +/+ littermates.

The gene steel (Sl^J) was introduced onto the genetic background of strain 129 (Stevens and Mackensen, 1961) because in the homozygous state it affects the development of primordial germ cells. It was hoped that in the heterozygous viable state it would change the primordial germ cells in such a way as to make them more susceptible to teratocarcinogenesis. A new inbred subline of strain 129, strain 129/Sv-Sl^J C P, was established and Sl^J/+ males had twice as many teratomas as their +/+ littermates. The Sl^J gene strongly influences susceptibility to teratocarcinogenesis, but we do not yet know how. About 5% of the males of this subline had spontaneous teratomas, and we used it to perform the developmental study to be described below.

Another gene, as yet unidentified, is also involved in susceptibility to teratocarcinogenesis. A pair of mice of a subline of strain 129 produced 38 male offspring, and 8 of them had testicular teratomas. This is a high incidence when compared with 3% for the closely related 129 mice. A colony was established from the descendants of this pair in which about one-third of the males have spontaneous teratomas (Stevens, 1973). This new inbred subline was designated 129/terSv.

B. Development of Teratomas from Primordial Germ Cells

The relatively high incidence of teratomas in strain 129/Sv-Sl^J C P mice made it feasible to conduct a developmental study of these tumors. The tissues in tumors of young mice appeared immature. For example, in mice 4 days old immature neuroepithelium could be recognized (Fig. 3). In mice 2 or 3 days old, the tumors were composed of disorganized ectodermal and endodermal epithilia with mesodermal cells between and around them (Fig. 4) (Stevens, 1959). In 18-day fetuses (the day before birth), the tumors were composed of clumps of undifferentiated embryonic cells (Fig. 5). There were cavities lined by layers of cells that could not yet be identified as ectoderm or endoderm.

The earliest recognizable teratomas were observed in 15-day fetuses. They were composed of small clusters of embryonic cells and were located within the seminiferous tubules (Fig. 6). This suggested that the tumors were derived from a component of the seminiferous tubules, and the primordial germ cells seemed more likely to be the cell of origin than the cells that would normally mature into Sertoli cells (Stevens, 1962).

We were able to obtain genetic evidence to support the hypothesis that primordial germ cells gave rise to testicular teratomas, but not until after we had developed a method of experimentally inducing them.

C. Experimental Production of Teratomas from Primordial Germ Cells

By comparing the size of teratomas in 15- to 19-day fetal mice, we estimated that they originated at about 12 days of gestation. Many unsuccessful attempts were made experimentally to induce teratomas in offspring of females that were subjected to various chemical and physical stimuli during the twelfth day of gestation.

Genital ridges were dissected from 12-day fetuses and subjected directly to various experimental conditions. One of these conditions resulted in the induction of teratomas. When strain 129 male genital ridges were dissected from 12-day fetuses and grafted to the testes of adults, they developed into testes, most of which had teratomas. When these genital ridges were grafted into sites that did not descend into the scrotum, they developed into testes, and few of them had teratomas (Stevens, 1964; 1970a). All available evidence indicates that it is the lower temperature in the host scrotum that induces teratomas in testes that develop from genetically susceptible grafted genital ridges.

About 7 days after grafting a 12-day genital ridge, small clumps of embryonic cells were found within the seminiferous tubules indicating that they, like the spontaneous tumors, originated from either primordial germ cells or from Sertoli cells. These cells proliferated, ruptured the seminiferous tubules, and spread through the interstitial regions of the testes. They formed the primary germ layers which gave rise to many types of immature tissues. The development of experimentally induced teratomas resembled in every way the development of spontaneous teratomas.

This simple method of experimentally inducing teratomas made it feasible to attack the question of the cell of origin by genetic means (Stevens, 1967). Mice homozygous at the steel (Sl^J) locus are sterile because they have very few or no primordial germ cells. Matings of $Sl^J/+$ x $Sl^J/+$ mice produce sterile (Sl^J/Sl^J) and fertile $(Sl^J/+$ and $+/+)$ offspring. When genital ridges from 12-day sterile fetuses were grafted to the testes of adults, they developed into testes without teratomas. When genital ridges from 12-day fertile fetuses were similarly grafted, they developed into testes, and 75% of them had teratomas. This genetic evidence supports the morphological evidence that teratomas originate from primordial germ cells.

The activation of male germ cells and their development into teratomas closely parallels the parthenogenetic activation of ovarian eggs to be described below. Both result in the formation of embryonic primary germ layers which differentiate into many of the kinds of tissues found in immature and adult animals. It will be seen that early stages of ovarian teratomas actually resemble early stages of normal embryonic development including cleavage stages, blastocysts, egg cylinders, and occasionally primitive streak stages. Early stages of testicular teratomas do not mimic normal development so closely, but when they are maintained as transplantable tumors they form structures that morphologically resemble egg cylinders. These structures have been called embryoid bodies.

III. EMBRYOID BODIES DERIVED FROM TRANSPLANTABLE TERATOMAS

Peyron (1939) first discovered structures (boutons embryonaire) in human teratomas that resembled early embryos. Since then many workers have identified embryoid bodies in teratomas of the human ovary and testis (see Marin-Padilla, 1965; and Gaillard, 1974).

Embryoid bodies occur in some transplantable testicular teratomas in mice (Pierce and Dixon, 1959a, 1959b; Pierce, 1961; Stevens, 1959, 1960). When solid transplantable teratomas are grafted intraperitoneally, some of them produce hundreds of thousands of free floating structures that resemble normal mouse embryos. They may be derived from sloughing from the partially necrotic surface of peritoneal implants of teratocarcinoma or they may arise from preexisting embryoid bodies that divide in half.

During early transplant generations, embryoid bodies consist of many types of cells. They all have an outer envelope of embryonic endoderm. Within the envelope there may be a layer of ectoderm. Other cell types in embryoid bodies include undifferentiated embryonic cells, neural, mesenchymal, hematopoietic, and muscle cells.

After several transplant generations the embryoid bodies usually become more simple in structure. They may be composed of a layer of endodermal cells surrounding a layer of ectodermal or undifferentiated cells (Fig. 7) and bear a strong resemblance to normal 6-day embryos (Stevens, 1959). When these embryoid bodies are grafted singly into the anterior chamber of the eye (Stevens, 1960) or other sites in adults, they develop into teratomas with several tissue types. This demonstrates that they resemble normal embryos in their developmental potency as well as in their morphology.

Kleinsmith and Pierce (1964) isolated single cells from embryoid bodies and grafted them into mice. From more than 1,700 single cell grafts, 44 clonal lines were obtained, and 43 of them were teratocarcinomas composed of as many as 14 differentiated benign somatic tissues in addition to embryonal carcinoma. This demonstrated that single embryonal carcinoma cells are pluripotent like normal embryonic cells.

98 LEROY C. STEVENS

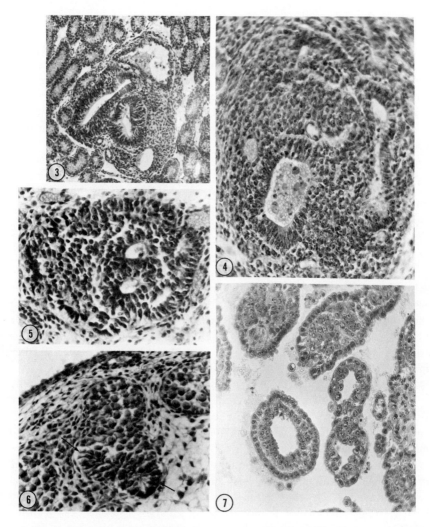

Fig. 3. Immature neuroepithelium in a teratoma of a 4-day mouse.

Fig. 4. Disorganized ectodermal (mid-left) and endodermal (right) epithelia and undifferentiated pluripotent embryonic cells between and around them.

Fig. 5. Undifferentiated cells with primitive epithelium in teratoma of an 18-day fetus.

Fig. 6. Teratoma (arrows) within seminiferous tubule of a 16-day fetus.

Fig. 7. Embryoid bodies in ascitic fluid of a mouse with a transplantable teratoma. Note outer layer of endodermal cells and inner undifferentiated cells.

Pierce and Beals (1964) and Pierce et al. (1967) examined the ultrastructure of embryonal carcinoma cells from embryoid bodies and early stages of primary teratomas, and found them to be similar to that of primordial germ cells in 15-day fetuses. This supports the evidence presented above that embryonal carcinoma of the testis is derived from primordial germ cells in mice.

IV. SPONTANEOUS OVARIAN TERATOCARCINOGENESIS

Fortuitously, we found that spontaneous parthenogenesis occurs remarkably frequently in the ovaries of an obscure inbred strain of mice (Stevens and Varnum, 1974). About half of strain LT females have ovarian teratomas at three months of age, and they are derived from ovarian eggs that behave as if they were fertilized. All stages from cleavage, blastocyst, egg cylinder, to primitive streak stages were observed in the ovary (Figs. 8–12). Like the uterine implanted parthenotes to be described here and in the next chapter by Dr. Tarkowski, most ovarian parthenotes become disorganized at about 6 days of gestation. However, rather than being aborted as are uterine parthenotes, they remain in the ovary as disorganized growths (Fig. 13). Undifferentiated cells derived from these parthenotes spread through the ovary and either continue to proliferate or differentiate into normal tissues of many varieties such as: nerve, muscle, many types of epithelia, cartilage, bone, lens, and teeth.

The stem cells of ovarian as well as testicular teratomas are undifferentiated proliferating embryonic cells. In most cases, they finally differentiate, and the tumors become benign. Occasionally, some cells remain undifferentiated and the tumors continue to grow. These may be maintained indefinitely as transplantable ovarian teratocarcinomas.

It has been shown that teratomas originate from male and female germ cells that develop into disorganized populations of cells resembling those in normal 5- to 6-day embryos. This raised the question - what would be the developmental fate of normal embryonic cells if they were artificially disorganized?

V. EMBRYO-DERIVED TERATOMAS

When 2- to 6-day mouse embryos are grafted into ectopic sites in adults, they become disorganized and develop into growths composed of many kinds of tissues (Stevens, 1968, 1970; Solter et al., 1970; Skreb et al., 1972; Damjanov and Solter, 1974). In about half of the grafts, some of the cells remain undifferentiated for remarkably long periods of time. These cells are pluripotent and they may continue to proliferate indefinitely and serve as stem cells of transplantable teratocarcinomas. These tumors can be trans-

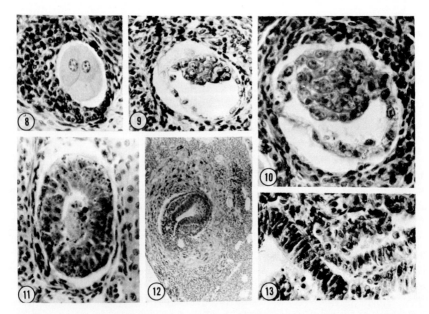

Fig. 8. Four-cell egg in the ovary of an LT mouse.

Fig. 9. Blastocyst in ovary of an LT mouse.

Fig. 10. Blastocyst in the ovary of an LT mouse. Note endoderm forming on ventral surface of the inner cell mass.

Fig. 11. Egg cylinder in the ovary of an LT mouse. Note cell debris in the proamniotic cavity.

Fig. 12. Unusually well developed embryo in ovary of an LT mouse. Note trophoblastic giant cells surrounding Reichert's membrane, proximal endoderm, mesoderm, and ectoderm. The primitive streak is at the upper pole of the embryo, the ectoplacental cone at the lower.

Fig. 13. Disorganized growth of ectodermal cells and undifferentiated cells in an ovarian teratoma of an LT mouse.

planted serially, and they behave like transplanted teratomas derived from primary testicular and ovarian teratomas.

Teratomas originate from disorganized populations of pluripotent embryonic cells. For spontaneous gonadal teratomas this population is derived from germ cells. For embryo-derived tumors this population is derived from grafted embryos.

The development of teratomas from grafted early embryos may be explained in two different ways. They may develop directly from undetermined embryonic cells. Alternatively, the grafted embryos may give rise to cells which develop characteristics of primordial germ cells, and it is these that give rise to teratomas. If the tumors are derived from cells like primordial germ cells, they should all be male, since teratomas have never been induced from grafted female primordial germ cells. If, on the other hand, undifferentiated embryonic cells are directly involved in a neoplastic change, half of the tumors should be male and half female. The chromosomal sex was determined in embryo derived teratomas, and both sexes were found (Dunn and Stevens, 1970).

It should be emphasized that embryo-derived teratomas were produced by artificial means, and there is no evidence that disorganized populations of pluripotent embryonic cells emanate from cells other than those that would normally give rise to eggs or spermatozoa. In females, the first stage of teratocarcinogenesis is spontaneous parthenogenesis within the ovary. It was of interest to find that even after ovulation, spontaneous parthenogenesis occurred in a small percentage of strain LT mice.

VI. SPONTANEOUS PARTHENOGENESIS

Spontaneous parthenogenesis in mice, or any other mammal, would be extremely difficult to detect unless it were a commonly occurring event. A parthenogenetic individual would be hemizygous if haploid, or homozygous if diploid, and deleterious recessive genes affecting viability would express themselves. The most likely place to look for spontaneous parthenogenesis would be in inbred animals which are assumed to be homozygous at all genetic loci except those on the sex chromosomes of males. During inbreeding, recessive lethal genes are eliminated by the inevitable selection for viability and fertility. If all recessive lethals were eliminated, spontaneous parthenogenetic development might be expected to produce viable individuals in inbred mice. In fact, until recently, the only reports on parthenogenetic development of mouse eggs have been those discussed by Dr. Tarkowski in this symposium, and they all involved activation by artificial means. The most advanced of these induced parthenotes were at about the same stage of development as a normal 8-day embryo, and most of the others were several days younger. Apparently even parthenotes in inbred mice have a limited life span. The cause of death is not yet known. Possibly balanced lethal genes are involved.

Solter et al. (1974) found several ultrasturctural differences between cleavage stages of normal and parthenogenetic mouse eggs. They suggested that the reasons for developmental arrest of parthenotes are numerous and complex and exist from the very beginning.

Even though parthenogenetically derived mouse embryos have a limited life span as individuals, their component cells are capable of prolonged survival. If parthenogenesis occurs in the ovary, for example, the resulting growths survive as teratomas for the life span of the host. In fact, some ovarian teratomas have been established as transplantable tumors (Stevens and Varnum, unpublished), and they have an indefinite life span. This capacity for continuous growth demonstrates that parthenogenetically derived cells have the genetic equipment necessary for growth and survival.

A. Implantation of Spontaneous LT Parthenotes

During post-mortem examination of virgin LT females (caged in groups of 5), about 10% of them are found to be pregnant (Stevens and Varnum, unpublished). Usually, only one or two decidual swellings are observed in these cases of spontaneous parthenogenesis. Most LT parthenotes develop to the stage of about 6 days of gestation (Figs. 14-16), but then they become disorganized and die.

Occasionally, parthenotes survive beyond the 6-day stage (Figs. 17-20). They form trophoblast, Reichert's membrane, yolk sac (without blood islands), ectoderm, primitive streak, and amnion. The most advanced we have observed had died a day or so before fixation, but a wavy neural tube and several somites could be recognized.

Strain LT females were superovulated and mated with vasectomized males. The day after ovulation, eggs were flushed from the oviducts and examined. Most females had at least one or two 2-cell parthenotes. The parthenotes were cultured in Whitten's medium (1971) until the blastocyst stage.

Nine cultured blastocysts were transferred to the uterus of a pseudopregnant host. Six days later the host was killed, and there were 7 decidual swellings. The decidua were dissected from the uterus and examined histologically. One contained only a remnant of Reichert's membrane, but the

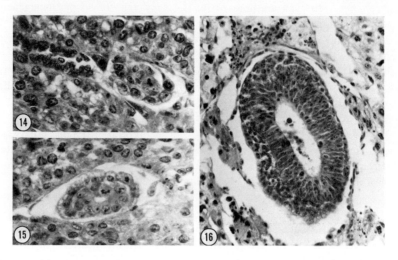

Figs. 14-16. Parthenotes implanting in the uteri of LT mice. The primitive streak is developing at the top of Fig. 16.

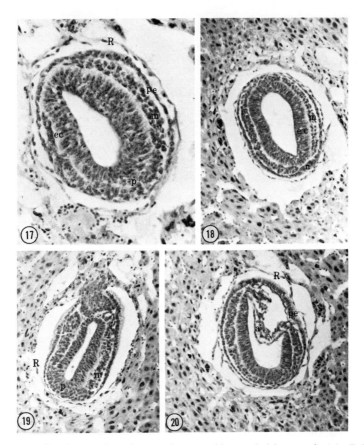

Figs. 17-20. Post-implantation parthenotes. Note primitive streak (p), Reichert's membrane (R), proximal endoderm (pe), ectoderm (ec), notochord (n), mesoderm (m), and amnion (a).

others contained healthy appearing embryos of about 7 days of gestation. There were trophoblastic giant cells attached to Reichert's membrane. Proximal endoderm was continuous with distal endoderm in the region of the ectoplacental cone. Mesodermal cells had migrated from the primitive streak between the ectoderm and endoderm. The ectodermal cells were oriented around a proamniotic cavity. There appeared to be a deficiency of ectoplacental cone and trophoblastic cells.

It is possible that the prolonged survival of this comparatively large litter of parthenotes was due to a uterine environment more favorable than when only one or two parthenotes were present.

The development of teratomas from parthenogenetically activated ovarian mouse eggs could not have been demonstrated until an inbred strain with a high incidence of these tumors was found. These tumors are extremely rare in other strains, and it would have been impossible to identify all stages from cleaving eggs, to egg cylinders, to teratoma formation. The chance finding that strain LT mice had this uncommon characteristic made the developmental study possible.

VII. SUMMARY AND CONCLUSIONS

Teratocarcinogenesis occurs spontaneously in the testes of inbred strain 129 mice. At least three genes are involved. Teratomas originate during the twelfth day of gestation from primordial germ cells that begin to develop spontaneously. The earliest identifiable stage is a disorganized population of pluripotent undifferentiated embryonic cells. These cells give rise to more cells like themselves and also to the three primary germ layers that differentiate into many kinds of immature and adult tissues.

Teratomas may be experimentally induced in some strains of mice by grafting 12-day male genital ridges to sites that descend into the scrotum. The male grafts develop into testes and most of them have teratomas. Apparently the low temperature in the scrotum initiates teratocarcinogenesis.

Teratomas may also be experimentally produced by grafting 2- to 6-day embryos to the testes and other sites in adults.

The early stages in the development of testicular teratomas does not closely mimic normal embryonic development. However, some transplantable testicular teratomas produce thousands of embryoid bodies that resemble normal mouse embryos. They are composed of an outer layer of endoderm and a core of pluripotent undifferentiated embryonic cells.

Teratomas occur spontaneously in the ovaries of about half of strain LT females. They are derived from ovarian eggs that begin to develop parthenogenetically. They closely mimic normal embryonic development as far as the egg cylinder stage, but then most of them become disorganized, and embryonal cells invade the ovary. Occasionally they attain the primitive streak stage.

About ten per cent of the eggs of strain LT virgin mice begin to develop parthenogenetically after ovulation. They cleave in the oviduct and implant in the uterus. Most of them develop quite normally for 5 or 6 days, but then they become disorganized and are aborted. Primitive streak stages and somite formation have been observed, but they are rare.

Parthenogenesis is the first stage of teratocarcinogenesis in females. Testicular teratomas are also derived from germ cells by a process that may be akin to parthenogenesis.

ACKNOWLEDGMENTS

I am deeply indebted to Don S. Varnum for his expert professional assistance. The work reported here was supported in part by NIH Research Grant CA 02662 from the

National Cancer Institute; by the Ladies Auxiliary to the Veterans of Foreign Wars, Cancer Society of Greenville County (S.C.); and by the Al Rose Cancer Aid Society. The Jackson Laboratory is fully accredited by the American Association for Accreditation of Laboratory Animal Care.

REFERENCES

Beatty, R. A. (1967). Parthenogenesis in vertebrates. *In* "Fertilization" (C. B. Metz and A. Monroy, eds.), Vol. 1, pp. 413–440. Academic Press, New York.

Damjanov, I. and Solter, D. (1974). Host-related factors determine the outgrowth of teratocarcinomas from mouse egg cylinders. *Z. Krebforsch.* **81**, 63–69.

Dunn, G. R. and Stevens, L. C. (1970). Determination of sex of teratomas derived from early mouse embryos. *J. Nat. Cancer Inst.* **44**, 99–105.

Gaillard, J. A. (1974). Differentiation and organization in teratomas. *In* "Neoplasia and Cell Differentiation" (G. V. Sherbet, ed.), pp. 319–349. Karger, Basel.

Heston, W. E. (1965). Genetic factors in the etiology of cancer. *Cancer Res.* **25**, 1320–1326.

Kleinsmith, L. J. and Pierce, G. B. (1964). Multipotentiality of single embryonal carcinoma cells. *Cancer* **24**, 1544–1552.

Marin-Padilla, M. (1965). Origin, nature and significance of the "embryoids" of human teratomas. *Virchows Arch. path. Anat.* **340**, 105–121.

Peyron, A. (1939). Faits nouveaux relatifs a l'origine et a l'hitogenese des embryomes. *Bull. Ass. p. et. du Cancer* **28**, 658–681.

Pierce, G. B. (1961). Teratocarcinomas, a problem in developmental biology. *Canadian Cancer Conference* **4**, 119–137.

Pierce, G. B. and Beals, T. F. (1964). The ultrastructure of primordial germinal cells of the fetal testes and of embryonal carcinoma cells of mice. *Cancer* **24**, 1553–1567.

Pierce, G. B. and Dixon, F. J. (1959a). Testicular teratomas I. Demonstration of metamorphosis of multipotential cells. *Cancer* **12**, 573-583.

Pierce, G. B. and Dixon, F. J. (1959b). Testicular teratomas II. Teratocarinomas as an ascites tumor. *Cancer* **12**, 584–589.

Pierce, G. B., Stevens, L. C., and Nakane, P. K. (1967). Ultrastructural analysis of the early development of teratocarcinomas. *J. Nat. Cancer Inst.* **39**, 755–773.

Skreb, N., Damjanov, I., and Solter, D. (1972). Teratomas and teratocarcinomas derived from rodent egg-shields. *In* "Cell Differentiation" (R. Harris, P. Alin, and D. Viza, eds.), pp. 151–155. Munksgaard, Copenhagen.

Solter, D., Damjanov, I., and Skreb, N. (1971). Teratocarcinogenesis as related to the age of embryos grafted under the kidney capsule. *Wilhelm Roux' Archiv* **167**, 288–290.

Stevens, L. C. (1958). Studies on transplantable testicular teratomas of strain 129 mice. *J. Nat. Cancer Inst.* **20**, 1257–1275.

Stevens, L. C. (1959). Embryology of testicular teratomas in strain 129 mice. *J. Nat. Cancer Inst.* **23**, 1249–1295.

Stevens, L. C. (1960). Embryonic potency of embryoid bodies derived from a transplantable testicular teratoma of the mouse. *Develop. Biol.* **2**, 285–297.

Stevens, L. C. (1962). Testicular teratomas in fetal mice. *J. Nat. Cancer Inst.* **28**, 247–267.

Stevens, L. C. (1964). Experimental production of testicular teratomas in mice. *Proc. Nat. Acad. Sci. U.S.* **52**, 654–661.

Stevens, L. C. (1967). Origin of testicular teratomas from primordial germ cells in mice. *J. Nat. Cancer Inst.* **38**, 549–552.

Stevens, L. C. (1968). The development of teratomas from intratesticular grafts of tubal mouse eggs. *J. Embryol. Exp. Morph.* **20,** 329–341.

Stevens, L. C. (1970a). Environmental influences on experimental teratocarcinogenesis in testes of mice. *J. Exp. Zool.* **174,** 407–414.

Stevens, L. C. (1970b). The development of transplantable teratocarcinomas from intratesticular grafts of pre- and postimplantation mouse embryos. *Develop. Biol.* **21,** 364–382.

Stevens, L. C. (1973). A new inbred subline of mice (129/terSv) with a high incidence of spontaneous congenital testicular teratomas. *J. Nat. Cancer Inst.* **50,** 235–242.

Stevens, L. C. and Hummel, K. P. (1957). A description of spontaneous congenital testicular teratomas in strain 129 mice. *J. Nat. Cancer Inst.* **18,** 719–747.

Stevens, L. C. and Little, C. C. (1954). Spontaneous testicular teratomas in an inbred strain of mice. *Proc. Nat. Acad. Sci. U.S.* **40,** 1080–1087.

Stevens, L. C. and Mackensen, J. A. (1961). Genetic and environmental influences on teratocarcinogenesis in mice. *J. Nat. Cancer Inst.* **27,** 443–453.

Stevens, L. C. and Varnum, D. S. (1974). The development of teratomas from parthenogenetically activated ovarian mouse eggs. *Develop. Biol.* **37,** 369–380.

Webster (1961). See Webster's Third New International Dictionary.

Whitten, W. K. (1971). Nutrient requirements for the culture of preimplantation embryos *in vitro. Adv. Biosci.* **6,** 129–139.

Induced Parthenogenesis in the Mouse

Andrzej K. Tarkowski

Department of Embryology
Institute of Zoology
University of Warsaw, 00-927/1
Warszawa, Poland

I. Introduction 107
II. Initiation of Parthenogenesis 108
  A. Types of Reaction of Eggs
    to Parthenogenetic Agents 108
  B. Methods of Parthenogenetic Activation 111
  C. The Efficiency of Parthenogenetic Activation and the
    Frequency of Various Types of Reaction 111
  D. Efficiency of Activation 112
    1. Activating Agent 112
    2. Post-ovulatory Age of Eggs 113
    3. Genetic Constitution 113
    4. Origin of Eggs: Spontaneous vs. Induced Ovulation 118
  E. Frequency of Various Types of Reaction 118
    1. Activating Agent 118
    2. Post-ovulatory Age of Eggs 119
    3. Genetic Constitution 119
    4. Origin of Eggs: Spontaneous vs. Induced Ovulation 120
III. Concluding Remarks 120
IV. Development of Parthenogenetic Embryos 121
  A. Preimplantation Development 121
    1. Cortical and Zona Reactions 121
    2. The Pronucleate Stage and Early Development 122
  B. Post-implantation Development 123
V. Possible Causes of Death of Parthenogenetic Embryos 125
  A. Mortality After Implantation 126
  References 127

I. INTRODUCTION

Interest in artificial activation of mammalian eggs to induce parthenogenetic development dates back to the 1930's when Pincus and his co-workers began to experiment with rabbit oocytes and early embryos. This pioneering work culminated in articles by Pincus and Shapiro reporting the

107

birth of a number of parthenogenetic rabbits. In the decade 1947-1957 the subject was again taken up by Chang in the USA and by Thibault in France both of whom confirmed the production of parthenogenetic rabbit blastocysts but they failed to obtain development beyond implantation. At the same time Austin and Braden in Australia, and subsequently in Great Britain, carried out similar experiments on rat, mouse, and hamster eggs but did not observe development beyond the 4-cell stage. The work carried out up until the late fifties has been fully and critically reviewed by Beatty (1957, 1967, 1972), Austin and Walton (1960), and Graham (in press) and will not be described in detail in this review.

After 1957 practically no work was done in this field for over a decade. The fading interest was probably due to the lack of success in obtaining more advanced rodent embryos and to the failure in obtaining postimplantation development in the rabbit. Graham (1970) and Tarkowski et al., (1970) reported success in attempts to produce parthenogenetic mouse blastocysts. Tarkowski and co-workers showed in addition that parthenogenetic blastocysts can implant and occasionally develop until mid-pregnancy, while Graham obtained from such blastocysts transplanted to the testis "growths" composed in one case of diploid and in the other of haploid cells. These two papers were followed during the next four years by nearly twenty reports all concerned with experimentally induced parthenogenesis in the mouse. This renewed interest in mammalian parthenogenesis is inspired, at least in part, by the possible applications of this material for developmental biology and genetics and for somatic cell genetics. In particular, parthenogenetic embryos may prove helpful in analyzing the essence of the activation process, in studying the effect of haploidy and highly homozygous diploidy on embryonic development and, finally, as a source of haploid cells for genetic studies.

The aim of this paper is to review the most recent work on experimental parthenogenesis in the mouse. Whenever necessary, studies on other species will also be discussed, but no attempt will be made to include in this review all references to earlier papers.

II. INITIATION OF PARTHENOGENESIS

A. Types of Reaction of Eggs to Parthenogenetic Agents

All the types of reaction that occur in response to parthenogenetic agents may also occur spontaneously as a result of fertilization. Of course, apart from normal reaction, i.e., extrusion of the second polar body and formation of a haploid female pronucleus, the incidence of these parthenogenetic type reactions is extremely low after normal fertilization (Braden, 1957).

The egg can react to artificial activation in one of the following ways (Fig. 1):

1. Extrusion of the second polar body and formation of a haploid pronucleus. This is a normal and most common reaction.

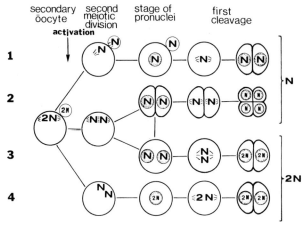

Fig. 1. Types of reaction of the egg to artificial activation. Numbers 1, 2, 3, 4 denote types of reaction described in the text. (From Tarkowski, 1971, modified).

2. Division of the oocyte into two cells of more or less equal size, each containing a haploid pronucleus. One cell corresponds to the egg and the other to the second polar body. Following Braden and Austin (1954) this phenomenon has been referred to in the literature as "immediate cleavage".

3. Suppression of the second polar body and formation of two haploid pronuclei. Eggs reacting along this route may develop either into diploid or into haploid embryos. In the first case the two pronuclei contribute chromosomes to one metaphase plate of the first cleavage (Witkowska, 1973a and unpublished observations; Kaufman, 1973c; Graham and Deussen, 1974), in the second oocyte undergoes, after a few hours delay, immediate cleavage as described under 2, which leads to haploidy (Graham, 1971).

4. Suppression of the second polar body and formation of a diploid pronucleus. Each of these routes of development may be accompanied by abnormalities (Fig. 2) such as the following: *a*. Formation besides the pronucleus (pronuclei) of tiny nuclei (subnuclei) derived from one or a few chromosomes which had detached from the main group. *b*. Unequal distribution of chromatids between the second polar body and a pronucleus or between the two pronuclei in the case of suppression of the second polar body or of immediate cleavage. *c*. Unequal division of the cytoplasm between the two cells in the process of immediate cleavage (often accompanied also by uneven distribution of chromosomes). Finally, formation of many nuclei of various sizes due to scattering of chromosomes.

Among treated eggs one usually obtained in addition to activated eggs, non-activated eggs with the plate of metaphase II intact or with scattered chromosomes, and degenerated eggs showing fragmentation, lysis, etc. The incidence of the two latter groups can be correlated with the strength of the stimulus — the stronger it is or the longer it acts the higher will be the

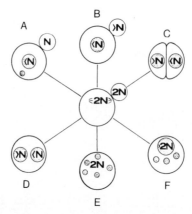

Fig. 2. Nuclear abnormalities observed in artificially activated eggs. *a*. Extrusion of the second polar body and formation, apart from a pronucleus, of a small subnucleus (subnuclei) derived from one or a few chromatids. The pronucleus is hypo-haploid unless the subnucleus originates from a set of chromatids which have moved to the second polar body. *b*. Uneven distribution of chromatids between the second polar body and pronucleus. The pronucleus may be either hypo-haploid or hyper-haploid. *c*. Uneven distribution of chromatids between the two pronuclei in the case of immediate cleavage. The resulting embryo is a hypo-haploid – hyper-haploid mosaic. *d*. Suppression of the second polar body and uneven distribution of chromatids between the two pronuclei. Diploidy is likely to be restored at first cleavage. *e*. Suppression of the second polar body and formation of many nuclei of varying chromatid content due to scattering of chromosomes of the second metaphase plate. *f*. Suppression of the second polar body and formation of hypo-diploid pronucleus and subnucleus (subnuclei).

percent of damaged eggs and the lower will be the percent of non-activated eggs. As "activated" one should consider only those eggs in which reformation of a pronucleus (pronuclei) has occurred. Eggs which extrude the second polar body but do not reform the pronucleus, or eggs undergoing division into many fragments should not be included in this category.

Most of the abnormalities which concern the nuclear apparatus (formation of subnuclei, unequal distribution of chromosomes) pass unnoticed when eggs are inspected in the living state and can be detected only in fixed and stained material or with the help of karyological examination. Knowledge of the character and the extent of these abnormalities is very important for the correct interpretation of the development of parthenogenetic embryos and, for this reason, karyological studies are indispensable.

Uneven distribution of chromosomes may not be of much importance when all products of the second meiosis remain in the egg as there is a chance that diploidy will be restored at the first cleavage. It will, however, result in aneuploidy – strictly speaking in hypo- and hyper-haploidy – when the second polar body is extruded or the egg undergoes immediate cleavage.

Hypo-haploidy is probably lethal in early cleavage and should be taken into account as one of the possible causes of mortality of parthenogenetic embryos in the preimplantation period (see section V).

B. Methods of Parthenogenetic Activation

The methods of activation which so far have proved effective and are in current use include: electric shock applied to the ampullar region of the oviduct (Tarkowski et al., (1970), heat shock applied to the oviduct in vivo (Braden and Austin, 1954) or in vitro (Komar, 1973), and removal of cumulus cells with hyaluronidase followed by culture in vitro (Graham, 1970). This last method can be combined with a hypotonic shock (Graham, 1972). (Only those papers in which a given method was employed for the first time are referred to). Activation of eggs inside the oviduct with electric current permits one to avoid culturing the eggs in vitro which – especially as far as the one-cell stage is concerned – remains very difficult. The disadvantage of this procedure is that the reaction of the eggs to the stimulus cannot be observed and consequently the original ploidy of the embryos is unknown. The method of heating the oviduct of an anesthetized mouse by taking it out of the body and immersing it in a water bath has the same disadvantage; in addition it is difficult to maintain precise control of the temperature. The latter difficulty can be overcome by applying the temperature shock in vitro (Komar, 1973).

The main advantage of all in vitro methods of activation is the possibility of observing the immediate reaction of eggs to the activating agent and of segregating the eggs according to the type of reaction. This makes it possible to compare the development of eggs displaying various types of reaction, for instance, haploids versus diploids. Eggs activated in vitro can be cultured in vitro to the blastocyst stage or, after a few hours when selection of those activated is possible, transplanted to the oviducts of recipient females. If the culture of one-cell eggs to blastocysts can not be mastered, then the latter procedure is recommended.

C. The Efficiency of Parthenogenetic Activation and the Frequency of Various Types of Reaction

Both the efficiency of parthenogenetic activation and the frequency of various types of reaction depend on four factors: 1. the type of parthenogenetic agent itself, 2. postovulatory age of eggs, 3. genetic constitution of eggs, and 4. type of ovulation, i.e., spontaneous versus induced with gonadotrophins.

The efficiency of activation is measured simply by the percent of eggs with a pronucleus or pronuclei. However, although simple, such an estimate is not very accurate as it does not necessarily express the ability of eggs to undergo further development. A much better measure of the efficiency of the parthenogenetic agent is the percent of treated eggs which develop to blastocysts. Unfortunately very little information of this kind is available.

Unless all but one of the above mentioned factors are the same, comparison of results of various experiments is difficult. Experimental conditions are often so different that speculations regarding the superiority or inferiority of any method in comparison with others is open to criticism, especially if the efficiency of a particular method is evaluated at the one-cell stage. With these reservations in mind the importance of each of the four factors mentioned above is discussed below, not only with reference to its efficiency of activation but also as to the types of reaction.

D. Efficiency of Activation

1. Activating agent. Table 1 presents data on the percent of eggs activated by means of various techniques. Only those experiments were selected in which the number of treated eggs approached or exceeded one hundred. Efficiency is expressed by the percent of pronucleate eggs. From the comparison of these data it can be concluded that the efficacy of all techniques is satisfactory since activation usually exceeded 50%. The very high or very low rate of activation obtained in some experiments may be due to the age factor or to genetic constitution rather than to the activating agent itself. The same agent may be potent in one strain and very inefficient in the other, or it may be effective only if aging oocytes are used.

There is some disagreement regarding the effect of hypotonic treatment on the rate of activation. According to Kaufman and Surani (1974) when eggs collected 20-21 hr post HCG are treated for 2 hr with culture medium diluted 4:1 or 3:2 with distilled water, the incidence of activated eggs remains the same as it is in control medium. Contrary to this are the results of the work of Graham and Deussen (1974) which show that hypotony may increase the percent of activated eggs and that eggs of various strains differ in this respect. The basis for the difference in results obtained by these two groups may be in the use of genetically different material; the use of slightly different culture media; to different pH, and last, but not least, to differences in the age of the eggs. The eggs used by Kaufman and Surani were collected as late as 20-21 hr post HCG and it may be that the maximal incidence of activation had already occurred in response to the hyaluronidase treatment itself.

Effectiveness as measured by the percent of multicellular morulae and blastocysts is more difficult to evaluate because of additional experimental variants, such as activation and development *in vivo* versus *in vitro*, contrasted with activation *in vitro* followed by transplantation to the oviduct. The highest percent of eggs developing parthenogenetically to blastocysts and the highest rate of implantation was observed when eggs were activated with electric current *in situ* and allowed to remain in the reproductive tract (Tarkowski *et al.,* 1970; Witkowska, 1973a) or activated *in vitro* with

INDUCED PARTHENOGENESIS IN THE MOUSE   113

hyaluronidase and hypotony and transplanted back into the oviduct (Kaufman and Gardner, 1974) (Tables 2 and 3). The low incidence of blastocysts in the experiments by Komar (1973), in which whole oviducts after being submitted to heat shock were kept in organ culture, can be partly explained by suboptimal culture conditions. This conclusion finds support in the fact that under the same culture conditions only 30.7% of fertilized eggs reached the blastocyst stage.

   2. Postovulatory age of eggs. The most precise data on the role of postovulatory age on the rate of activation have been presented by Kaufman (1973a) who activated eggs with hyaluronidase *in vitro*. He showed that eggs treated 14 hr after ovulatory injection of HCG do not react at all, and that the percent of activated eggs that had been treated at 16, 18, and 20 hr rises from 27 to 76.1 and to 83.9%, respectively. These figures express in precise terms the experience of other authors who note that activation becomes easier with aging of the eggs. The importance of the factor of age finds support also in observations by Kaufman and Surani (1974) who found that when cumulus-free eggs are subjected to hypotony almost none are activated at 15.25 hr while at 20-21 hr the activation rate is as high as 75%. Graham and Deussen (1974) do not present exact data on this topic but they state that 70 to 90% of F1C57BL/CBA and CBA eggs subjected to hypotony are activated at 16 hr post HCG or later, while at earlier times only 20 to 50% are activated. However, the older the eggs are the greater is the chance of their fragmentation, lysis or their inability to develop beyond the first cleavages. Newly ovulated eggs (13-16 hr post HCG) require stronger stimuli to become activated, e.g., electric shock (Tarkowski *et al.,* 1970; Witkowska, 1973a,b; Mintz and Gearhart, 1973), heat shock (Komar, 1973) or hyaluronidase treatment followed by osmotic shock in acid medium (Graham and Deussen, 1974).

   3. Genetic constitution. The relevant comparative data are provided by experiments carried out in the author's laboratory on spontaneously ovulated eggs of A and CBA origin activated by electric current *in situ* (Tarkowski *et al.,* 1970; Witkowska, 1973a) and by *in vitro* experiments of Graham and Deussen (1974). In the first case no differences between these two strains could be observed in the percent of pronucleate eggs. On the other hand, the observations of Graham and Deussen clearly show that the same treatment (hyluronidase with or without hypotony) applied to eggs of the same age produce strikingly different results depending on the strain (Table 1). For instance, a 2 hr incubation of C3H eggs in 3/5 White's medium led to 100% activation (62.5% of eggs subsequently underwent lysis) as compared to 78.6% activation of ICR/A eggs, and 15.3% of AKR eggs. Thus it seems that genetic differences play an important role in the susceptibility of eggs to artificial activation, or at least to some activating agents.

TABLE 1

Efficiency of various techniques of activation expressed as a percent of pronucleate eggs

Treatment	Strain	Type of ovulation I–induced S–spontaneous	No. of treated eggs	Overall % activation	Age of eggs post HCG or time of activation	References	Remarks
Hyaluronidase, culture in vitro	F_1 C57BL/ CBA $T_6 T_6$	I	397	66.5	not specified probably 24 – 29h	Graham, 1971	
	F_1 C57BL/ A2G	I	137	83.9	30	Kaufman, 1973a	only the age group with highest overall % activation is cited here
	C3H	I	122	100.0			5.3% of eggs subsequently lyse
Hyaluronidase, $^4/_5$ hypotonic medium for 2h, culture in vitro	C57BL	I	139	28.7	$15^{1/2}$ – 17		$^4/_5$ hypotonic medium = culture medium diluted 4:1 with distilled water
	AKR	I	120	5.0		Graham and Deussen, 1974	
Hyaluronidase, $^3/_5$ hypotonic medium, culture in vitro	ICR/A	I	215	78.6	$15^{1/2}$ – 17		$^3/_5$ hypotonic medium = culture medium diluted 3:2 with distilled water
	AKR	I	98	15.3			

Method	Strain		No.	%	Age/timing	Reference	Remarks
Hyaluronidase, culture *in vitro*	F₁ C57BL/ A2G	I	737	74.0			
Hyaluronidase, $^4/_5$ hypotonic medium, culture *in vitro*		I	107	73.8	20 – 21	Kaufman and Surani, 1974	
Hyaluronidase, $^3/_5$ hypotonic medium, culture *in vitro*		I	1633	75.2			
Electric shock *in vivo*	A	S	171	53.2	8:00–11:00 a.m. on the day of vaginal plug	Tarkowski *et al*, 1970	Combined results of activation with 30, 40, and 50V
	CBA	S	94	55.3		Witkowska, 973a	
	CBA	I	154	38.3	14 – 16	Tarkowski *et al*, 1970	
Heat shock (44 – 44.5°C) *in vitro*	Swiss albino A	I and S	121	57.0	14½–17½ or 8:00–11:00 a.m. on the day of vaginal plug	Komar, 1973	
Heat shock 44.0°C *in vitro*, Cytochalasin B, culture *in vitro*	Swiss albino	I	1057	73.0	13 – 16½	Balakier and Tarkowski, in preparation	

115

TABLE 2

Efficiency of various techniques of activation expressed as a percent of morulae and blastocysts

Treatment	Strain	Type of ovulation I–induced S–spontaneous	No. of treated eggs	Age of eggs post HCG(hrs) or time of activation	Day of examination	Morulae (≥8 cell) and blastocysts %	Blastocysts %	References
Electric shock *in vivo*	CBA	S	89[1] 112[2]	8:00–11:00 a.m. on the day of vaginal plug	4th	60.7 48.2	2.2 1.8	Witkowska, 1973a
	CBA	I	238[1] 277[2]		5th[3]	46.2 39.7	26.0 22.4	
	CBA	I	252[1] 315[2]	14 – 16	5th[3]	7.5 6.0	3.2 2.5	
Heat shock (44.5°C), *in vitro,* oviduct cultured *in vitro*	Swiss albino and A	S	145[1]	8:00–11:00 a.m. on the day of vaginal plug	5th	6.2	4.1	Komar, 1973
	Swiss albino	I	207[1]	14½ – 17½	5th	7.2	1.4	
Hyaluronidase, ⅗ and ⅘ hypotonic medium for 2h, transfer to oviduct within 6–9h of activation[4]	F₁ C57BL/ A2G	I	308[5] 410[6]	20 – 21	4 – 4½th	54.9 41.2	22.4 16.8	Kaufman and Gardner, 1974

[1] Number of eggs recovered.

[2] Presumed number of eggs at the time of activation calculated by multiplying the number of oviducts by the mean number of eggs per oviduct found in females autopsied on the first day of pregnancy. This calculation was carried out in view of the fact that the number of eggs per oviduct was lower on the 4th and 5th day than it was shortly after ovulation, thus suggesting that a number of eggs originally present in the oviduct must have undergone degeneration and were not recovered.

[3] Eggs were recovered from ligated oviducts.

[4] Only eggs with two pronuclei (potentially diploid) were transferred.

[5] Number of transferred eggs.

[6] ... number of treated eggs was calculated as follows: number of eggs transplanted $\times \dfrac{100}{75.1}$ (75.1 is an overall percent of activation in

116

TABLE 3

Efficiency of activation expressed as a percent of implantation

Treatment	Strain	Type of ovulation of donors I–induced S–spontaneous	No. of treated eggs	Age of eggs post HCG (hrs) or time of activation	Implantations %	Reference
electric shock *in vivo*	CBA	S	618[1]	8:00–11:00 a.m. on the day of vaginal plug	21.4	Witkowska, 1973 a and b
	A	S	230[1]		13.5	Tarkowski *et al.* 1970, Nowicka and Tarkowski, unpublished results
hyaluronidase, 3/5 and 4/5 hypotonic medium, transfer to recipients within 6–9h of activation	F[1] C57BL/A2G	I	509[2]	20 – 21	24.4[3]	Kaufman and Gardner 1974; Kaufman and Surani, 1974

[1] Presumed number of eggs at the time of activation calculated as follows: number of uterine horns x mean number of eggs per oviduct found in females autopsied on the 1st day of pregnancy.

For rationale of this calculation see foot-note 2 in Table 2.

[2] Presumed number of treated eggs calculated as explained in foot-note 6 in Table 2.

[3] Percentage of eggs able to implant is here underestimated because of the inevitable loss of some eggs as a result of transplantation.

117

4. Origin of eggs: spontaneous versus induced ovulation. Most experiments have been carried out on eggs resulting from induced ovulation, the only exception being the work from the author's laboratory. The observations of Tarkowski *et al.,* (1970) and Witkowska (1973a) speak clearly to the disadvantage of using eggs from hormonally induced ovulations. Fewer eggs undergo activation and, what is more important, very few develop to blastocysts (Tables 1 and 2). However, these observations are open to criticism in that hormonally treated females were not pseudopregnant as were those ovulating spontaneously. Thus one cannot exclude the possibility that the poor development of parthenogenetic embryos was due to improper oviducal conditions rather than to the inferior quality of the eggs. Roblero (1973) has recently shown that in superovulated mice ovariectomized 34 hr after HCG the rate of cleavage is significantly retarded and that this effect can be partly mitigated by daily injections of progesterone. On the other hand, it is known that mouse zygotes can develop to blastocysts in oviducts of prepuberal mice (Beyer and Zeilmaker, 1973) or in oviducts cultured *in vitro* irrespective of whether they were obtained from mated or unmated females (Whittingham, 1968). In view of these conflicting observations the question of whether eggs of inferior quality result from induced ovulations remains open.

E. Frequency of Various Types of Reaction

The type of reaction of the egg to parthenogenetic activation probably depends on the four above mentioned factors even more clearly than does the incidence of activation.

1. Activating agent. The evidence for the role of this factor comes from the following observations: *First* – electric shock versus heat shock of 44-45.5°C applied to spontaneously ovulated eggs of A strain (Tarkowski *et al.,* 1970; Komar, 1973). Regular immediate cleavage was observed among eggs activated with heat shock in as many as 36% so treated, whereas after electric shock cleavage occurred in only 6.6%. *Second* – CBA eggs induced to ovulate, electric shock versus activation *in vitro* with hyaluronidase with or without hypotony (Tarkowski *et al.,* 1970; Graham and Deussen, 1974). The results are very similar when electric shock is compared with normal medium and with $\frac{4}{5}$ hypotonic culture medium (in hypotonic medium the incidence of eggs with one diploid pronucleus is significantly higher than after electric stimulation) and completely dissimilar in comparison with $\frac{3}{5}$ hypotonic medium, which suppresses formation of the second polar body in 96.6% of activated eggs with 84.3% having one diploid pronucleus. *Third* – in comparison to all other techniques activation induced by submitting cumulus-free eggs to hypotony causes in a very high percent of cases, suppression of the second polar body thus leading to potentially diploid embryos (Graham, 1972; Graham and Deussen, 1974; Kaufman and Surani, 1974). However, this effect is prominent only in F1C57BL/CBA, CBA, and F1C57BL/A2G eggs which suggests that the genotype of eggs plays an important role here. While in the experiments of Graham and Deussen suppression of the second polar body was very often accompanied by formation of one diploid pronucleus rather than two haploid pronuclei, only eggs with two pronuclei were observed by Kaufman and Surani. However, it is not clear whether this difference is due to different conditions of activation (different media and different pH) or to different genotypes and/or ages of eggs.

2. *Postovulatory age of eggs.* This factor has been investigated most precisely by Kaufman and Surani (1974). Eggs collected from the oviduct up to 20 hr after HCG administration react predominantly by extrusion of the second polar body and by the formation of one pronucleus. In eggs aged 25 hr immediate cleavage becomes the dominant reaction. Suppression of the second polar body and formation of two pronuclei also slightly increases with age. The same tendency has been noted by Graham and Deussen (1974) in F1C57BL/CBA eggs harvested at 13-16, 17-20, and 21-24 hr post HCG and submitted to hypotonic shock. Changes in relative frequencies of various types of reaction depending on the age of oocytes were also observed in eggs stimulated electrically. Activation performed shortly after ovulation, i.e., 14-16 hr after HCG, yielded 25 eggs with the second polar body and one pronucleus among 43 activated (Tarkowski *et al.,* 1970). Out of 21 eggs activated (38.5% of recovered eggs) which received a shock 24-26 hr after HCG (12-14 hr after ovulation) only one had the second polar body and one pronucleus; 3 underwent immediate cleavage and in all others the second polar body was suppressed, the most common reaction being formation of two pronuclei (Sierajewska and Tarkowski, unpublished observations). The increasing frequencies of eggs undergoing immediate cleavage and eggs with two haploid pronuclei are probably due to migration of the spindle toward the center of the egg and/or its inability to rotate following activation so that its position in relation to the egg surface does not change from tangential to perpendicular. Migration of the spindle was in fact observed in aging mouse eggs by Szollosi (1971).

3. *Genetic constitution.* Comparison of data presented by Tarkowski *et al.,* (1970) and Witkowska (1973a) regarding activation of A and CBA strains suggests that eggs of the A strain react to electric shock more often by suppression of the second polar body and formation of two pronuclei than do CBA eggs. However, since the two groups of eggs were not very large (60-91) the differences may be due to chance.

Clear cut differences regarding the incidence of various types of reaction have been observed in experiments in which eggs of CBA, C3H, C57BL, ICR/A, AKR, and F1C57BL/CBA were submitted to both hyaluronidase treatment and hypotonic medium (Graham and Deussen, 1974). The most interesting finding was that CBA and F1C57BL/CBA eggs reacted much more frequently than did the eggs of other strains investigated by the suppression of the second polar body and by the formation of two haploid pronuclei or one diploid pronucleus. Since the latter type of reaction is exceptionally common in CBA eggs, very rare in C57BL eggs, and of intermediate frequency in F1C57BL/CBA eggs, genetic factors are clearly involved. This conclusion is also supported by the finding of Kaufman and Surani (1974) that when C57BL are crossed with A2G rather than CBA the formation of a diploid pronucleus occurs only very rarely (1.5%).

4. *Origin of eggs: spontaneous versus induced ovulation.* Very little information is available regarding the role of this factor. According to Tarkowski *et al.,* (1970) and Witkowska (1973a) CBA eggs from induced ovulations react more often by suppression of the second polar body and by the formation of two pronuclei or many nuclei than do spontaneously ovulated eggs. However, because of the small number of eggs examined, sixty in each group, the significance of this difference is not certain.

III. CONCLUDING REMARKS

From the above considerations it follows that all of the above factors may influence the rate of activation, type of reaction, and survival time of parthenogenetic embryos. However, very often it is impossible to evaluate precisely the significance of each factor because experiments of various authors differ not only in the techniques of activation but also in other variables such as age and genotype of eggs, etc. In the view of the present author the optimal experiment would consist of the following steps: First, activation *in vitro* with a moderately strong shock shortly after ovulation; second, culture of eggs *in vitro* for several hours until the types of reaction can be clearly defined and the eggs segregated accordingly; third, transplantation to the oviducts of pseudopregnant recipients. Step one needs a comment. It is beyond doubt that susceptibility of eggs to artificial activation increases with age and that shortly after ovulation only very strong stimuli, such as electric current, heat, or hypotony, are effective. Such strong shocks irreversibly damage a fraction of treated eggs, which subsequently degenerate, and one cannot exclude the possibility that sublethal changes are also produced in activated eggs. For this reason the ability of eggs to develop to blastocysts is a much more precise criterion of the value of a technique than is the percent of eggs activated, as observed a few hours after the shock. On the other hand, the disadvantage of using older eggs which require only a mild shock to become activated, is that aging processes may have already begun that will subsequently result in the death of the embryos. For this reason it is advisable to test the developmental potential of the aging eggs by fertilizing them and studying their ability to develop until term, or at least until the later postimplantation stages. The only control experiments of this type were performed on spontaneously ovulated eggs activated between 8:00 and 10:00 A.M. Females mated during this period (delayed mating) produced litters of normal size which proved that eggs artificially activated at this time were fully viable (Tarkowski *et al.*, 1970).

IV. DEVELOPMENT OF PARTHENGENETIC EMBRYOS

A. Preimplantation Development

1. Cortical and zona reactions. These are the first reactions of the egg to the contact with the spermatozoan and it is generally believed that changes in the egg membrane initiate a chain of reactions which lead to activation. Cortical and zona reactions have been studied in parthenogenetic eggs in direct and indirect ways.

Observations with the light and electron microscopes by Austin (1956) and Longo (1974b) on hamster eggs undergoing spontaneous activation, and ultrastructural studies on rabbit eggs activated with cold-shock (Flechon and Thibault, 1964; Longo, 1974a), and observations on mouse eggs activated with hyaluronidase (Solter *et al.*, 1974) provide evidence that cortical reactions in these eggs did not proceed normally, i.e., rupture of cortical granules either did not take place or was significantly impaired. The recent work of Gwatkin *et al.*, (1973) showed, however, that in electrically stimulated hamster eggs cortical granules disappeared rapidly, thus rendering the zona pellucida impermeable to sperm. Thus, it seems that various parthenogenetic agents differ in their ability to evoke the cortical reaction and that one should be cautious in generalizing results obtained by using only one technique.

According to Chalmel (1969) rabbit eggs activated by cold shock can be subsequently fertilized, a fact which provides additional evidence that the block to polyspermy is not operating. Austin and Braden (1954) have also observed that rat eggs activated by cold shock and ether or nitrous oxide anesthesia can undergo fertilization. Similar experiments have not yet been performed in the mouse. The result of such experiments would be critical for evaluation changes that occur in the cell membrane of artificially activated eggs.

Failure of the cortical reaction should also be reflected in a subnormal zone reaction or in its absence. Activation of the egg by the spermatozoan changes properties of the zona so that penetration of additional sperm is prevented or at least reduced; there are species differences and in the mouse also strain differences in the efficiency of the zona reaction. (Braden *et al.*, 1954; Austin and Braden, 1956). The question of whether or not parthenogenetic activation in the mouse brings about a zona reaction has been approached indirectly by Mintz and Gearhart (1973) who took advantage of the fact discovered by Smithberg (1953) in the mouse, and Chang and Hunt (1956) in the rat and the rabbit, that the zona pellucida of fertilized eggs becomes more resistant to digestion with proteolytic enzymes. Mintz and Gearhart compared the times required for dissolution of the zona in weak solution of pronase of fertilized, unfertilized, and artificially activated eggs. The eggs were activated by the electric shock method according to Tarkowski *et al.,* (1970). The results were not clear cut but the authors conclude that "...the zonas of parthenogenetic embryos in stages ranging from 2 to 14 cells varied in their lysis times in pronase and overlapped with those of unfertilized and fertilized egg zonas. As a population the zonas of parthenogenones had intermediate lysis times...". It is worth recalling in this connection, however, that according to Gwatkin *et al.,* (1973), whose work was referred to earlier in the present paper, electrical stimulation of hamster eggs causes not only disappearance of cortical granules but results also in a zona reaction as shown by the failure of capacitated sperm to adhere to the zona pellucida. These authors have also found that, when unstimulated mouse oocytes were treated with the material released from the cortical granules of hamster eggs, fertilization was prevented.

From the comparison of the above results it follows that, depending on the parthenogenetic agent, cortical and zona reactions may or may not be evoked and that absence of these reactions does not preclude initiation of parthenogenic development. Whether failure of the cortical reaction has any adverse effect on survival of parthenogenetic embryos remains unknown.

Evidence that parthenogenetic agents can also produce other changes in the egg surface which normally follow penetration of the spermatozoan has recently been provided by Pienkowski (1974) who studied agglutinability of mouse eggs in the presence of Concanavalin A. Eggs activated by hyaluronidase agglutinate at the same very low concentration of Con A (10 μg/ml) as do fertilized eggs, whereas unfertilized eggs become agglutinable only at concentrations as high as 2000 μg/ml.

2. The pronucleate stage and early development. Graham and Deussen (1974) studied DNA synthesis in parthenogenetic eggs and found that the majority of pronuclei started to synthesize DNA between 3½ and 6 hours after activation, the conclusion being that this process is not grossly different in timing from that of fertilized eggs (Luthardt and Donahue, 1973). The duration of the one-cell stage was precisely timed by Kaufman (1973b,c) in fertilized control eggs and in parthenogenetic haploid eggs. According to this author parthenogenetic haploids are by no means retarded in comparison with control diploids and may even enter first cleavage slightly earlier. The next cleavages have not been timed so precisely, but it seems that up to the 8-cell stage the rate of cleavage is not grossly delayed (CBA eggs developing *in vivo* after electric shock — Witkowska, 1973a). The work of Witkowska (1973a) presents the most detailed account to date of the fate of parthenogenones in the preimplantation period because the embryos were studied day by day in the living state as well as in permanent and in air-dried preparations. The following remarks are based mainly on her results.

Starting with the two-cell stage some eggs contain small cytoplasmic fragments in the perivitelline space. As these fragments are not seen up to 7 hours after activation they are probably extruded during the first cleavage. Witkowska put forward a suggestion that this might reflect a subnormal cortical reaction in some of the parthenogenones. Another abnormality already observed during early cleavage but occurring only in eggs from induced ovulations is the occurrence of binucleate blastomeres which must arise because of the absence of cytokinesis. Retardation in the development of parthenogenetic embryos first becomes evident on the 4th day; some embryos are arrested in cleavage and the average cell number in the developing embryos is slightly below that of the controls. The majority of developing embryos are morulae rather than blastocysts. However, it should be born in mind that in CBA strains, which were used in the experiments, the preimplantation development is slow, and in control material examined at the same time only 37.4% of embryos are blastocysts. On the 5th day of development, however, the average cell number is quite high and well above the average cell number in one-day-younger controls, indicating that the delay in the development of parthenogenones does not exceed half a day. The same conclusion can be drawn from the experiments of Kaufman and Gardner (1974) who transferred diploid parthenogenetic eggs to the oviduct and on the 4th day recovered 69 blastocysts and 100 morulae. Thus, the blastocysts were 40% of the embryos. Morphological and karyological studies of Tarkowski *et al.,* (1970) and Witkowska (1973a) on late preimplantation embryos have shown in addition that: *first*, many morulae do not transform into blastocysts despite the fact that they are composed of more than 32 cells (at this stage cavitation usually begins) and occasionally of as many as 70-80 cells; *second*, the embryos are haploid, haploid-diploid mosaics, or diploid; *third*, no correlation can be detected between ploidy and normalcy of embryos as regards their morphology; *fourth*, the cell number appears to depend on ploidy — haploid embryos were built of the greatest number of cells, diploid embryos of the smallest, mosaic embryos being intermediate. It is also worth noting that much more normal development, both in quantitative and qualitative terms, was observed when eggs originated from spontaneous rather than from induced ovulations.

   B. Postimplantation Development. Postimplantation development of parthenogenetic mouse embryos has been studied by Tarkowski *et al.,* (1970), Graham (1972), Witkowska (1973b), Mintz and Gearhart (1973) and Kaufman and Gardner (1974). The most advanced embryo obtained so far was at the stage of 8 somites; macroscopically it looked perfectly normal but unfortunately was destroyed prior to microscopical and karyological examination (Tarkowski *et al.,* 1970; Witkowska, 1973b). Thus far, implanted parthenogenetic embryos have not been studied karyologically; the implantations have been examined only histologically or by dissection under the dissecting microscope. From the thorough study of Witkowska (1973b), who collected as many as 152 implantations, the following conclusions can be drawn. *First*, some embryos although they evoke a decidual reaction are not able to destroy the uterine epithelium and consequently do not establish direct contact with the mucosa. These embryos, which include irregular forms, multicellular morulae which have not cavitated, and a number of morphologically normal blastocysts, die at the time of implantation. *Second*, embryos which establish contact with the uterine mucosa transform into egg-cylinder which, however, are usually small and retarded in development for approximately one day. It seems that the majority of these embryos die during the seventh day. *Third*, the most successful embryos die at the stage of egg-cylinder, a characteristic of the 7th day of normal development, i.e., at the stage when the proamniotic cavity appears and the ectoderm becomes subdivided into embryonic and extra-embryonic parts. Mintz and Gearhart (1973) have also examined a number of females on day 6, probably corresponding to the 7th day in the terminology used by Witkowska, and found empty decidual swellings as well as small and underdeveloped egg-cylinders.

When electric shock is applied *in situ*, as it was in the studies of Witkowska and Mintz (1973b) and Gearhart (1973) the ploidy of each particular embryo that has implanted is unknown. However, from the karyological studies of Witkowska (1973a) on preimplantation embryos it is known that the majority (60%) of embryos were haploid and that of the remainder 21.6% were haploid-diploid mosaics, 1.7% were tetraploid, and only 16.7% were diploid. The very high mortality of parthenogenetic embryos at, and shortly after, implantation could be accounted for by the lethality of haploids and at least some of the mosaics. However, if diploidy were the only condition for the survival of parthenogenones beyond implantation, one might expect that 1 in 6 implantations should contain diploid embryos and that these should be alive at mid-term. On the 9th and 10 days Witkowska collected 86 implantations altogether but of that only two contained live embryos (2.3%). These observations show that the majority of diploid parthenogenetic embryos must also die before the 9th day and that diploidy by itself is not a sufficient condition for survival of parthenogenones. This conclusion is corroborated by the studies of Graham (1972) who, by suppressing formation of the second polar body, obtained diploid eggs and transplanted them to pseudopregnant recipients. None of the implanted embryos survived beyond the 7th day. A similar experiment was recently carried out by Kaufman and Gardner (1974) but the content of the decidual swellings collected on the 6th and 7th days was not described. Over 50% of potentially diploid eggs implanted and this figure corresponds well with the percentage of morulae and blastocysts recovered in another series of experiments checked on the 4th day. Kaufman and Gardner also transplanted haploid eggs – those with the second polar body and one pronucleus, as well as those which had undergone immediate cleavage. About $\frac{1}{3}$ of each type of haploid egg produced a decidual reaction but again a histological description of the content of the decidual swellings was not presented. Finally, diploid eggs obtained by heat shock followed by culture in the medium containing cytochalasin B also failed to develop beyond the egg-cylinder stage (Balakier and Tarkowski, in preparation).

V. POSSIBLE CAUSES OF THE DEATH OF PARTHENOGENETIC EMBRYOS

Eggs artificially activated die at all developmental stages investigated, beginning with the one-cell stage and ending with the egg-cylinder stage. The causes of this mortality are undoubtedly manifold and by no means clear. Some of the factors which may be involved are discussed below (see also Graham, 1971 and in press; Tarkowski, 1971.

1. The parthenogenetic agent does not fully reproduce the activating action of the spermatozoan. Subnormal cortical and zona reactions may serve as an example. This is probably true in the majority of eggs but it is not clear whether this is also the case in the most successful embryos.

2. The parthenogenetic agent produces damage at the ultrastructural and/or molecular level which does not express itself immediately but leads to the death of the embryo at later stages. As all parthenogenetic agents act on the basis of sublethal damage, there is no doubt that some of the eggs are irreversibly injured. Solter et al., (1974) have recently analyzed the ultrastructure of parthenogenetic mouse embryos developing in vitro and have described manifold abnormalities, including the persistence of primary nucleoli, the persistence of nucleoli on mitotic chromosomes, the presence of subnuclei and of cytoplasmic fragments, etc. However, since fertilized eggs cultured under the same conditions were not studied, it cannot be excluded that many of these abnormalities were simply due to suboptimal culture conditions rather than to parthenogenetic activation. Many of the parthenogenones will undoubtedly show abnormalities in ultrastructure as they do not look perfectly normal even under the dissecting microscope. The question which should be answered is, "Do the parthenogenetic embryos developing under optimal conditions, i.e. in vivo, and displaying normal morphology

differ from control embryos?" In other words, the problem is not how bad they can be but whether the best ones are as normal as the controls.

3. There may be constituents of the spermatozoan, apart from the nucleus, which are necessary for the normal development of zygote. Such a possibility cannot be excluded although it appears unlikely in view of the fact that at least some of the parthenogenetic embryos develop as far as the egg-cylinder stage.

4. Artificial activation may lead to aneuploidy in two ways: *first*, by formation on subnuclei which may not contribute chromosomes to the metaphase plate at first cleavage, and *second*, by uneven distribution of chromatids at the second meiotic division (Graham and Deussen, 1974). Both phenomena have been observed and it seems that their frequency depends mainly on the technique of activation. Taking into account that haploid blastocysts have been obtained on several occasions, it does not seem likely that aneuploidy in the form of hypo-diploidy or hyper-haploidy would be lethal in the preimplantation period. Hypo-haploidy, however, is probably lethal very early in cleavage and may be considered as one of the least questionable factors responsible for the death of parthenogenetic embryos before implantation.

A. Mortality after Implantation

1. The possible role of asynchrony between the embryos and the uterus resulting from retarded development of parthenogenetic embryos was investigated by Witkowska (1973b) in two ways. *First*, females were mated with vasectomized males in the morning rather than at night (delayed mating) and the eggs were activated *in situ* shortly afterwards. *Second*, morulae and blastocysts recovered on the 4th day of pseudopregnancy. Neither of these two procedures has increased the survival rate of embryos. Thus there is no evidence that the asynchrony might be a major cause of embryonic mortality at the time of implantation. Re-transplantation of parthenogenetic morulae and blastocysts to one day younger recipients was also carried out by Kaufman and Gardner (1974) but their paper does not make clear whether this procedure had any effect on the postimplantation survival of embryos.

2. All aneuploids which have survived until implantation are likely to die at or shortly after implantation. Although there is no direct information regarding the fate of haploid embryos it seems that they also die soon after implantation. This conclusion can be indirectly deduced from experiments of Tarkowski *et al.*, (170) and Witkowska (1973a,b) who observed heavy mortality soon after implantation and a preponderance of haploids among late preimplantation embryos. When haploid eggs were transplanted by Kaufman and Gardner (1974), about one third of them implanted thus proving beyond doubt that haploids can induce a decidual reaction and initiate implantation.

Unfortunately no information has so far been presented regarding the morphology of these embryos and the stage at which they die. Early death of haploids can also be inferred from the fact that implanted haploid embryos have never been observed in mammals and that this condition is known to be lethal in other vertebrates.

3. The most puzzling problem is the death of diploid parthenogenones at the egg-cylinder stage. To explain this phenomenon Graham (1971) favored the theory of exposure of recessive lethal genes due to the high homozygosity of parthenogenetic embryos. This view was accepted with some doubt by other authors (Tarkowski, 1971; Witkowska 1973b; Mintz and Gearhart, 1973) who find difficulty in reconciling it with the origin of the eggs from highly inbred strains − presumably already nearly homozygous. It should also be bourne in mind that, because of recombination at meiosis parthenogenetic embryos arising as a result of suppression of the second polar body are never fully homozygous.

Brown and Chandra (1973) have recently formulated a hypothesis according to which the number of active X chromosomes in mammals conforms to the number of maternally derived sets of autosomes. If this supposition is correct, then in diploid mammalian parthenogenones both X chromosomes should be genetically active which might have a deleterious effect on embryonic development. The hypothesis explains satisfactorily the observed behavior of X chromosomes in various aberrant karyotypes but as yet it has not been tested on parthenogenetic material. It is worth recalling in this connection, however, that digynic triploid mouse embryos which, according to the hypothesis of Brown and Chandra, should also have two active X chromosomes, do develop beyond the stage at which diploid parthenogenones die (Wroblewska, 1971).

REFERENCES

Austin, C. R. (1956). Cortical granules in hamster eggs. *Exp. Cell. Res.* **10**, 533–540.

Austin, C. R. and Braden, A. W. H. (1954). Induction and inhibition of the second polar division in the rat egg and subsequent fertilization. *Aust. J. Biol. Sci.* **7**, 195–210.

Austin, C. R. and Braden, A. W. H. (1956). Early reactions of the rodent egg to spermatozoan penetration. *J. Exp. Biol.* **33**, *358–365.*

Austin, C. R. and Walton, A. (1960). Fertilization. *In*, "Marshall's Physiology of Reproduction.", Vol. I, Part 2, (A. S. Parkes, ed.), Longmans Green and Co., Ltd. London.

Beatty, R. A. (1957). Parthenogenesis and polyploidy in mammalian development, Cambridge University Press.

Beatty, R. A. (1967). Parthenogenesis in vertebrates. *In*, "Fertilization", Vol. 1, (C. B. Metz and A. Monroy, eds.), pp. 413–440. Academic Press, New York.

Beatty, R. A. (1972). Parthenogenesis and heteroploidy in the mammalian egg. *In*, "Oogenesis", (J. D. Biggers and A. W. Schuetz, eds.), pp. 277–299. University Park Press, Baltimore.

Beyer, G. and Zeilmaker, G. H. (1973) Development of mouse and rat zygotes following transfer to non-synchronized rat and mouse oviducts. *J. Reprod. Fert.* **33**, 141–143.

Braden, A. W. H. (1957). Variation between strains in the incidence of various abnormalities of egg maturation and fertilization in the mouse. *J. Genet.* **55**, 476–486.

Braden, A. W. H. and Austin, C. R. (1954). Reactions of unfertilized mouse eggs to some experimental stimuli. *Exp. Cell Res.* **7**, 277–280.

Braden, A. W. H., Austin, C. R., and David, H. A. (1954). The reaction of the zona pellucida to sperm penetration. *Aust. J. Biol. Sci.* **7**, 391–409.

Brown, S. W. and Chandra, H. S. (1973). Inactivation system of the mammalian X chromosome. *Proc. Nat. Acad. Sci., U.S.,* **70**, 195–199.

Chang, M. C. and Hunt, D. M. (1956). Effects of proteolytic enzymes on the zona pellucida of fertilized and unfertilized mammalian eggs. *Exp. Cell. Res.* **11**, 497–499.

Chalmel, M. C. (1962). Possibilité de fécondation des oeufs de lapine activés parthénogénétiquement. *Ann. Biol. Anim. Biochem. Biophys.* **2**, 279–297.

Flechon, J. E. and Thibault, C. (1964). Modifications ultrastructurales de l'ovocyte de la Lapine au cours de l'activation. *J. Microsc.* **3**, 34.

Graham, C. F. (1970). Parthenogenetic mouse blastocysts. *Nature, (London)* **226**, 165–167.

Graham, C. F. (1971). Experimental early parthenogenesis in mammals. *Adv. Biosci.* **6**, 87–97.

Graham, C. F. (1972). Genetic manipulation of mouse embryos. *Adv. Biosci.* **8**, 263–273.

Graham, C. F. (1974). The production of parthenogenetic mammalian embryos and their use in biological research. *Biol. Rev.* (in press).

Graham, C. F. and Deussen, Z. A. (1974). *In vitro* activation of mouse eggs. *J. Embryol. Exp. Morph.* **31**, 497–512.

Gwatkin, R. B. L., Williams, D. T., Hartmann, J. F., and Kniazuk, M. (1973). The zona reaction of hamster and mouse eggs: production *in vitro* by a trypsinlike protease from cortical granules. *J. Reprod. Fert.* **32**, 259–265.

Kaufman, M. H. (1973a). Parthenogenesis in the mouse. *Nature, (London)* **242**, 475–476.

Kaufman, M. H. (1973b). Timing of the first cleavage division of the mouse and the duration of its component stages: a study of living and fixed eggs. *J. Cell. Sci.* **12**, 799–808.

Kaufman, M. H. (1973c). Timing of the first cleavage division of haploid mouse eggs, and the duration of its component stages. *J. Cell. Sci.* **13**, 553–566.

Kaufman, M. H. and Gardner, R. L. (1974). Diploid and haploid mouse parthenogenetic development following *in vitro* activation and embryo transfer. *J. Embryol. Exp. Morph.* **31**, (in press).

Kaufman, M. H. and Surani, M. A. H. (1974). The effect of osmolarity on mouse parthenogenesis. *J. Embryol. Exp. Morph.* **31**, 513–526.

Komar, A. (1973). Parthenogenetic development of mouse eggs activated by heat-shock. *J. Reprod. Fert.* **35**, 433–443.

Longo, F. J. (1974a). Ultrastructural analysis of parthenogenetic rabbit eggs. *Anat. Rec.* **178**, 404–405a.

Longo, F. J. (1974b). An ultrastructural analysis of spontaneous activation of hamster eggs aged *in vivo. Anat. Rec.* **179**, 27–56.

Luthardt, F. W. and Donahue, R. P. (1973). Pronuclear DNA synthesis in mouse eggs. An autoradiographic study. *Exp. Cell Res.* **82**, 143–151.

Mintz, B. and Gearhart, J. D. (1973). Subnormal zona pellucida change in parthenogenetic mouse embryos. *Dev. Biol.* **31**, 178–184.

Pienkowski, M. (1974). Study of the growth regulation of preimplantation mouse embryos using concanavalin A. *Proc. Soc. Exp. Biol. Med.* **145**, 464–469.

Roblero, L. (1973). Effect of progesterone *in vivo* upon the rate of cleavage of mouse embryos. *J. Reprod. Fert.* **35**, 153–155.

Smithberg, M. (1953). The effect of different proteolytic enzymes on the zona pellucida of mouse ova. *Anat. Rec.* **117**, 544a.

Solter, D., Biczysko, W., Graham, C., Pienkowski, M. and Koprowski, H. (1974). Ultrastructure of early development of mouse parthenogenones. *J. Exp. Zool.* **181**, 1–23.

Szollosi, D. (1971). Morphological changes in mouse eggs due to aging in the fallopian tube. *Am. J. Anat.* **130**, 209–226.

Tarkowski, A. K. (1971). Recent studies on parthenogenesis in the mouse. *J. Reprod. Fert.,* **Suppl. 14**, 31–39.

Tarkowski, A. K., Witkowska, A., and Nowicka, J. (1970). Experimental parthenogenesis in the mouse. *Nature, (London)* **226**, 162–165.

Witkowska, A. (1973a). Parthenogenetic development of mouse embryos *in vivo*. I. Preimplantation development. *J. Embryol. Exp. Morph.* **30**, 519–545.

Witkowska, A. (1973b). Parthenogenetic development of mouse embryos *in vivo*. II. Preimplantation development. *J. Embryol. Exp. Morph.* **30**, 547–560.

Whittingham, D. G. (1968). Development of zygotes in cultured mouse oviducts. I. The effect of varying oviductal conditions. *J. Exp. Zool.* **169**, 391–398.

Wroblewska, J. (1971). Developmental anomaly in the mouse associated with triploidy. *Cytogenetics* **10**, 199–207.

III. Early Development

Genetic and Biochemical Activities in Preimplantation Embryos

Cole Manes

Department of Pediatrics
University of Colorado Medical Center
Denver, Colorado

I.	Introduction	133
II.	Genetic Activities: Patterns of Gene Expression	136
	A. Identification of the Products of Transcription	137
	1. Transfer RNA	137
	2. Ribosomal RNA	140
	3. Messenger RNA	143
	B. Analysis of the Products of Translation	146
	C. The Relative Roles of Maternal and Embryonic Gene Products During the Immediate Post-fertilization Period	148
	1. Evidence from the Use of Inhibitors of Transcription	148
	2. Genetic Evidence	150
III.	Biochemical Activities: An Open System	151
	A. Membrane Composition and Function	152
	B. The Role of the Mitochondrion	153
	C. Steroid Metabolism	155
	D. The Interaction between the Blastocyst and the Uterus	155
	1. The Role of the Reproductive Tract	156
	2. Models of Tissue Interaction	157
IV.	Summary and Conclusions	157
	References	158

I. INTRODUCTION

There are simple as well as complex ways of viewing the early development of mammals. In one perspective, the embryo possesses all the geometrical elegance and simplicity of a colonial organism akin to *Pandorina* or *Volvox*. Cell numbers during cleavage are few. Blastocyst formation results in only two cell types: trophoblast and embryoblast (Fig. 1). The tasks required of the embryo during the preimplantation interval are seemingly straightforward: increase the copies of the genome; segregate the early descendants of the zygote into the two necessary cell types; make contact with the uterine

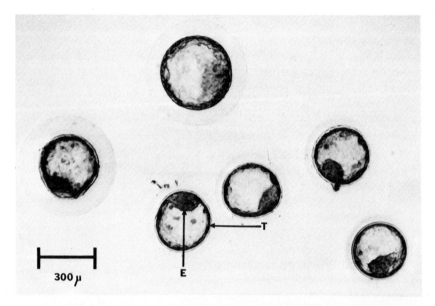

Fig. 1. Early blastocysts of the rabbit approximately 72 hours after egg fertilization, showing the segregation of embryonic cells into trophoblast (T) and embryoblast (E).

lining. The complexities of organogenesis and the finer aspects of cell differentiation lie ahead. Simplicity and success seem to be the keynotes of this period.

The perspectives provided by the cuvette and the scintillation counter, however, tend to be complex. Charles Epstein recently presented a graph illustrating much of the available information regarding enzyme activities in the preimplantation mouse embryo (Fig. 2). He intentionally did not identify the multiple curves included in the graph, for the purpose was to impress upon all of us that many things are occurring simultaneously in these early embryos, that values for enzyme activity do not fall into any obvious and simple pattern, and that such information may not readily lend itself to "explaining" mammalian embryogenesis (Epstein, 1973).

It is difficult in the early stages of any science to predict what kinds of data will be illuminating and what kinds will be merely confusing. The field of early mammalian embryology is encountering its own growth problems, and Paul Weiss has cautioned us to distinguish between true growth and obesity (Weiss, 1969). A determined effort will be made in this review to heed this caution. The accumulating data will be focused around two major questions; namely, (1) to what extent, and in what manner, does the embryonic genome participate in the events of early embryogenesis? And (2), what are the major avenues of interaction between the blastocyst and the uterus that allow the

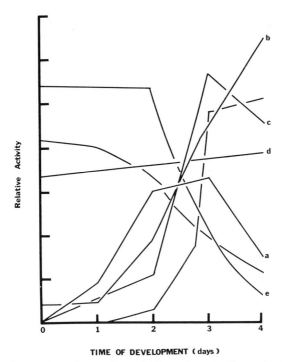

TIME OF DEVELOPMENT (days)

Fig. 2. The activity of several metabolic systems in the preimplantation mouse embryo as a function of embryonic age. (From Epstein, C. J., in "The Regulation of Mammalian Reproduction," 1973. Courtesy of Charles C. Thomas, Publisher, Springfield, Illinois.)

initiation of implantation? Given the current motivation of concerned citizens to intervene either positively or negatively in human reproduction, these are not idle questions. If transcription of the embryonic genome is indeed essential for certain phases of preimplantation development, then the "necessary conditions" for such genetic expression become of critical importance (Graham, 1973). Alternatively, if preimplantation development is largely supported by gene products synthesized during oogenesis, then a greater research effort must be directed toward identifying such synthetic activity during maturation of the mammlian oocyte. Furthermore, the observed capacity of the mammalian uterus to regulate mitotic activity in the developing blastocyst (McLaren, 1973) is a phenomenon relevant not only to problems of human fertility but also to the fundamental biological question of the regulation of cell replication in Metazoa.

It should be remarked right away that virtually all the data regarding genetic and biochemical activities in the preimplantation embryo have been derived from studies on the mouse and the rabbit. In addition, many of these studies depend heavily upon the use of hormonal superovulation and upon the maintenance of living embryos under more-or-less defined conditions outside the reproductive tract. Some of these studies also rely upon the ability of the embryo to transport, or at least to allow, radioactive molecules into its cytoplasm and to incorporate these molecules into identifiable metabolites. It therefore seems appropriate for someone to suggest periodically that we must temper our enthusiasm for extrapolating these data to "all mammals" – or even to all normally-ovulated mouse and rabbit embryos developing happily *in vivo* in a non-radioactive environment.

That *caveat* having been posed, however, it must be admitted that these studies do in fact constitute a logical beginning in the attempt to understand the preimplantation development of mammals. Not only are the mouse and the rabbit *bona fide* mammals which lend themselves rather readily to the work of the average laboratory, but by virtue of the differences they present they may well span a major portion of the entire spectrum of metabolic patterns which will be encountered as more mammalian embryos are investigated. The diminutive mouse embryo, one of the smallest mammalian embryos known, might well be expected to accomplish its preimplantation tasks a bit differently from the large rabbit embryo, in spite of the evolutionary proximity of rodents and lagomorphs. The rodent embryo activates its nucleolus early, sheds the *zona pellucida* before implantation, and grows scarcely at all. The rabbit embryo activates its nucleolus relatively late, retains an acellular coating up to the moment of implantation, and undergoes logarithmic growth as a blastocyst. The rodent blastocyst is also capable of prolonged periods of reversible growth arrest, whereas the rabbit blastocyst quickly dies if its growth is arrested. It may be predicted that the final picture of "the mammalian embryo" will be a composite, and that some of the noted differences will turn out to be trivial, while others will be more fundamental and significant. It also appears reasonable to expect that, since both mouse and rabbit embryos will continue to develop normally when replaced in the reproductive tract after periods of *in vitro* maintenance, much of the information derived from these studies will apply equally well to the undisturbed embryo *in vivo*.

II. GENETIC ACTIVITIES: PATTERNS OF GENE EXPRESSION

In order to approach the question regarding the participation of the embryonic genome in the events of preimplantation development, it becomes relevant to ask whether transcription is in fact occurring and, if so, to attempt to identify the products of transcription. It is worth noting that the first published work to provide a glimpse of what the embryo is "up to" in terms of genetic expression is scarcely ten years old. In 1964, Mintz published autoradiographs of preimplantation mouse embryos which had been exposed *in vitro* to labeled precursors of nucleic acids and proteins. She showed quite clearly that these materials were incorporated into acid-insoluble products by the early embryo and that there were potentially interesting changes in the pattern of this incorporation during the preimplantation period (Mintz, 1964a). It is difficult to escape the conclusion that her findings opened the door to a great deal of work which has taken place during the ensuing decade, and that Mintz brought the mammalian embryo within the reach of the molecular biologist.

Four years after Mintz's paper, Ellem and Gwatkin reported on the MAK column analysis of radiolabeled RNA synthesized *in vitro* by the preimplantation mouse (Ellem and Gwatkin, 1968), and over the next two years Woodland and Graham (1969) and Piko (1970) published sucrose gradient analyses of these same products. Manes (1971) reported the electrophoretic analysis of radiolabeled RNA synthesized by the preimplantation rabbit embryo *in vitro*, and Greenslade *et al.* (1972) have reported the MAK-binding properties of RNA synthesized by the rabbit embryo *in vivo*. Even though the detection of radiolabeled materials is an extremely sensitive business, the degree of animal husbandry involved in some of these experiments is impressive. To gather up enough biological material for an analysis of the RNA synthetic pattern in the 2-cell mouse embryo, for example, Woodland and Graham (1969) were called upon to provide some 900-1000 embryos for each experiment. Perhaps it is not amiss to remind both those investigators who plan to explore genetic expression in the early mammalian embryo, as well as the granting agencies who may express astonishment at the budget requests for experimental animals, that such logistics are sometimes involved.

A. Identification of the Products of Transcription

1. Transfer RNA. One of the salient points which has emerged from these studies of transcription in preimplantation embryos is that at all periods examined, no matter how early, transcription is in fact occurring (see review by Schultz and Church, in *Biochemistry of Animal Development,* in press). The immediate post-fertilization period in both the mouse and rabbit embryo, however, reveals a pattern of transcription that is somewhat rudimentary and unlike that found in growing cells. Ribosomal RNA synthesis is conspicuous by its absence. The one newly synthesized RNA which is readily identifiable in sucrose gradients and polyacrylamide gels is a homogeneous species having a molecular weight of 25,000 – 30,000 daltons (Fig. 3). Woodland and Graham (1969) found that this low molecular weight RNA, synthesized by the 2-cell mouse embryo, elutes from a G-100 Sephadex column ahead of the transfer RNA marker, and concluded that it is probably the tRNA precursor. At the 4-cell stage, both embryonic and marker tRNA elute coincidentally. I have

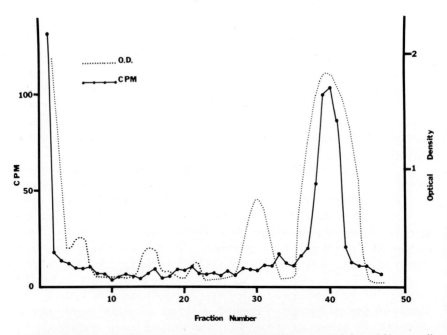

Fig. 3. Electrophoretic separation of cytoplasmic RNA extracted from 141 two-cell rabbit embryos. The embryos were exposed to uridine-5-H³ (200 microcuries/ml) for 4 hours *in vitro*. Electrophoresis was carried out in 7% polyacrylamide gels, with migration of the RNA sample from left to right. The optical density marker (dotted line) is liver cytoplasmic RNA extracted from an adult rabbit.

found that the low molecular weight RNA synthesized by the 2-cell rabbit embryo electrophoreses identically with marker tRNA obtained from adult rabbit liver, and so far have obtained no evidence of a precursor tRNA in the rabbit embryo (Manes, 1971, and unpublished observations). However, to refer to this low molecular weight RNA with the functional title of "transfer RNA" requires more justification.

Circumstantial evidence as to its identity has been accumulating. It has also been necessary to distinguish synthesis of the entire molecule from turnover labeling of the –CCA segment at the 3'-hydroxyl terminus of the presumptive tRNA molecule. Although the radioactive precursor offered to the embryo was uridine-5-H³, after four hours approximately 30% of the radioactivity can be recovered from 2-cell rabbit embryos as cytidine. Thus, it was significant to find that the radioactive precursor incorporated into rabbit low molecular weight RNA is resistant to removal by *Crotalus* venom phosphodiesterase under conditions which almost completely degrade single-stranded RNA (Manes, unpublished observations). This finding reveals that the molecule has considerable double-stranded structure, as well as that the radioactive label is not confined to the 3'-hydroxyl terminus (Zubay and Takanami, 1964). Further characterization of the material has shown that it is methylated (Woodland and Graham, 1969; Manes and Sharma, 1973) and contains the expected proportion of pseudouridine (Woodland and Graham, 1969). The final proof that this RNA is indeed transfer RNA is based upon its "acceptance" of amino acids, and such amino acid acceptance can be demonstrated for some of the newly-synthesized RNA in the 2-cell rabbit embryo (Fig. 4). Aminoacylation followed by phenoxyacetylation will alter the elution properties only of transfer RNA when applied to BD-cellulose (Gillam *et al.*, 1967). The shift in the elution pattern of labeled, 2-cell embryonic RNA following this treatment is compared in Figure 4 with the elution behavior of late blastocyst RNA both before and after aminoacylation

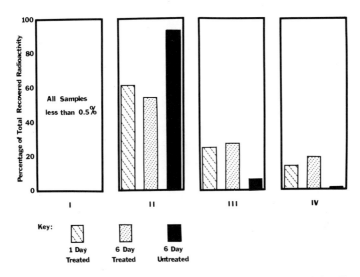

Fig. 4. Elution of radiolabeled cytoplasmic RNA extracted from 1-day (2- to 4-cell) and 6-day (0.8 – 1.0 x 10^5 cells) rabbit embryos and applied to BD-cellulose. One day embryos and 6-day embryos were exposed to uridine-5-H^3 (100 microcuries/ml and 50 microcuries/ml respectively) for 4 hours *in vitro*. The RNA was applied to the BD-cellulose column in 0.02 M sodium acetate (pH 5.0) containing 0.01 M $MgCl_2$ and 0.25 M NaCl. Stepwise elution was carried out using the following salt solutions in the acetate $MgCl_2$ buffer: 1.0 M NaCl (fraction II), 1.0 M NaCl and 10% ethanol (fraction III), and finally 2.5 M NaCl with 30% ethanol (fraction IV). tRNA which is not aminoacylated elutes in fraction II. Fraction II RNA was aminoacylated by using a crude tRNA-aminoacyltransferase from adult rabbit liver and a mixture of 15 amino acids, followed by phenoxyacetylation (Gillam *et al.*, 1967). It was reapplied to the BD-cellulose column and eluted by the stepwise procedure described above. Only RNA which can be aminoacylated and subsequently phenoxyacetylated – namely, transfer RNA – now elutes in fraction IV. A portion of the untreated fraction II RNA from the 6 day embryo was reapplied to the column as a control. (Manes, in preparation)

and phenoxyacetylation. It now seems justifiable, on the basis of this evidence, to state without equivocation that the 2-cell rabbit embryo is synthesizing transfer RNA (Manes, in preparation).

Parenthetically, it should be noted that the embryonic origin of this RNA was, for a time, in doubt. During the course of fine structural studies on early rabbit embryos (Van Blerkom *et al.*, 1973), it was discovered that a few normal-appearing follicle cells remain attached to the rabbit *zona pellucida* for two to three days after fertilization, and it was realized that a synthetic contribution from these cells might well be mistaken for a contribution from an embryo which itself consists of only a few cells. Removing the *zona pellucida* and the attached follicle cells with pronase following the labeling period did not alter the synthetic pattern seen in the denuded embryos (Manes, in preparation). Isolated follicle cells were found to incorporate uridine into low molecular weight RNA, however, so that a potential contribution from these cells must be considered in studies of the synthetic activities in very early embryos.

Some questions still remain. Is this tRNA a population unique to the early embryo? What is its stability? With regard to the last question, preliminary evidence in our laboratory shows that the tRNA in the early cleaving rabbit embryo has a half-life in the range of 24-30 hours, which is in contrast to the half-life of five days reported for tRNA in adult rat liver (Hanoune and Agarwal, 1970). A fairly rapid turnover of early embryonic RNA is to be expected, in fact. The total RNA content of these embryos – allowing for a probable 10-20% range of measurement error – does not appear to increase whatever during early cleavage (Reamer, 1963; Manes, 1969).

With regard to the first question, there is as yet no direct evidence to characterize the isoaccepting tRNA species present at any given time during the preimplantation period. It may well be of importance, however, to the qualitative control of protein synthesis in these embryos. In some situations the translation of messenger RNA templates is dependent upon rather subtle changes in tRNA availability (Anderson and Gilbert, 1969). Perhaps the most striking demonstration of this phenomenon is the regulation of hemoglobin synthesis in the chick embryo, which is utterly dependent upon a minor species of alanine tRNA (Wainwright, 1971). Total tRNA of the cleaving rabbit embryo is more extensively methylated than that of the blastocyst (Manes and Sharma, 1973), which may indicate significant differences in its participation in the events of translation.

2. Ribosomal RNA. The embryonic nucleolus shortly after egg fertilization is not functional in ribosome biosynthesis, and the few ribosomes visible in the embryonic cytoplasm are evidently of maternal origin (Fig. 5a). The nucleolus assumes a condensed, fibrillar, spherical configuration during oocyte maturation, as it simultaneously becomes quiescent, and then undergoes

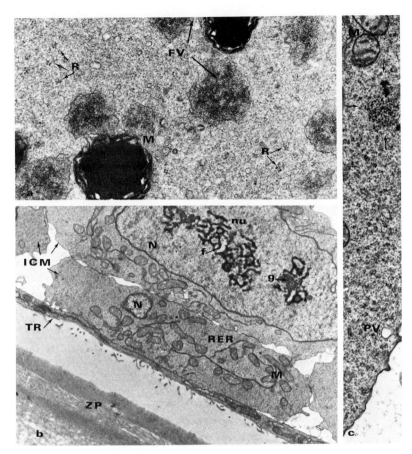

Fig. 5. Changes in the fine structure of the embryonic cytoplasm during preimplantation development in the rabbit. (a) The cytoplasm of the 1-day embryo contains spheroidal mitochondria (M) having an electron-dense matrix and flattened cristae. Flocculent vesicles (FV) are prominent, while ribosomes (R) are relatively scarce and are seen in small clusters of 3 to 6. (X 38,000) (b) The cytoplasm of the 4.5-day embryo, in contrast, contains elongated mitochondria with numerous transverse cristae. The cytoplasm now contains a dense population of ribosomes as well as a rough-surfaced endoplasmic reticulum (RER). The inner cell mass (ICM), trophoblast (TR), and *zona pellucida* (ZP) are identified (X 18,200). (c) At higher magnification, the abundance of cytoplasmic ribosomes and polysomes in the 4.5-day embryo is more evident. Large clusters of apparently free ribosomes are frequently seen (arrows). Pinocytotic vesicle (PV). (X 41,000)

"reactivation" during early embryogenesis, with reappearance of the pars granulosa, on a schedule characteristic of the species (Fig. 6). In the mouse, the nucleolus is fully functional by the 4-cell stage (Woodland and Graham, 1969), and may begin functioning as early as the late 2-cell stage (Knowland

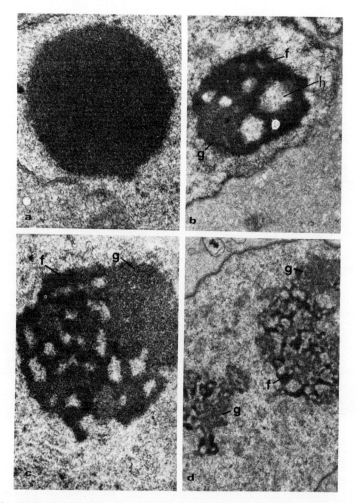

Fig. 6. Fine structure of the embryonic nucleolus during preimplantation development in the rabbit. (a) The compact, spherical, exclusively fibrillar nucleolus seen at the
1- and 2-cell stage. (b) At the 8-cell stage, granular elements (g) begin to appear, but they
are segregated from the fibrillar elements (f). The nucleolus is becoming more irregular in
outline, and it contains "holes" (h). (c) At the 16- to 32-cell stage there is an increase in
the proportion of granular elements as well as a general loosening or reticulation of the
nucleolus. (d) The embryonic nucleolus at the early blastocyst stage (64- to 128-cells),
when newly-synthesized cytoplasmic ribosomes are readily detected. (Reproduced with
permission from Van Blerkom et al., 1973).

and Graham, 1972). The rabbit is somewhat more leisurely in reactivating its
nucleolus, which becomes fully functional only at the late morula, or 64- to
128-cell stage (Manes, 1971; Fig. 7). It is worth noting that, in spite of the
differences in cell number, full nucleolar reactivation in both mouse and

rabbit embryos occurs between 48 and 60 hours following egg fertilization. The ultrastructural differentiation of the nucleolus closely parallels its bio-synthetic activity in both the mouse and rabbit embryo (Hillman and Tasca, 1969; Van Blerkom *et al.*, 1973), as it has been found to do in other developing systems (Hay and Gurdon, 1967; Dumont *et al.*, 1970).

3. Messenger RNA. The initial identification of RNA having a DNA-like base composition in the preimplantation embryo was that of Ellem and Gwatkin (1968), who reported that 15-20% of the labeled RNA recovered from the 8-cell mouse embryo after a 5-hour exposure to uridine-H[3] behaved as "tenaciously-bound" material on the MAK column. Greenslade *et al.* (1972) reported similar findings in rabbit blastocysts labeled *in vivo*. Schultz *et al.* (1973a) have used the presence of polyadenylic acid (poly(A)) sequences at the 3'-hydroxyl end of messenger RNA to identify presumptive messengers in the early rabbit embryo by the Millipore filter binding assay. Poly(A)-containing material was detectable at all stages examined, the earliest being the 16-cell cleaving embryo. Furthermore, in cleaving embryos as well as in blastocysts, it comprised about 20% of the heterogeneous RNA synthesized. Schultz (1973) has more recently reported that poly(A)-containing RNA is intimately associated with polysomes as early as the 16-cell stage in the rabbit. He finds, in the rabbit blastocyst, that this RNA decays according to kinetics which suggest two populations of messenger, one with a half-life of about 7 hours and the other with a half-life of about 18 hours. The longer-lived

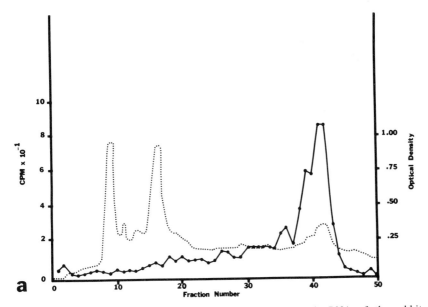

Fig. 7. Electrophoretic analysis of radiolabeled cytoplasmic RNA of the rabbit embryo during the preimplantation period. The patterns shown are from (a) 2-cell (b)

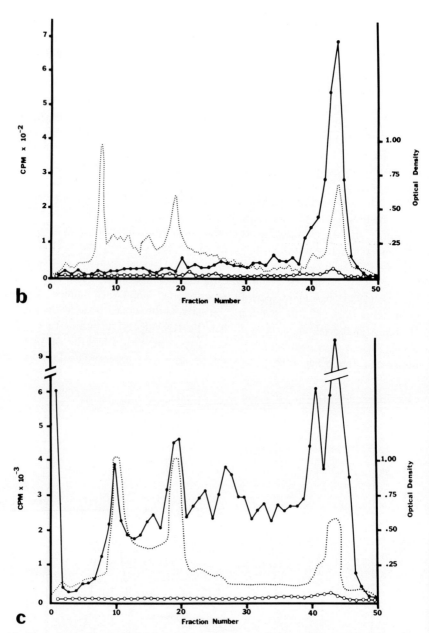

16-cell, and (c) 64- to 128-cell embryos exposed to uridine-5-H³ *in vitro*. The optical density marker (dotted line) is adult rabbit liver cytoplasmic RNA, and migration of the RNA sample was from left to right. (Reproduced with permission of the Wistar Press from Manes, 1971).

component comprises some 60% of the total poly(A)-containing RNA (Schultz, in press). These messenger stabilities are similar to those reported for HeLa cells (Singer and Penman, 1973).

Since the templates for most proteins are transcription products of unique, or few-copy DNA sequences, it is possible to assay for presumptive messenger RNA synthesis in embryos by examining specifically the transcription of this unique copy DNA. Schultz *et al.* (1973b) approached this problem by removing repetitive sequences from total, labeled rabbit DNA (defined as DNA having a Cot value less than 500 in 0.12M phosphate buffer) on hydroxyapatite. Total, unlabeled embryonic RNA from the late rabbit blastocyst was then allowed to reanneal with the unique sequence DNA to saturation. It could be shown that approximately 1.8% of the unique sequence DNA was represented by RNA transcripts in the embryo just before implantation (Fig. 8). This amount of unique sequence DNA can be calculated as capable of specifying as many as 60,000 different proteins, each having a molecular weight of 50,000. This figure obviously does not allow one to conclude that the embryo is, in fact, synthesizing such a vast array of proteins, but it does provide a glimpse of the transcriptional complexity of these relatively "simple" embryonic cells.

By reversing the procedure, and allowing labeled non-ribosomal and non-transfer RNA to hybridize with total DNA, Schultz *et al.* (1973b) were

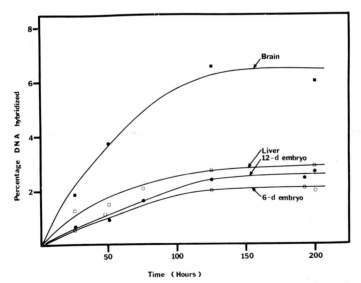

Fig. 8. Hybridization of total, unlabeled rabbit RNA obtained from the tissues indicated with tritiated, unique sequence rabbit DNA. The reactions reached saturation plateaus in each instance except with 12-day embryo RNA. (Reproduced with permission of Plenum Publishing Corp. from Schultz *et al.*, 1973).

able to demonstrate that approximately 70% of this heterogeneous material hydridized with unique sequence DNA and the remaining 30% with repetitive sequence DNA at all stages of preimplantation rabbit development, from the 16-cell cleaving embryo to the late blastocyst. This information reveals that not only are both types of DNA sequences being transcribed during early rabbit development, but also that they are transcribed in proportion to their prevalence in the genome. Such findings are obviously compatible with some current models of gene regulation in eukaryotes (Britten and Davidson, 1969). There is as yet no information regarding what proportion of this heterogeneous RNA is confined to the nucleus. Neither is it known whether all newly-synthesized messenger RNA directly enters the translation apparatus of the polysome, or whether a significant portion is delayed in transit in the form of sub-ribosomal ribonucleoprotein particles.

B. Analysis of the Products of Translation

It has been known since the report of Mintz (1964a) that the preimplantation embryo is not only engaged in transcription shortly after fertilization but is also synthesizing proteins. Several subsequent reports have indicated that there is a substantial increase in the total protein synthetic capacity of the embryo at about the time of blastocyst formation (Monesi and Salfi, 1967; Manes and Daniel, 1969; Tasca and Hillman, 1970; and Karp et al., in press). The only detailed study of the qualitative, stage-specific aspects of protein synthesis in the preimplantation mammal has been carried out on rabbit embryos (Van Blerkom and Manes, in press). The technique of autoradiography of thin, SDS-polyacrylamide slab gels can be used to resolve approximately one hundred discrete embryonic protein bands (Fig. 9). Several interesting facts have emerged from this study. One is that the pattern of protein synthesis is complex at all stages, the newly-synthesized proteins ranging in size from less than 10,000 to about 250,000 daltons. Another is that the major qualitative changes in the pattern of protein synthesis occur *prior* to blastocyst formation, rather than at a time when the embryonic cells are visibly differentiating into trophoblast and embryoblast. A third fact is that development is characterized not only by the sequential addition of new protein species to the cellular repertoire, but also by the stage-specific *loss* of certain protein synthetic capacities.

Several tentative facts can also be derived from this study. If one examines the cytoplasmic ultrastructure of the early cleaving rabbit embryo (Fig. 5a), scattered, small clusters of 3 to 6 ribosomes – which are presumed to be polysomes – are all that can be found. By analogy with the reticulocyte polysome (Warner et al., 1963; Scherrer and Marcaud, 1968), one would expect polysomes of this size to be synthesizing peptides in the range of 15,000-30,000 daltons. Since the early embryo is evidently synthesizing proteins much larger than would be predicted from polysome size, it appears that the

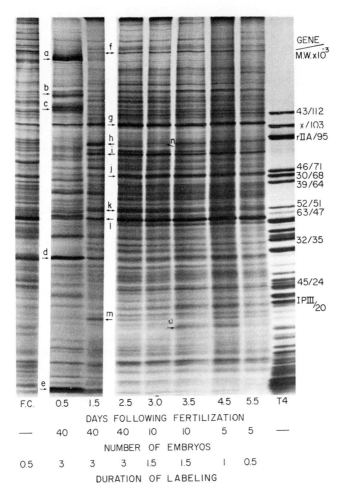

Fig. 9. Electrophoretic analysis of radiolabeled embryonic proteins during preimplantation development in the rabbit. Embryos were exposed to methionine-S[35] *in vitro* for the times indicated (in hours). Total embryonic protein, which had been dissolved in a solution containing 2-mercaptoethanol and sodium dodecyl sulfate (SDS), was heated prior to molecular weight determination using thin slab gels of polyacrylamide containing SDS. The figure is an autoradiogram of the resulting separation. Patterns of protein synthesis in rabbit follicle cells (F.C.) and in pre-replicative bacteriophage T4 are shown for comparison and to provide molecular weight markers. (Reproduced with permission of Academic Press from Van Blerkom and Manes, in press.)

available messenger templates are not in fact "saturated" with ribosomes (Latham and Darnell, 1965; Kuff and Roberts, 1967). During the period of development prior to activation of the embryonic nucleolus, ribosome availability is quite conceivably the rate-limiting factor in protein synthesis

(Kennell and Magasanik, 1962). It has been found in both mouse and rabbit embryos that amino acid incorporation increases remarkably as embryonic ribosomes become available (Tasca and Hillman, 1970; Karp *et al.*, in press). It is also found that following the activation of the embryonic nucleolus, cytoplasmic ribosomes are much more numerous and the polysomes visible by electron microscopy are much larger (Fig. 5, b and c).

Another derived fact is that the changing patterns of protein synthesis seen in rabbit embryos through the 8-cell stage (1.5 days following egg fertilization) probably occur in all embryonic cells simultaneously, since the blastomeres of the 8-cell rabbit embryo have been shown to be developmentally totipotent (Moore *et al.*, 1968). It is possible that protein synthetic patterns in presumptive trophoblast and embryoblast diverge shortly after the 8-cell stage, but such a divergence would not be evident in this study of whole embryos.

A third derived fact is that embryonic protein synthesis, if it is affected or regulated in any way by the maternal reproductive tract, is subject primarily to quantitative, rather than qualitative, controls. The changing patterns seen during cleavage, and the establishment of the "blastocyst pattern" of protein synthesis by 3.5 days following egg fertilization, occur equally well in embryos developing from the 2-cell stage to blastocysts *in vitro* (Van Blerkom and Manes, in press). With the apparent exception of two minor protein bands (Bands "N" and "O", Fig. 8), no other *new* proteins are synthesized between blastocyst formation and implantation, the developmental period when the embryo is in the uterus and is most susceptible to growth regulation (McLaren, 1973).

Since many of the protein bands seen in these gels can be resolved into multiple bands when two-dimensional separations are used, it is not possible to equate each band with a single protein. Nonetheless, many of the early bands are found to be labeled in ovulated, but unfertilized eggs (Van Blerkom and Manes, in preparation), a fact which provides presumptive evidence for maternal support of at least some protein synthesis during cleavage.

C. The Relative Roles of Maternal and Embryonic Gene Products During the Immediate Post-Fertilization Period

1. Evidence form the Use of Inhibitors of Genetic Transcription. Since it is evident that the preimplantation embryo is engaged in both RNA and protein synthesis, it becomes pertinent to ask how the two are related. Is expression of the embryonic genome required to sustain the observed protein synthesis, or is much of it supported by gene products synthesized during oogenesis? Have mammals essentially abandoned the technique of pre-programming which is characteristic of early embryogenesis in invertebrates and the lower vertebrates? Answers to these questions have classically relied upon ablating the embryonic genome itself, and assaying the effect of this ablation upon subsequent development. Since physical enucleation of the mammalian egg has proved to be technically formidable, most investigators have resorted to the use of inhibitors of genetic transcription to accomplish the "chemical enucleation" of the embryo.

Actinomycin D, the inhibitor traditionally used to block RNA synthesis, has a number of effects upon mammalian cells. In addition to its reported depression of oxidative phosphorylation (Honig and Rabinovitz, 1965; Laszlo et al., 1966) and its direct interference with phospholipid synthesis (Pastan and Friedman, 1968), it also appears to cause degradation of free polysomes (Sarma et al., 1969), and to inhibit the initiation of translation (Soeiro and Amos, 1966; Singer and Penman, 1972). Thus, its use in the investigation of the direct requirements for genetic transcription during early mammalian embryogenesis is fraught with hazards. In concentrations sufficient to inhibit uridine incorporation by at least 90%, it blocks further cleavage of both mouse and rabbit embryos (Mintz, 1964a; Thomson and Biggers, 1966; Tasca and Hillman, 1970; Monesi et al., 1970; Manes, 1973). These same studies show that the drug has a delayed suppressive effect on amino acid incorporation.

Because of the distinct possibility that Actinomycin D might be affecting the embryos at several levels, I have examined the effects of another inhibitor – α-amanitin – which for the moment appears to be a "specific" inhibitor of genetic transcription. It was found that relatively high concentrations of this poison would virtually abolish uridine incorporation in the early cleaving rabbit embryo, while allowing limited further cleavage and a level of protein synthesis somewhat higher than that seen in Actinomycin D (Manes,

While the requirement for genetic intervention may not be as immediate and continuous as the Actinomycin D studies suggested, the evidence to date nonetheless indicates that the mammalian embryo, even during the immediate post-fertilization period, does indeed rely upon its own genome for continued normal development. The nature of the required transcription product(s) is not immediately obvious. Amino acid incorporation is detectable in the unfertilized mouse egg (Brinster, 1971a), and we have found that the ovulated, but unfertilized, rabbit egg is involved in a complex pattern of protein synthesis (Van Blerkom and Manes, in preparation). Since a complete protein synthetic apparatus is apparently present and functioning in the egg prior to fertilization, we may assume that contributions are required from the embryonic genome to bring about important *qualitative* changes in the pattern of protein synthesis.

The embryonic contribution during this early period is evidently not new ribosomes. Although RNA with the properties of messenger is synthesized during cleavage (Ellem and Gwatkin, 1968; Schultz et al., 1973a), and some of this RNA is associated with polysomes (Schultz, 1973), the evidence does not necessarily allow the conclusion that it is the required embryonic contribution. Messenger RNA is synthesized and incorporated into polysomes by cleaving sea urchin embryos (Kedes and Gross, 1969), yet these embryos carry out the activities of cleavage and blastula formation in its total absence (Gross, 1964). Furthermore, some messages associated with polysomes may not be translated (Shafritz, 1974).

It should be recalled that one early and prominent gene product in the cleaving embryo is transfer RNA. The decline in the rate of amino acid incorporation in early rabbit embryos maintained in totally inhibitory concentrations of α-amanitin is almost identical to the observed rate of decay of newly-synthesized transfer RNA (Manes, 1973). Thus, it appears probable that one essential embryonic contribution during this period is transfer RNA, and such a possibility would help to explain much of the genetic data to be presented below.

2. Genetic Evidence. If cleavage of the mammalian embryo can be arrested by the essentially total chemical blockade of genetic expression, can it also be arrested by a more selective blockade? Can single embryonic genes be shown to be involved in supporting cleavage? The answer is: apparently not. The search for mutations in mammals which affect early development has centered around the T-locus in the mouse (Bennett, 1964; Mintz, 1964b), and the earliest point at which development is clocked is at the late morula stage. No single genetic locus has yet been identified whose mutation leads to faulty cleavage. To date, the earliest expression of a specific protein known to be coded in the paternal genome has been detected only at the morula stage in the mouse (Brinster, 1973a).

Indirect evidence also suggests that, if gene expression is indeed required during cleavage, it may be of a somewhat unusual kind. "Lethal hybrids" can be produced in rabbits by interspecific matings (Adams, 1957; Chang, 1960); these hybrid embryos appear to cleave normally, but die at the early blastocyst stage. Parthenogenetic embryos, and embryos with highly abnormal chromosome complements, also cleave in an apparently normal manner but go awry at the blastocyst stage or later (see Graham, 1973). Chemical mutagens and X-irradiation have no effect short of lethality on cleaving embryos, whereas abnormalities in blastocysts are readily produced (Adams *et al.*, 1961; Fisher and Smithberg, 1973).

Can this evidence be reconciled with the apparently absolute requirement for genetic expression demonstrated when "chemical enucleators" are used? It should be recalled that the eukaryotic genome is composed of both repetitive and unique sequence DNA (Britten and Kohne, 1968), and that both varieties of DNA are transcribed by the preimplantation embryo (Schultz *et al.*, 1973b). Most specific proteins are encoded by unique DNA sequences, whereas each transfer RNA species may be encoded by up to 700 DNA sequences (Roufa and Axelrod, 1971). Single-hit, or few-hit mutational events could eliminate a unique sequence gene product entirely while sparing at least some of the sequences present in multiple copies. Total inhibition of transcription, on the other hand, would obviously eliminate the expression of repetitive sequences as well. Since it has been found that transfer RNA is a prominent gene product during cleavage, this notion would resolve the apparent contradiction between genetic data and data obtained with chemical inhibitors of genetic transcription, and would lend support to the concept that protein synthesis and normal development in the immediate post-fertilization period in mammals is transfer RNA-dependent.

Another potential embryonic contribution during the cleavage period which is consistent with multiple copy gene expression is histone message. These DNA sequences may be present in up to 1200 copies (Weinberg et al., 1972), and newly-synthesized histone message has been identified in the cleaving sea urchin embryo (Kedes et al., 1969). Histone synthesis, and its dependence on support from the embryonic genome, is an unexplored area of preimplantation mammalian development.

III. BIOCHEMICAL ACTIVITIES: AN OPEN SYSTEM

The metabolic and biochemical activities which have been discovered in preimplantation embryos, and which are not limited to the direct pathway of genetic transcription and translation, have been extensively reported elsewhere and will not be treated in detail here (see, for example, Fridhandler, 1968; Biggers, 1971; Brinster, 1971b; Lutwak-Mann, 1971; Whitten, 1971; Brinster, 1973b). It would appear more to the purpose of this review to stress only certain aspects of this metabolic activity, some of which are assuming increasing importance as potential avenues of interaction between the uterus and the developing blastocyst.

One property of the preimplantation embryo which bears repetition is that it is an *open system*, in the sense of exchanging organic molecules with its immediate environment. In this regard, the embryo resembles a mammalian cell in culture more than it resembles the *closed system* embryos of the sea urchin, the frog, or the chick. In the nutritional requirements exhibited by the preimplantation embryo, a distinction must be made between the cleaving embryo and the blastocyst. Both mouse and rabbit zygotes will cleave in very simple media – much simpler than the media required for tissue culture. As to organic substrates, the mouse requires only pyruvate (or an immediately derivable metabolite, such as lactate, phosphoenolpyruvate, or oxaloacetate) for apparently normal cleavage (Cholewa and Whitten, 1970), whereas the rabbit requires only an amino nitrogen source (Kane, 1972). Normal blastocyst formation, growth, and attachment, however, require a more complex nutritional support (Pincus, 1941; Gwatkin, 1966; Hsu *et al*., 1974), and the cells of the blastocyst thus exhibit growth properties very similar to cultured embryonic or adult cells.

The nutritional requirements placed by the preimplantation mammalian embryo upon the maternal reproductive tract allow for the possibility of growth regulation, and the phenomenon of "embryonic diapause" (Enders, 1963) attests to the fact that such regulation can most certainly occur. It seems probable that the more complex nutritional requirements of the blastocyst increase its sensitivity to such regulation (McLaren, 1973), since the diapause, when it occurs, affects not the cleaving embryo but the blastocyst. In fact, Psychoyos (1971) reports that the 2-cell rat embryo develops normally to the blastocyst stage in a rat uterus in ovariectomy delay, a report which underscores the differences in growth characteristics during the pre-implantation period.

A. Membrane Composition and Function

Holley (1972) has proposed that the growth of all mammalian cells is regulated by the availability, inside the cell, of certain relatively common low molecular weight nutrients and ions. In his view, the role of macromolecular "growth factors" is to regulate the intracellular availability of the low molecular weight nutrients by regulating their transport across the cell surface. Such a concept is appealing by virtue of its simplicity and by virtue of the fact that it is supported by a great deal of experimental information. It may also explain some aspects of the growth regulation exerted by the mammalian uterus on the developing blastocyst.

Gwatkin (1966) found that omission of arginine and leucine from the culture medium would cause a reversible growth arrest in mouse blastocysts very similar to that seen in normal diapause. Examination of uterine washings during ovariectomy delay in the mouse, however, did not indicate that these two amino acids were missing, or that amino acid deprivation was the mechanism for normal diapause (Gwatkin, 1969). These findings may still be reconciled in the light of Holley's proposal, since it is not the concentration of nutrients in the uterine washings which is critical, but their concentration *inside the cells* of the blastocyst which determines whether those cells grow or do not grow. And the intracellular concentration of these nutrients in turn depends not only upon their availability outside the cell but also upon the activity of membrane transport systems. This very important aspect of preimplantation embryonic development is just beginning to receive the attention it deserves.

Borland and Tasca (1974) have investigated amino acid transport in the developing mouse embryo. The uptake of methionine and leucine is reported to be sodium-independent in the 4-cell embryo, but totally sodium-dependent in the blastocyst. This finding implies not only that sodium ion concentrations can be growth regulatory to blastocysts *in vivo*, but also that membrane transport systems display characteristic developmental patterns. Preliminary data from our laboratory indicate that the uptake of a mixture of amino acids by preimplantation rabbit embryos is likewise governed by different transport systems as development proceeds. Agents which complex with free sulfhydryl groups are strongly inhibitory to amino acid uptake in rabbit blastocysts, but have little effect on the cleaving embryo (Manes, unpublished observations).

Changes in the chemical composition of cell surfaces during the preimplantation period may also be correlated with changes in cell growth properties, much as they are in neoplastic or virally-transformed cells (Wallach, 1969). Pinsker and Mintz (1973) report an increase in the synthesis of higher molecular weight cell surface glycoproteins after blastocyst formation in the mouse, a change which has also been reported to occur in rapidly growing or virally-transformed mammalian cells (Buck *et al.*, 1971). Similarly, sialic acid becomes prominent on the cell surface of the mouse blastocyst following the estrogen surge on day 4 (Holmes and Dickson, 1973), a property which is shared with at least some transformed cells (Defendi and Gasic, 1963). It is tempting to attempt to correlate these changes with the invasive properties required of the trophoblast for implantation.

B. The Role of the Mitochondrion

Mammalian embryos require oxygen for development after fertilization (Pincus, 1941; Auerbach and Brinster, 1968), although oxygen can be toxic above certain concentrations (Whitten, 1971). It seems reasonable to postulate that this oxygen requirement reflects the necessary functioning of the mitochondrial electron transport system in the early embryo. As early as 1941, Pincus reported that cleavage of the rabbit egg could be inhibited by respiratory poisons. Thomson (1967) found that the development of 2-cell mouse embryos was arrested by cyanide at 10^{-3} M or by 2,4-dinitrophenol at 10^{-4} M. To this list can now be added Antimycin A, oligomycin, rotenone, and barbiturates, all of which block cleavage of the rabbit egg (Manes, unpublished observations).

This evidence of a requirement for unimpeded electron transport in the mitochondrial "respiratory chain" appears to be at variance with the ultrastructural appearance of embryonic mitochondria during early cleavage. If examined shortly after fertilization, mitochondria in both mouse and rabbit embryos are seen to be nearly spherical, devoid of cristae, and filled with a dense matrix (Fig. 5a). With blastocyst formation they begin to appear much more "functional", having elongated contours and conspicuous transverse cristae (Fig. 5b and c). As expected, this change in mitochondrial ultrastructure is paralleled by an increase in oxygen consumption (Fridhandler, 1961). The biochemical evidence, however, forces one to conclude that, in spite of their appearance, mitochondria and their electron transport systems are essential to early cleavage. The absolute requirement of the mouse embryo for pyruvate (Whitten, 1971), and the demonstration that carbon dioxide produced by the 2-cell mouse embryo is much more readily labeled when the ^{14}C-containing pyruvate is labeled in the C-1 rather than the C-2 position, provide further evidence for the functioning of the Krebs cycle (Wales and Whittingham, 1973).

It has been found that glucose alone will not serve as a single carbon energy source during early cleavage in either the mouse (Brinster, 1965) or the rabbit (Daniel, 1967), and that the majority of glucose metabolized by the early rabbit embryo is utilized via the hexose monophosphate pathway (Fridhandler, 1961). It therefore seems reasonable to conclude that ATP production is absolutely required by the cleaving embryo, that the chief source of this ATP is the electron transport system fueled by the Krebs cycle, and that the Embden-Meyerhof glycolytic pathway is not fully functional until somewhat later. That the glycolytic pathway can serve as a significant source of ATP after blastocyst formation is suggested by the finding that rabbit blastocysts display high rates of *aerobic* glycolysis (Fridhandler, 1961), and by Thomson's report (1967) that the mouse blastocyst is not as sensitive to electron transport inhibitors as is the cleaving embryo.

The mitochondria carrying out this necessary function are apparently all maternal (Szollosi, 1965). Furthermore, they are almost certainly a heterogeneous population (Wagner, 1969) which is not replicating (Piko, 1970; Manes, unpublished observations). Taken together, these facts allow for some interesting possibilities. They suggest that maternal, cytoplasmic inheritance of mitochondrion-associated metabolic properties may be found in mammals if carefully looked for. They further suggest that the segregation of a potentially heterogeneous population of mitochondria can occur during cleavage, and that clones of cells derived from the early blastomeres may inherit significantly different metabolic capacities. It should be recalled that marked biochemical differences have indeed been identified in the offspring of the nine-banded armadillo (*Dasypus novemcinctus*), which characteristically gives birth to monozygous quadruplets (Storrs and Williams, 1968). A heterogeneity in growth properties among cell populations in single individuals is also suggested by the occurrence of hemi-hypertrophy in humans, as well as by unilateral sensitivity to teratogens, as in the so-called "Poland anomaly" (David, 1972).

Other than the provision of an electron transport system, do mitochondria perform any other necessary function during the preimplantation period? Piko and Chase (1973) have recently reported that normal preimplantation development to the blastocyst stage occurs in the mouse under conditions in which the mitochondrial genome is allowed neither to replicate nor to synthesize RNA. Blastocysts grown in the presence of appropriate concentrations of ethidium bromide possess mitochondria which look scarcely more functional than those seen during early cleavage. These blastocysts, however, were found to implant and yield apparently normal offspring when transfered to the uteri of pseudopregnant hosts (Piko and Chase, 1973). Earlier work by Piko (1970) had suggested that DNA replication in mitochondria does not normally occur during the entire preimplantation period of the mouse at any event. It may be that mitochondrial replication is directly related to embryonic growth, since the mouse embryo, as noted earlier, does not actually grow prior to implantation. By way of contrast, the rabbit embryo grows considerably following blastocyst formation; we have not been able to demonstrate thymidine incorporation into mitochondrial DNA in the cleaving rabbit embryo, but such incorporation is unmistakably present in the rabbit blastocyst (Manes, unpublished observations). Whether this replication of mitochondrial DNA is *required* for normal rabbit blastocyst growth and differentiation is not yet known.

C. Steroid Metabolism

It has generally been assumed that the preimplantation embryo plays a passive role within a reproductive tract which is responding to ovarian steroid hormones. It is therefore of some interest to discover that the embryo's role may not be entirely passive. Huff and Eik-Nes (1966) showed some time ago that the late rabbit blastocyst can incorporate radio-labeled acetate into material which appears to be progesterone. More recently, Dickmann and Dey (1974) have demonstrated, by histochemical methods, the presence of Δ^5-3β-hydroxysteriod dehydrogenase (HSD) activity in rat morulae and early blastocysts, again suggesting that the preimplantation embryo is capable of progesterone biosynthesis. The production of steroid hormones by the embryo might well account for the reported differences in uterine secretions seen between normally pregnant and pseudopregnant uteri (Renfree, 1972; Beier and Kühnel, 1973). The reciprocal influence of the preimplantation embryo upon the uterus is an aspect of mammalian reproduction which will undoubtedly receive increasing attention.

D. The Interaction between the Blastocyst and the Uterus

One seemingly reasonable obligation imposed upon the investigator of early mammalian embryogenesis is the extrapolation of the information gleaned from *in vitro* studies to the understanding of normal and abnormal mammalian development and reproduction. The fact that the preimplantation embryo is an open system renders it all the more necessary to include the immediate and changing environment of the embryo in the total embryogenic process. There is every indication that biochemical activities in these embryos are responsive to conditions in the maternal reproductive tract; just *how* they are responsive becomes a critical question.

1. The Role of the Reproductive Tract. It has been claimed (Kirby, 1965) that at least a brief exposure to the hormonally-primed uterine milieu is a requirement for normal embryonic development past the early blastocyst stage. In view of the recent reports from Hsu and his collaborators (Hsu *et al.*, 1974) that mouse embryos can develop normally *in vitro* from the 2-cell stage to post-implantation organogenetic stages, it is difficult to maintain that the uterus is really necessary. Although the proportion of 2-cell embryos which can be successfully carried through to post-implantation stages is about 3%, and although the culture conditions are by no means defined, an absolute requirement for "special uterine factors" in preimplantation embryogenesis appears to be ruled out. Similar *in vitro* development of rabbit embryos (or others) has not yet been achieved, but it seems unlikely that the mouse will prove to be unique in this regard.

If the uterus cannot be assigned a promotional role in embryogenesis, it must nonetheless be accorded a sometime inhibitory role. Dating from the initial studies of Chang (1950), the concept of "synchrony" between embryo and reproductive tract has become firmly established. This concept simply gives recognition to the fact that two precisely-timed but separate chains of events are initiated at the time of ovulation, if it is followed by egg fertilization. The first sequence is that of embryogenesis itself, the second that of metabolic changes within the reproductive tract in response to post-ovulatory hormonal stimuli. That these two sequences must proceed in parallel for successful embryogenesis was shown by introducing artificial asynchrony in rabbits by means of embryo transfer to pseudopregnant recipients (Chang, 1950). Embryos out of synchrony with the reproductive tract by more than 24 hours showed low survival rates compared with those in synchrony. The same phenomenon has been demonstrated also in rodents, although here the embryo which is "older" than the recipient reproductive tract may slow or arrest its development until the events in the reproductive tract "catch up" (Dickmann and Noyes, 1960; Doyle *et al.*, 1963).

The uterus deprived of hormonal support by ovariectomy is directly lethal to the rabbit blastocyst (Adams, 1958), but results in an embryonic arrest in the rodent blastocyst which can be reversed by hormonal supplementation after periods as long as ten days (Weitlauf and Greenwald, 1968). Thus, judged by these two representatives, it appears that the response of the mammalian blastocyst to the non-synchronous uterus is either a rapidly irreversible growth arrest or a rather lengthy diapause during which the growth arrest is quite reversible. Although it is still controversial whether the uterus exerts its inhibitory action by failing to supply the embryo with certain necessary growth requirements or by inflicting upon the embryo some growth-inhibitory "substance" (McLaren, 1973), the growth-regulatory capacity of the uterus poses an intriguing question in mammalian reproduction.

2. Models of Tissue Interaction. Since growth regulation of mammalian cells is not a phenomenon unique to blastocysts and uteri, it is reasonable to expect that information derived from other instances of growth regulation may provide clues to illuminate this aspect of mammalian reproduction. One obvious system with potential relevance is that of monolayer cell culture, where contact inhibition or its absence can be studied. It should come as something of a reassurance to the embryologist to realize that, even under the more controllable conditions of tissue culture, controversy still surrounds the phenomenon of "contact inhibition". There is evidence that both nutrient depletion (Cunningham and Pardee, 1969; Vasilieu *et al.*, 1970) and directly inhibitory substances (Yeh and Fisher, 1969; Pariser and Cunningham, 1971) may be involved. Nonetheless, the study of the reversible inhibition of replication of mammalian cells in culture must be considered relevant to embryology until proven otherwise.

Of perhaps equal relevance, since the tissue interactions involved are heterotypic, is the analogy between embryo-uterine and epitheliomesenchymal interaction (Kirby and Cowell, 1968). Here the roles of extracellular macromolecules and the frequently inverse relationship between cell division and differentiation have been reasonably well established (Grobstein, 1967). The interplay between the growth-inhibitory properties of mucopolysaccharides (Lippman, 1965) and the growth-promoting properties of protein "growth factors" (Holley, 1972) may well have implications for the behavior of blastocysts in uteri. The mammalian uterus, during the secretory phase of the ovarian cycle, secretes large quantities of mucopolysaccharides (Zachariae, 1958; Endo and Yosizawa, 1973) and "uterine specific protein" (Krishnan and Daniel, 1967) into the immediate environment of the blastocyst. This hormone-specific response of the uterus must again be assumed to have relevance to embryonic development and the phenomenon of synchrony until proven otherwise.

IV. SUMMARY AND CONCLUSIONS

The initial stages of mammalian embryogenesis evidently require the transcription of the embryonic genome for their normal progression. Prior to the activation of the embryonic nucleolus, the transcription products include both transfer and messenger RNA. A model has been presented in an attempt to reconcile the fact that complete inhibition of genetic transcription will ultimately arrest cleavage with the fact that genetic lesions are not expressed until the morula or blastocyst stage. The model postulates that the contribution required of the embryonic genome for normal cleavage is the product of multiple-copy genes, and potential candidates for this role would include both transfer RNA species and histone messenger RNA.

The preimplantation embryo has also been considered as an open system, exchanging organic substrates with its environment in an obligatory manner. The nutritional requirements of the blastocyst appear to be more complex than those of the cleaving embryo. The ability of the mammalian uterus to arrest — either reversibly or irreversibly — the growth of the blastocyst, is quite possibly related to these nutritional requirements, and to the acquisition by the blastocyst of membrane transport systems which can be influenced by the uterine milieu.

An attempt has also been made to point out what is *not* known about preimplantation embryos as well as the contributions which molecular and membrane biology have already made to this area. Above all, it is suggested that a determined search for relatively simple, integrative mechanisms will be more fruitful in the near future than the continued collection of data. It seems highly unlikely, to use Robert Frost's' phrase, that "accumulated fact will, of itself, take fire and light the world up" (1963).

ACKNOWLEDGMENTS

The work discussed from the author's laboratory was supported by grants from the United States Public Health Service, National Institute of Child Health and Human Development. I thank Dr. Jonathan Van Blerkom for his criticisms and suggestions during the preparation of this manuscript.

REFERENCES

Adams, C. E. (1957). An attempt to cross the domestic rabbit (*Oryctolagus cuniculus*) and hare (*Lepus europaeus*). *Nature (London)* **180**, 853.

Adams, C. E. (1958). Egg development in the rabbit: the influence of postcoital ligation of the uterine tube and of ovariectomy. *J. Endocrin.* **16**, 283–293.

Adams, C. E., Hay, M. F., and Lutwak-Mann, C. (1961). The action of various agents upon the rabbit embryo. *J. Embryol. Exp. Morph.* **9**, 468–491.

Anderson, W. F., and Gilbert, J. M. (1969). tRNA-dependent translation control of *in vitro* hemoglobin synthesis. *Biochem. Biophys. Res. Commun.* **36**, 456–462.

Auerbach, S., and Brinster, R. L. (1968). Effect of O_2 concentration on development of two-cell mouse eggs. *Nature (London)* **217**, 465–466.

Beier, H. M., and Kühnel, W. (1973). Pseudopregnancy in the rabbit after stimulation by human chorionic gonadotropin. *Hormone Res.* **4**, 1–27.

Bennett, D. (1964). Abnormalities associated with a chromosome region in the mouse. *Science* **144**, 260–267.

Biggers, J. D. (1971). New observations on the nutrition of the mammalian oöcyte and the preimplantation embryo. *In* "The Biology of the Blastocyst" (R. J. Blandau, ed.), pp. 319–327. University of Chicago Press, Chicago.

Borland, R. M., and Tasca, R. J. (1974). Activation of a Na^+-dependent amino acid transport system in preimplantation mouse embryos. *Develop. Biol.* **36**, 169–182.

Brinster, R. L. (1965). Studies on the development of mouse embryos *in vitro*. II. The effect of energy source. *J. Exp. Zool.* **158**, 59–69.

Brinster, R. L. (1971a). Uptake and incorporation of amino acids by the preimplantation mouse embryo. *J. Reprod. Fertil.* **27**, 329–338.

Brinster, R. L. (1971b). Mammalian embryo metabolism. *In* "The Biology of the Blastocyst" (R. J. Blandau, ed.), pp. 303–318. University of Chicago Press, Chicago.

Brinster, R. L. (1973a). Parental glucose phosphate isomerase activity in three-day mouse embryos. *Biochem. Gen.* **9**, 187–191.

Brinster, R. L. (1973b). Protein synthesis and enzyme constitution of the preimplantation mammalian embryo. *In* "The Regulation of Mammalian Reproduction" (S. J. Segal, ed.), pp. 302–316. Charles C. Thomas, Publisher, Springfield, Illinois.

Britten, R. J., and Kohne, D. E. (1968). Repeated sequences in DNA. *Science* **161**, 529–540.

Britten, R. J., and Davidson, E. H. (1969). Gene regulation for higher cells: a theory. *Science* **165**, 349–357.

Buck, C. A., Glick, M. C., and Warren, L. (1971). Glycopeptides from the surface of control and virus-transformed cells. *Science* **172**, 169–171.

Chang, M. C. (1950). Development and fate of transferred rabbit ova or blastocyst in relation to the ovulation time of recipients. *J. Exp. Zool.* **114**, 197–225.

Chang, M. C. (1960). Fertilization of domestic rabbit (*Oryctolagus cuniculus*) ova by cottontail rabbit (*Sylvilagus transitionalis*) sperm. *J. Exp. Zool.* **144**, 1–10.

Cholewa, J. A., and Whitten, W. K. (1970). Development of 2-cell mouse embryos in the absence of a fixed nitrogen source. *J. Reprod. Fertil.* **22**, 553–555.

Cunningham, D. D., and Pardee, A. B. (1969). Transport changes rapidly initiated by serum addition to "contact inhibited" 3T3 cells. *Proc. Nat. Acad. Sci. U.S.* **64**, 1049–1056.

Daniel, J. C., Jr. (1967). The pattern of utilization of respiratory metabolic intermediates by preimplantation rabbit embryos *in vitro*. *Exp. Cell Res.* **47**, 619–624.

David, T. J. (1972). Nature and etiology of the Poland anomaly. *New England J. Med.* **287**, 487–489.

Defendi, V., and Gasic, G. (1963). Surface mucopolysaccharides of polyoma virus transformed cells. *J. Cell Comp. Physiol.* **62**, 23–26.

Dickmann, Z., and Noyes, R. W. (1960). The fate of ova transferred into the uterus of the rat. *J. Reprod. Fertil.* **1**, 197–212.

Dickmann, Z., and Dey, S. K. (1974). Steroidogenesis in the preimplantation rat embryo and its possible influence on morula-blastocyst transformation and implantation. *J. Reprod. Fertil.* **37**, 91–93.

Doyle, L. L., Gates, A. H., and Noyes, R. W. (1963). Asynchronous transfer of mouse ova. *Fertil. Steril.* **14**, 215–225.

Dumont, J., Yamada, T., and Cone, M. (1970). Alteration of nucleolar ultrastructure in iris epithelial cells during initiation of Wolffian lens regeneration. *J. Exp. Zool.* **174**, 187–203.

Ellem, K. A. O., and Gwatkin, R. B. L. (1968). Patterns of nucleic acid synthesis in the early mouse embryo. *Develop. Biol.* **18**, 311–330.

Enders, A. C. (ed.) (1963). "Delayed Implantation". University of Chicago Press, Chicago.

Endo, M., and Yosizawa, Z. (1973). Hormonal effect on glycoproteins and glycosaminoglycans in rabbit uteri. *Arch. Biochem. Biophys.* **156**, 397–403.

Epstein, C. J. (1973). Discussion following: Protein syntehsis and enzyme constitution of the preimplantation mammalian embryo. *In* "The Regulation of Mammalian Reproduction" (S. J. Segal, ed.), pp. 317–318. Charles C. Thomas, Publisher, Springfield, Illinois.

Fisher, D. L., and Smithberg, M. (1973). *In vitro* and *in vivo* X-irradiation of preimplantation mouse embryos. *Teratology* **7**, 57–64.

Fridhandler, L. (1961). Pathways of glucose metabolism in fertilized rabbit ova at various preimplantation stages. *Exp. Cell Res.* **22**, 303–316.

Fridhandler, L. (1968). Intermediary metabolic pathways in preimplantation rabbit blastocysts. *Fertil. Steril.* **19**, 424–434.

Frost, R. (1963). *Selected Poems of Robert Frost.* Holt, Rinehart, & Winston, Inc., New York, p. 293.

Gillam, I., Millward, S., Blew, D., Tigerstrom, M., Wimmer, E., and Tener, G. M. (1967). The separation of soluble ribonucleic acids on benzoylated diethylaminocellulose. *Biochemistry* **6**, 3043–3056.

Graham, C. F. (1973). The necessary conditions for gene expression during early mammalian development. In "Genetic Mechanisms of Development" (F. H. Ruddle, ed.), pp. 202–224. Academic Press, New York.

Greenslade, F. C., McCormack, J. J., and Hahn, D. W. (1972). Embryonic and uterine RNA synthesis during early gestation of the rabbit. *Biol. Reprod.* **7**, 142.

Grobstein, C. (1967). Mechanisms of organogenetic tissue interaction. *Nat. Cancer Inst. Monogr.* **26**, 279–299.

Gross, P. R. (1964). The immediacy of genomic control during early development. *J. Exp. Zool.* **157**, 21–38.

Gwatkin, R. B. L. (1966). Amino acid requirements for attachment and outgrowth of the mouse blastocyst *in vitro. J. Cell Physiol.* **68**, 335–344.

Gwatkin, R. B. L. (1969). Nutritional requirements for post-blastocyst development opment in the mouse. *Int. J. Fertil.* **14**, 101–105.

Hanoune, J., and Agarwal, M. K. (1970). Studies on the half-life time of rat liver transfer RNA species. *F.E.B.S. Letters* **11**, 78–80.

Hay, E., and Gurdon, J. (1967). Fine structure of the nucleolus in normal and mutant *Xenopus* embryos. *J. Cell Sci.* **2**, 151–162.

Hillman, N., and Tasca, R. J. (1969). Ultrastructural and autoradiographic studies of mouse cleavage stages. *Am. J. Anat.* **126**, 151–174.

Holley, R. W. (1972). A unifying hypothesis concerning the nature of malignant growth. *Proc. Nat. Acad. Sci. U.S.* **69**, 2840–2841.

Holmes, P. V., and Dickson, A. D. (1973). Estrogen-induced surface coat and enzyme changes in the implanting mouse blastocyst. *J. Embryol. Exp. Morphol.* **29**, 639–645.

Honig, G. R., and Rabinovitz, M. (1965). Actinomycin-D inhibition of protein synthesis unrelated to effect on template RNA synthesis. *Science* **149**, 1504–1506.

Hsu, Y., Baskar, J., Stevens, L. C., and Rash, J. E. (1974). Development *in vitro* of mouse embryos from the two-cell egg stage to the early somite stage. *J. Embryol. Exp. Morph.* **31**, 235–245.

Huff, R. L., and Eik-Nes, K. B. (1966). Metabolism *in vitro* of acetate and certain steroids by six-day-old rabbit blastocysts. *J. Reprod. Fertil.* **11**, 57–63.

Kane, M. T. (1972). Energy substrates and culture of single cell rabbit ova to blastocysts. *Nature (London)* **238**, 468–469.

Karp, G., Manes, C., and Hahn, W. E. (1974). Ribosome production and protein synthesis in the preimplantation rabbit embryo. (in press).

Kedes, L. H., and Gross, P. R. (1969). Synthesis and function of messenger RNA during early embryonic development. *J. Mol. Biol.* **42**, 559–575.

Kedes, L. H., Gross, P. R., Cognetti, G., and Hunter, A. (1969). Synthesis of nuclear and chromosomal proteins on light polyribosomes during cleavage in the sea urchin embryo. *J. Mol. Biol.* **45**, 337–351.

Kennell, D., and Magasanik, B. (1962). The relation of ribosome content to the rate of enzyme synthesis in *Aerobacter aerogenes. Biochim. Biophys. Acta* **55**, 139–151.

Kirby, D. R. S. (1965). The role of the uterus in the early stages of mouse development. *In* "Preimplantation Stages of Pregnancy" (G. E. W. Wolstenholme and M. O'Connor, eds.), pp. 325–339. Little, Brown & Co., Boston, Massachusetts.

Kirby, D. R. S., and Cowell, T. P. (1968). Trophoblast-host interactions. *In* "Epithelial-Mesenchymal Interactions" (R. Fleischmajer and R. E. Billingham, eds.), pp. 64–77. Williams and Wilkins, Baltimore, Maryland.

Knowland, J., and Graham, C. (1972). RNA synthesis at the two-cell stage of mouse development. *J. Embryol. Exp. Morph.* **27**, 167–176.

Krishnan, R. S., and Daniel, J. C., Jr. (1967). "Blastokinin": Inducer and regulator of blastocyst development in the rabbit uterus. *Science* **158**, 490–492.

Kuff, E. L., and Roberts, N. E. (1967). *In vivo* labeling patterns of free polyribosomes: relationship to tape theory of messenger ribonucleic acid function. *J. Mol. Biol.* **26**, 211–225.

Laszlo, J., Miller, D. S., McCarty, K. S., and Hochstein, P. (1966). Actinomycin D: Inhibition of respiration and glycolysis. *Science* **151**, 1007–1010.

Latham, H., and Darnell, J. E. (1965). Distribution of mRNA in the cytoplasmic polyribosomes of the HeLa cell. *J. Mol. Biol.* **14**, 1–12.

Lippman, M. (1965). A proposed role for mucopolysaccharides in the initiation and control of cell division. *Trans. N. Y. Acad. Sci.* **27**, 342–360.

Lutwak-Mann, C. (1971). The rabbit blastocyst and its environment: Physiological and biochemical aspects. *In* "The Biology of the Blastocyst" (R. J. Blandau, ed.), pp. 243–260. University of Chicago Press, Chicago.

Manes, C. (1969). Nucleic acid synthesis in preimplantation rabbit embryos. I. Quantitative aspects, relationship to early morphogenesis and protein synthesis. *J. Exp. Zool.* **172**, 303–310.

Manes, C. (1971). Nucleic acid synthesis in preimplantation rabbit embryos. II. Delayed synthesis of ribosomal RNA. *J. Exp. Zool.* **176**, 87–96.

Manes, C. (1973). The participation of the embryonic genome during early cleavage in the rabbit. *Develop. Biol.* **32**, 453–459.

Manes, C., and Daniel, J. C., Jr. (1969). Quantitative and qualitative aspects of protein synthesis in the preimplantation rabbit embryo. *Exp. Cell Res.* **55**, 261–268.

Manes, C., and Sharma, O. K. (1973). Hypermethylated tRNA in cleaving rabbit embryos. *Nature (London)* **244**, 283–284.

McLaren, A. (1973). Blastocyst activation. *In* "The Regulation of Mammalian Reproduction" (S. J. Segal, ed.), pp. 321–328. Charles C. Thomas, Publisher, Springfield, Illinois.

Mintz, B. (1964a). Synthetic processes and early development in the mammalian egg. *J. Exp. Zool.* **157**, 85–100.

Mintz, B. (1964b). Gene expression in the morula stage of mouse embryos as observed during development of t^{12}/t^{12} lethal mutants *in vitro*. *J. Exp. Zool.* **157**, 267–272.

Monesi, V., and Salfi, V. (1967). Macromolecular synthesis during early development in the mouse embryo. *Exp. Cell Res.* **46**, 632–635.

Monesi, V., Molinaro, M., Spalletta, E., and Davoli, C. (1970). Effect of metabolic inhibitors on macromolecular synthesis and early development in the mouse embryo. *Exp. Cell Res.* **59**, 197–206.

Moore, N. W., Adams, C. E., and Rowson, L. E. A. (1968). Developmental potential of single blastomeres of the rabbit egg. *J. Reprod. Fertil.* **17**, 527–531.

Pariser, R. J., and Cunningham, D. D. (1971). Transport inhibitors released by 3T3 mouse cells and their relation to growth control. *J. Cell Biol.* **49**, 525–529.

Pastan, I., and Friedman, R. M. (1968). Actinomycin D: Inhibition of phospholipid synthesis in chick embryo cells. *Science* **160**, 316–317.

Piko, L. (1970). Synthesis of macromolecules in early mouse embryos cultured *in vitro*: RNA, DNA and a polysaccharide component. *Develop. Biol.* **21,** 257–279.

Piko, L., and Chase, D. G. (1973). Role of the mitochondrial genome during early development in mice. *J. Cell Biol.* **58,** 357–378.

Pincus, G. (1941). The control of ovum growth. *Science* **93,** 438–439.

Pinsker, M. C., and Mintz, B. (1973). Change in cell-surface glycoproteins of mouse embryos before implantation. *Proc. Nat. Acad. Sci. U.S.* **70,** 1645–1648.

Psychoyos, A. (1971). Discussion following: Macromolecular synthesis and effect of metabolic inhibitors during preimplantation development in the mouse. *Adv. Biosci.* **6,** 119.

Reamer, G. R. (1963). Ph.D. Thesis, Boston University.

Renfree, M. B. (1972). Influence of the embryo on the marsupial uterus. *Nature (London)* **240,** 475–477.

Roufa, D. J., and Axelrod, D. (1971). The repeated transfer RNA genes of animal cells in culture. *Biochem. Biophys. Acta.* **254,** 429–439.

Sarma, D. S. R., Reid, I. M., and Sidransky, H. (1969). The selective effect of Actinomycin D on free polyribosomes of mouse liver. *Biochem. Biophys. Res. Commun.* **36,** 582–588.

Scherrer, K., and Marcaud, L. (1968). Messenger RNA in avian erythroblasts at the transcriptional and translational levels and the problem of regulation in animal cells. *J. Cell Physiol.* **72 (Suppl. 1),** 181–212.

Schultz, G. A. (1973). Characterization of polyribosomes containing newly synthesized messenger RNA in preimplantation rabbit embryos. *Exp. Cell Res.* **82,** 168–174.

Schultz, G. A. (1974). The stability of messenger RNA containing polyadenylic acid sequences in rabbit blastocysts. Exp. Cell Res., (in press)

Schultz, G., Manes, C., and Hahn, W. E. (1973a). Synthesis of RNA containing polyadenylic acid sequences in preimplantation rabbit embryos. *Develop. Biol.* **30,** 418–426.

Schultz, G. A., Manes, C., and Hahn, W. E. (1973b). Estimation of the diversity of transcription in early rabbit embryos. *Biochem. Gen.* **9,** 247–259.

Schultz, G. A., and Church, R. B. (1974). Transcriptional patterns in early mammalian development. *In* "The Biochemistry of Animal Development" in press.

Shafritz, D. A. (1974). Evidence for nontranslated messenger ribonucleic acid in membrane-bound and free polysomes of rabbit liver. *J. Biol. Chem.* **249,** 89–93.

Singer, R. H., and Penman, S. (1972). Stability of HeLa cell mRNA in Actinomycin. *Nature (London)* **240,** 100–102.

Singer, R. H., and Penman, S. (1973). Messenger RNA in HeLa cells: kinetics of formation and decay. *J. Mol. Biol.* **78,** 321–334.

Soeiro, R., and Amos, H. (1966). mRNA half-life measured by the use of Actinomycin D in animal cells – a caution. *Biochim. Biophys. Acta* **129,** 406–409.

Storrs, E. E., and Williams, R. J. (1968). A study of monozygous quadruplet armadillos in relation to mammalian inheritance. *Proc. Nat. Acad. Sci. U.S.* **60,** 910–914.

Szollosi, D. G. (1965). The fate of sperm middle-piece mitochondria in the rat egg. *J. Exp. Zool.* **159,** 367–378.

Tasca, R. J., and Hillman, N. (1970). Effects of Actinomycin D and cycloheximide on RNA and protein synthesis in cleavage stage mouse embryos. *Nature (London)* **225,** 1022–1025.

Thomson, J. L. (1967). Effects of inhibitors of carbohydrate metabolism on the development of preimplantation mouse embryos. *Exp. Cell Res.* **46,** 252–262.

Thomson, J. L., and Biggers, J. D. (1966). Effect of inhibitors of protein synthesis on the development of preimplantation mouse embryos. *Exp. Cell Res.* **41,** 411–427.

Van Blerkom, J., Manes, C., and Daniel, J. C., Jr. (1973). Development of preimplantation rabbit embryos *in vivo* and *in vitro*. I. An ultrastructural comparison. *Develop.
Biol.* **35**, 262–282.

Van Blerkom, J., and Manes, C. (1974). Development of preimplantation rabbit embryos
in vivo and *in vitro*. II. A comparison of qualitative aspects of protein synthesis.
Develop. Biol. (in press).

Vasiliev, J. M., Gelfand, I. M., Guelstein, V. I., and Fetisova, E. K. (1970). Stimulation of
DNA synthesis in cultures of mouse embryo fibroblast-like cells. *J. Cell Physiol.* **75**,
305–314.

Wagner, R. P. (1969). Genetics and phenogenetics of mitochondria. *Science* **163**,
1026–1031.

Wainwright, S. D. (1971). Stimulation of hemoglobin synthesis in developing chick
blastodisc blood islands by a minor alanine-specific transfer RNA. *Cancer Res.* **31**,
694–696.

Wales, R. G., and Whittingham, D. G. (1973). The metabolism of specifically labeled
lactate and pyruvate by two-cell mouse embryos. *J. Reprod. Fertil.* **33**, 207–222.

Wallach, D. F. H. (1969). Generalized membrane defects in cancer. *New England J. Med.*
280, 761–767.

Warner, J. R., Knopf, P. M., and Rich, A. (1963). A multiple ribosomal structure in
protein synthesis. *Proc. Nat. Acad. Sci. U.S.* **49**, 122–129.

Weinberg, E. S., Birnstiel, M. L., and Purdom, I. F. (1972). Genes coding for polysomal
9S RNA of sea urchins; conservation and divergence. *Nature (London)* **240**, 225–228.

Weiss, P. A. (1969). Living nature and the knowledge gap. *Saturday Review,* Nov. 29, pp.
19–22.

Weitlauf, H. H., and Greenwald, G. S. (1968). Survival of blastocysts in the uteri of
ovariectomized mice. *J. Reprod. Fertil.* **17**, 515–520.

Whitten, W. K. (1971). Nutrient requirements for the culture of preimplantation embryos
in vitro. Adv. Biosci. **6**, 129–139.

Woodland, H. R., and Graham, C. F. (1969). RNA synthesis during early development of
the mouse. *Nature (London)* **221**, 327–332.

Yeh, J., and Fisher, H. W. (1969). A diffusible factor which sustains contact inhibition of
replication. *J. Cell Biol.* **40**, 382–388.

Zachariae, F. (1958). Autoradiographic and histochemical studies of sulfomucopoly-
saccharides in the rabbit uterus, oviducts, and vagina. *Acta Endocrin.* **29**, 118–134.

Zubay, G., and Takanami, M. (1964). Observations on the configuration of nucleotides
near the 3'-hydroxy end of adapter RNA. *Biochem. Biophys. Res. Commun.* **15**,
207–213.

The Regulation of Enzyme Synthesis in the Embryogenesis and Germination of Cotton

Virginia Walbot, Barry Harris,
and L. S. Dure, III

Department of Biochemistry
University of Georgia
Athens, Georgia 30602

I. Introduction 165
II. The System 166
III. Cordycepin Experiments 169
IV. Isolation of a Ribonucleoprotein Particle
from Developing Cotyledons 173
V. Polyadenylation of Messenger RNA During Germination 179
VI. Conclusions 185
References 186

I. INTRODUCTION

Seed germination in higher plants is the most crucial stage in insuring the survival of the species. Thus, while the seed and embryo are developing on the vegetative plant, various reserve energy and information sources are laid down in the storage tissues, the cotyledons. During germination these stored materials such as proteins, lipids, starch, and phytic acid are metabolized to provide nutrition to the growing root-shoot axis. Thus, there is a temporal separation of the anabolism and catabolism of the stored reserves within the cotyledonary tissue.

It was noted a few years ago by Ihle and Dure (1969) that two of the enzymes involved in the degradation of the stored lipid and protein of cotton seeds were synthesized *de novo* during germination, whereas the messenger RNAs coding for these enzymes were apparently synthesized during embryogenesis. This suggests that, as part of the preparation for germination, mRNA for proteins unique to the germination process is also prepositioned. Thus, there is a temporal separation of the transcription and translation of the messenger RNAs involved in the breakdown of the stored reserve materials. The analogy to sea urchin and amphibian eggs, where the material genome

165

contributes a store of mRNA that is used to support the initial post-fertilization development of the zygote, is at once obvious to the developmental
biologist; however, it is important to remember that the seed is a multicellular
system composed of fully differentiated tissues. The phenomenon of stored
mRNA probably exists in other systems as well, as one developmental stage
provides an endowment of structural and macromolecular information that
will be utilized by a subsequent stage. How this separation is effected has
been one of the principle interests of this laboratory. Thus, our inquiry into
the formation, storage and subsequent use of the stored messenger RNA in
the cotton seed may prove useful in comprehending a phenomenon of general
importance to developmental biology.

II. THE SYSTEM

Cotton plants are maintained in a greenhouse so as to flower year-round
and flowers are tagged on the day of anthesis. Within each boll approximately
32 embryos develop synchronously. Figure 1 gives the developmental time-
table of the cotton embryo. During the initial 20 days post-anthesis there is
relatively little accumulation of wet or dry weight by the embryo, although
the endosperm of the seed and the boll itself reach their full size during this
period. During the subsequent 20 days the embryos increase rapidly in cell
number and weight. At approximately 28 days, when the embryo has
increased to 85 mg wet weight the vascular connection (finiculus) between
each ovule and the maternal tissue of the vegetative plant atrophies; concomitantly cell division ceases in the cotyledons and the mRNAs that are stored in

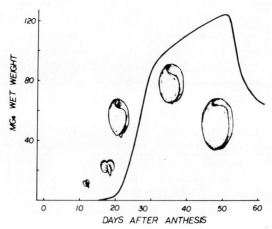

Fig. 1. Increase in wet weight of cotton embryos during embryogenesis, and a
depiction of the relative size of embryos during development (from Ihle & Dure, 1972c).

the cotyledons are first demonstratable (Ihle and Dure, 1969). The embryo continues to increase in wet and dry weight for an additional 10 days before the desiccation phase begins; during this period the reserve proteins and lipids continue to be synthesized.

An interesting characteristic of cotton and other higher plant embryos is that they are capable of precocious germination; that is, if embryos are removed from the surrounding ovular tissue and supplied with water they will germinate. Cotton embryos will germinate successfully, although slowly at the younger stages, from about 20 days post-anthesis onward. This phenonmenon of precocious germination has been used to test at what point the embryos form the mRNA required for germination in general and for the enzymes required to mobilize the stored reserves specifically. The two enzymes studied are carboxypeptidase C (Ihle and Dure, 1972a, b) and isocitratase. The carboxypeptidase activity develops in cotyledons in normal germination as shown in Fig. 2. Enzyme synthesis is seen to be sensitive to cycloheximide but insensitive to inhibition by Actinomycin D. This enzyme was purified from embryos germinated in radioactive amino acids to substantiate its *de novo* synthesis (Ihle and Dure, 1972a). Embryos precociously germinated before the 85 mg stage of development do not develop these enzyme activities in the presence of Actinomycin D, while embryos precociously germinated after the 85 mg stage do develop enzyme activity as shown in Fig. 2 (left panel).

This observation that the mRNAs coding for the two enzymes involved in the catabolism of the stored reserves are transcribed while these reserves are being deposited led to the question of how are these mRNAs prevented from being translated during the last stages of embryogenesis. That is, since dissected embryos readily germinate precociously, what prevents precocious germination *in vivo*? Vivipary, which in the realm of higher plants means precocious germination *in vivo*, is a genetic defect in some strains of cotton and other higher plants, and it is generally a lethal mutant since the embryos germinate while still in the arborea fruit. Thus, although much of the information required for germination seems to be contained in the embryos in the form of mRNA, it is not used until the proper developmental time, germination.

Fig. 2 summarizes the evidence presented by Ihle and Dure (1970) indicating that the plant hormone abscisic acid is responsible for suppressing the translation of the mRNAs coding for enzymes required during germination. The presence of this hormone becomes demonstrable in the ovular tissue surrounding each developing embryo at the time the vascular finiculus connecting the seed to the vegetative plant begins to degenerate about 3/5 through the development of the embryo. As is shown in Fig. 2 application of abscisic acid or an ovular extract to precociously germinating embryos

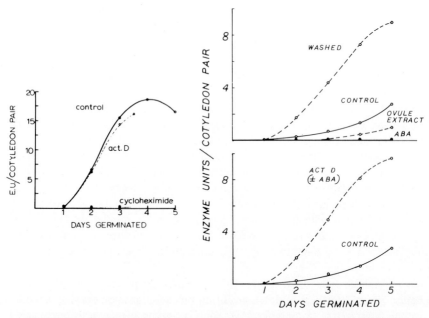

Fig. 2. *Left Panel*. Appearance of carboxypeptidase C activity in cotyledons during the normal germination of mature embryos and the effects of cycloheximide (1 mg/ml) and Actinomycin D (20 μg/ml) on the appearance of this activity. Embryos were imbibed and maintained in solutions containing the inhibitors for the entire incubation period. *Right Panel*. Appearance of carboxypeptidase C activity in cotyledons during the precocious germination of immature embryos weighing 95 mg wet weight, and the effects of various treatments on the appearance of this activity. *Top*. Control embryos were placed on wet filter paper directly from the ovule. Washed embryos were shaken for 30 minutes in distilled water prior to transfer to wet filter paper. Ovule extract refers to embryos initially washed, but placed on wet filter paper containing a concentrate of the material washed off the embryos in the washing procedure. ABA refers to embryos initially washed, but placed on wet filter paper containing 10^{-6} M ABA. *Bottom*. Act D (±ABA) refers to embryos washed in Actinomycin D (20 μg/ml) and transferred to filter paper saturated with distilled water or with a solution of 10^{-6} M ABA. Reprinted from Ihle & Dure (1972c).

prevents the appearance of the diagnostic enzyme activities. The enhancement of the appearance of enzyme activity in the washed embryos presumably is due to the removal of ovular abscisic acid adhering to the embryo surface in the process of being absorbed. The abscisic acid effect itself is mediated by the concomitant synthesis of RNA and protein, however, since the simultaneous application of Actinomycin D with the abscisic acid prevents abscisic acid mediated inhibition of enzyme appearance. Abscisic acid is only effective in mediating this translation inhibition during embryogenesis or precocious germination; the hormone is without inhibitory activity if applied during the normal germination of mature embryos. Thus, the control over the translation of preformed germination mRNAs is restricted to that period when their translation would bring about vivipary.

From the foregoing it can be seen that the developing cotton cotyledon provides a developmental system in which the transcription and translation of two enzymes required for germination can be followed. In addition, the cotton cotyledon offers some advantages over other systems in that large quantities of material of known physiological age can be obtained. Since the cotyledons cease mitosis after the 85 mg stage of embryogenesis and do not undergo cell division during the first 5 days of germination, the fate of the mRNA for the germination enzymes can be studied in a rather uncomplicated system. We have turned our attention, therefore, to studying what happens to these RNAs from the time of their synthesis to their translation. Basically we have taken two approaches: Firstly, we have attempted to isolate the storage form of the mRNAs and, secondly, we have attempted to follow the processing of these mRNAs during germination.

III. CORDYCEPIN EXPERIMENTS

The first indication that the putative stored mRNA may not be processed until germination came from experiments using the adenosine analogue cordycepin, 3'-deoxyadenosine, which has been used to inhibit the *in vivo* addition of polyadenylate chains to the 3' OH end of mRNAs in a variety of organisms (Penman, *et al.*, 1970). Poly(A) addition is considered to be one of the processing steps in the progression of an initial gene transcript to polysomal mRNA (Jelinek, *et al.*, 1973). Application of cordycepin to germinating cotton embryos results in a complete suppression of the process of germination whereas Actinomycin D treated embryos closely resemble controls after several days of germination. If the *in vivo* incorporation of radioactive amino acids into acid precipitable material is measured in antibiotic treated germinating cotyledons as is shown in Fig. 3, it can be seen that cordycepin is an effective inhibitor of this process as compared to Actinomycin D treated or control embryos. However, another nucleotide analogue, 3'-deoxycytidine, which could conceivably mimic the effect of cordycepin on RNA synthesis but leave poly(A) synthesis undisturbed, has no effect on the *in vivo* amino acid incorporation. Also shown in Fig. 3 is the development in germination of the diagnostic enzyme carboxypeptidase in the presence of the various inhibitors; as was found with amino acid incorporation into protein, cordycepin effectively suppresses the appearance of the enzymes we consider to be coded for by preformed mRNA, while 3'-deoxycytidine and, as shown before, Actinomycin D have no effect on enzyme appearance. If the 3'-deoxytriphosphate derivatives of these analogues are inserted into a growing

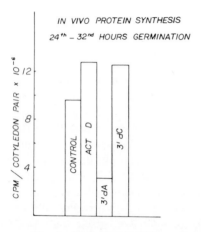
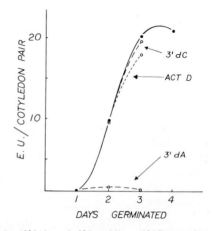

Fig. 3. *Right Panel.* Effect of cordycepin (3'dA) and 3'd-cytidine (3'dC) on the appearance of carboxypeptidase C activity in cotyledons during the germination of mature embryos compared with the effect of Actinomycin D. Embryos were imbibed and maintained in solutions of 5 mM nucleoside analogs or 20 µg/ml Actinomycin D. *Left Panel.* Effect of the same three compounds on the *in vivo* incorporation of [^{14}C] leucine and [^{14}C] valine into protein by germinating cotyledons during the indicated time period of germination. Embryos were imbibed and maintained in solutions of the compounds as given above and the radioactive amino acids introduced to the incubation solutions after 24 hours of germination. After 32 hours of germination the cotyledons were harvested and the incorporation of radioactivity into extractable protein measured by conventional means. Reprinted from Walbot *et al.* (in preparation).

polynucleotide chain they will act as chain terminators. The effect of cordycepin triphosphate on RNA and poly(A) polymerase *in vitro* supports this mechanism of inhibition (Horowitz, *et al.*, 1974). It appears from Fig. 3 that we had found an inhibitory effect specific to cordycepin and not a general effect of 3'-deoxynucleoside analogues. These observations are consistent with the hypothesis that the cordycepin sensitive step is the addition of poly(A) to the stored mRNAs, a step with which neither Actinomycin D nor 3'-deoxycytidine should directly interfere.

To further test this hypothesis we attempted to follow the time course of the cordycepin inhibition of carboxypeptidase appearance during germination. In Fig. 4 embryos were initially imbibed with water and then transferred to cordycepin at the times indicated. Little or no enzyme activity subsequently appears in embryos transferred at 6 hours; progressively more enzyme appears the longer the embryos were allowed to remain in water before transfer to the drug, and the control levels of enzyme activity are reached in embryos transferred after 30 hrs. Thus, it appears that the cordycepin sensitive step in enzyme appearance begins after the 6th hour of germination and is complete by the 30th hour. Also shown in Fig. 4 are the results of experiments in

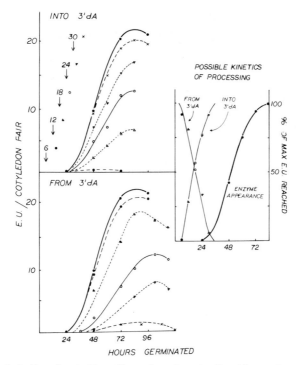

Fig. 4. *Left Top Panel.* The effect of cordycepin (5 mM) on the appearance of carboxypeptidase C activity in germinating cotyledons transferred to solutions of the inhibitor at various times after the commencement of germination. *Left Bottom Panel.* The appearance and increase of carboxypeptidase activity in germinating cotyledons after their removal from solutions of cordycepin (5 mM) at the times indicated. Reprinted from Walbot *et al.* (in preparation).

Right Panel. A time course of the processing of mRNA's in germinating cotyledons.

which embryos were first exposed to cordycepin and then transferred to water. In this case there is a progressive reduction in enzyme levels reached the later the embryos are removed from cordycepin. This finding suggests that the pool of stored mRNAs is depleted during incubation in cordycepin either through nuclease degradation or through inactivation by the poly(A) chain terminating action of 3' deoxy ATP. It is interesting to note that the apparent time course of mRNA decay in the presence of cordycepin is the same as the time course of presumed mRNA processing shown in Fig. 4. This, in turn, suggests that we are looking at a single process, presumably the addition of poly(A) to the stored mRNAs. If this process is aborted, as in the presence of cordycepin, the mRNAs are never used.

The data obtained from assays of carboxypeptidase in cotyledons transferred to cordycepin at intervals after germination is initiated is reflected by the extent to which the embryos visibly germinate. Fig. 5 shows embryos that were transferred to cordycepin solutions at intervals of 6, 12, 18, 24 and 30 hours after germination began. As can be seen, their visible germinative development reflects the point at which they were transferred to cordycepin. Embryos transferred to cordycepin after 30 hrs and embryos continuously in Actinomycin D for the 48 hour period show no gross morphological evidence of inhibition.

Since cordycepin is an analogue of adenosine it might be expected that it would have other effects on cell metabolism including possibly a general poisoning of the tissue by the interruption of ATP synthesis. However, we have found that the ATP pool size is the same in control and cordycepin treated embryos after 18 and 24 hours of continuous exposure to the drug. Cordycepin triphosphate could become incorporated into the 3' end of tRNA molecules through the CCA turnover, and inhibit protein synthesis in this fashion. Yet, we have found that the *in vitro* charging capacity of the tRNAs isolated from control and cordycepin treated embryos is identical for the 5

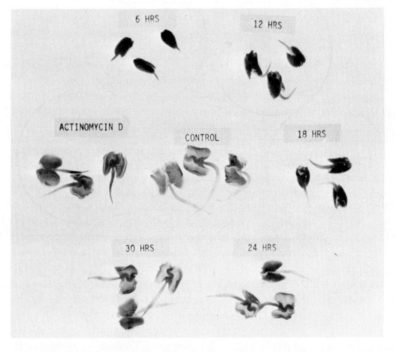

Fig. 5. Photograph of embryos after 48 hours of germination, some of which had been transferred to a solution of 5 m*M* cordycepin at the times indicated after the commencement of germination. Actinomycin D treated embryos were imbibed and maintained in Actinomycin D (20 μg/ml). Reprinted from Walbot *et al.* (in preparation).

amino acids tested. There are other conceivable sites of inhibition by cordycepin that could lead to an inhibition of protein synthesis and to the development of enzyme activity thought to come from stored mRNA, but our data at this point are at least suggestive of a cordycepin mediated inhibition of poly(A) addition. On this basis we continued to examine our system to test further the notion that the stored mRNA is not processed until the commencement of germination.

IV. ISOLATION OF A RIBONUCLEOPROTEIN PARTICLE FROM DEVELOPING COTYLEDONS

As mentioned in the Introduction, one approach to the study of the presumed stored mRNAs was to attempt to isolate the storage form of this class of mRNA. Although we assay for only two enzymes thought to be coded for by preformed mRNAs, the fact that germination proceeds normally for at least three days in the presence of large doses of Actinomycin D probably indicates that many enzyme activities required for early germination are coded for by this class of messengers. The transcription of this class of messengers is co-ordinate as shown by the fact that precocious germination of embryos after the 85 mg stage of embryogeny is insensitive to Actinomycin D.

Messenger RNAs and presumptive pre-mRNAs have now been isolated from a variety of tissue types from the nucleus, cytoplasm, and from polyribosomes. In all cases, the mRNA is found to be associated with proteins forming a ribonucleoprotein particle (RNP). Although the mRNAs are found in RNPs, it is not yet clear whether such particles are artifacts of the isolation procedures or the result of association of mRNAs or pre-mRNAs with specific proteins (Samarina, et al., 1973; Wilt, et al., 1973). Despite the uncertainty surrounding the in vivo state of mRNAs, it seemed reasonable to assume that attempts to isolate the storage form of the mRNAs found in cotton cotyledons should begin by the isolation of non-ribosomal RNP particles.

The presence of a non-ribosomal RNP particle in cotton was first surmised from experiments in which post-85 mg stage embryos were precociously germinated in the presence of ^{32}P. During the initial hours of incubation most of the label in sucrose gradients designed to display the ribosomes and ribosomal subunits was found coincident with the 40s small ribosomal subunit. This indicated either the preferential synthesis of the small ribosomal subunit or the synthesis of another component of similar sedimentation coefficient. The presence of two components in the 40s region of sucrose gradients having different densities can be demonstrated by neutral CsCl equilibrium density centrifugation analysis of particles fixed with glutaraldehyde. When the pooled fractions from the 40s region of sucrose gradients are treated with 0.01 M EDTA for 10 minutes at 0°C and then loaded onto a sucrose gradient, the contribution of the small ribosomal subunit to the A_{260} in the 40s region of the gradient is greatly reduced (Cammarano, et al. 1972). If the EDTA treatment is repeated, only one type of particle is found in the 40s region of the gradient, and it bands at a buoyant density of 1.41 gm/cm^3 in contrast to the small ribosomal subunit which bands at a buoyant density of 1.55 gm/cm^3 in a similar gradient as shown in Fig. 6. Particle fixation has been used routinely to assay for the purity of the non-ribosomal RNP found in the 40s region of sucrose gradients. Approximately 2% of the high molecular weight RNA from post-85 mg stage cotton cotyledons is found in this RNP particle.

In addition to the difference in buoyant density, the RNA extracted from the RNP fraction has a base composition different from the 18s or 25s cotton ribosomal RNA (Table 1). The proteins of the RNP particle have not been examined in detail, however, the buoyant density of the particle indicates that it is composed of approximately 80% protein in contrast to the ribosomal subunits which are approximately 60% protein and 40% RNA. The base composition data gives only an average of a variety of RNA molecules since gel electrophoresis of RNA from the RNP shows classes of RNA ranging from 10 to 18s (Fig. 7).

The RNP RNA does not contain long stretches of poly(A) as shown by the following tests. The RNP RNA does not bind to Millipore filters, an assay, which while not completely specific, does indicate the presence of poly(A)

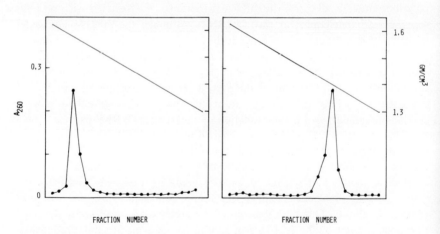

Fig. 6. Determination of the buoyant density of 40s ribosomal sub unit (left panel) and of RNP particles (right panel). The preparations were fixed in 2% glutaraldehyde and centrifuged in neutral CsCl gradients for 48 hr at 32,000 rpm and 25°C in a Spinco SW 39 rotor; fractions were collected and bouyant density determined by refractometry and A_{260} by spectrophotometry. Details will be published (Walbot and Dure, in preparation).

TABLE 1

*Molar Base Ratio of RNA Fractions as Determined by
the Method of Lane (1963)*

	A	G	C	U
18s rRNA	23.1	29.6	23.3	24.0
25s rRNA	21.7	32.4	24.1	23.5
RNP RNA	26.1	23.6	24.4	26.8

(Lee *et al.*, 1971). Furthermore, RNP RNA is completely sensitive to digestion by pancreatic and Tl RNAses (Edmonds and Carmela, 1969) indicating that there are no long stretches of poly(A) within the molecule. If the RNP particle does represent the form of the mRNAs presumed to be stored in the developing cotyledons, it would be expected, based on the data from the cordycepin experiments, that this RNA fraction would not contain long stretches of poly(A), since the addition of poly(A) is presumably the cordycepin sensitive processing step that occurs during germination.

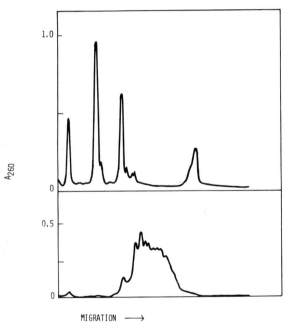

MIGRATION ⟶

Fig. 7. Gel electrophoresis of total RNA (top) and RNP RNA isolated from cotton embryos. Total RNA and RNP RNA were extracted from 85-90 mg stage cotton cotyledons and electrophoresed in 2.4% SDS-polyacrylamide gels, 6 mamp/gel, by the method of Loening (1967). Electrophoretograms were obtained by scanning the gels after a 15 min dialysis in distilled water in a Cary model 15 spectrophotometer.

The best proof that the RNP RNA is not degraded ribosomal RNA comes from competitive RNA-DNA hybridization experiments as shown in Fig. 8. It is difficult to label developing cotton embryo nucleic acids *in vivo* to sufficient specific radioactivity for hybridization experiments. Therefore, we have used *in vitro* iodination of cotton RNP and ribosomal RNAs following the procedure of Getz *et al.* (1972). Iodinated RNAs have been successfully used in both filter and liquid hybridization at 65°C reactions without detecting any loss of iodine from the iodinated RNA as long as care was taken to eliminate the heat labile 5 iodo-6 hydroxydihydropyrimidine product that is also formed during the iodination reaction (Getz *et al.*, 1972). ^{125}I labeled RNAs are routinely prepared with specific activities from 10^5 to 10^6 cpm/μg RNA.

The saturation kinetics of the hybridization of ^{125}I rRNA to filterbound cotton DNA are shown in Fig. 8, top left. Since each cotton cotyledon cell contains approximately 10×10^{-12} pg of DNA (Fisher and Jensen, 1972), it can be calculated that each cell contains about 2000 copies of each ribosomal cistron. The hybridization of ^{125}I rRNA is successfully competed with by unlabeled rRNA as shown in Fig. 8, top right; (the solid line indicates the theoretical competition curve at an increasing ratio of unlabeled rRNA and the solid circles indicates the experimental values.)

In contrast to the rRNA, ^{125}I RNP RNA does not show a true saturation value when hybridized to cotton DNA (Fig. 8, bottom left). Since the conditions for hybridization used would only allow the detection of RNA transcribed from repetitive DNA, we are not measuring the diversity of unique copy (possible mRNA) components in the RNP RNA. Our results do indicate, however, that some of the RNP RNA is homologous to repeated sequence DNA and that the RNP RNA is homologous to a relatively large proportion of the cotton genome (at least 1.3%) under these conditions. Further hybridization would be expected under conditions of higher C_0t. The ^{125}I RNP RNA is not competed with by an excess of unlabeled rRNA (Fig. 8, bottom right) indicating that there is no homology between these RNAs and the ribosomal RNAs. We are at present extending our investigation of the hybridization characteristics of the RNP RNA to include an analysis of the proportion of repetitive and unique transcripts in the RNP RNA, to determine if the unique and repetitive transcripts are found in the same RNA molecule, and to determine the amount of the cotton genome used in transcribing the RNP RNA.

The RNP particles represent a class of particles containing RNA with many of the characteristics expected of the putative stored mRNAs of cotton seeds. However, the definitive test that it indeed represents at least in part the stored mRNA would be the *in vitro* synthesis of a protein unique to the germination stage of development. We have not yet attempted such an analysis of the RNP RNA since it is likely that the stored mRNAs code for a very

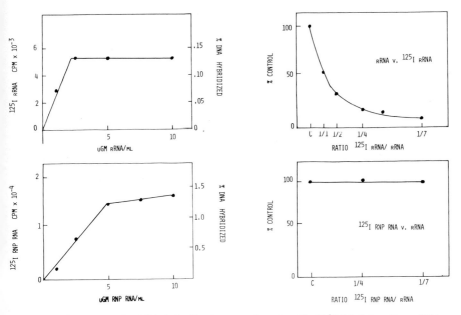

Fig. 8. Filter DNA-RNA hybridization experiments with ^{125}I labeled cotton rRNA and RNP RNA. The hybridizations were performed by the method of Gillespie & Spiegelman (1965) using 10 μg of CsCl purified cotton DNA per filter. [125]I labeled RNAs were prepared by modification of the procedure of Getz et al., (1972). The data are discussed in the text. Details of the purification of the DNA and RNA will be published (Walbot and Dure, in preparation).

large number of proteins. However, we have demonstrated, as shown in Fig. 9, that the RNP RNA functions as mRNA *in vitro* in directing the polymerization of amino acids into acid precipitable polypeptides. RNP RNA is shown to stimulate the incorporation of amino acids about two-fold over control reaction mixtures when RNP particles are added at one-tenth the concentration of the ribosomes. RNP directed amino acid incorporation is completely dependent on all of the components of the amino acid incorporating system found in the control reaction mixture including ribosomes. In reaction mixtures which have reached a plateau of incorporation at 20 minutes, the addition of RNP particles can re-initiate amino acid incorporation to approximately the same extent as the initial RNP directed incorporation as shown in the lower panel of Fig. 9. RNP directed incorporation is about 90% inhibited by cycloheximide and only about 20% inhibited by chloramphenicol, possibly indicating the proportion of cytoplasmic and organellar ribosomes in the reaction mixture.

To sum up this aspect of our work, we have found a RNP particle that is not a ribosomal subunit by several criteria: Different buoyant density, different RNA base composition, heterogeneous size of the RNA, lack of

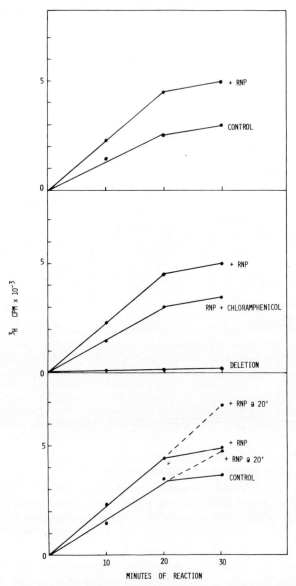

Fig. 9. Effects of RNP RNA on *in vitro* incorporation of [³H] leucine into polypeptides. The standard reaction mixture and characteristics of the cell-free incorporating system from cotton will be published (Walbot and Dure, in preparation). At the times indicated 0.1 ml aliquots of the 0.5 ml reaction mixtures were spotted on filter paper discs which were processed by the method of Bollum (1966) and their radioactivity determined. *Top Panel.* RNP particles were added to complete reaction mixtures at one-tenth the concentration of the cotton ribosomes (RNA:RNA). *Middle Panel.* Deletion refers to omission of any of the components found in the standard reaction mixture; chloramphenicol (10^{-4} M) was added to the reaction mixture at zero time; cycloheximide (10^{-4} M) added at zero time gave results similar to deletion experiments. *Bottom Panel.* Dashed lines give data obtained when RNP particles at one-tenth the concentration of ribosomes were added to reaction mixtures after 20 min of incubation; an additional 0.05 ml of the ATP generating system was also added at 20 min.

RNA homology by hybridization, and the insensitivity of the particle to degradation by EDTA. The RNA of the RNP particles does not contain significant amounts of poly(A), but is capable of supporting the *in vitro* incorporation of amino acids into polypeptides in a manner consistent with its possible functioning as messenger RNA.

V. POLYADENYLATION OF MESSENGER RNA DURING GERMINATION

Since the experiments with cordycepin indicated that a cordycepin sensitive step in the processing of the putative stored mRNAs occurs during the first 30 hours of germination, it seemed most likely that part of this processing involved the addition of poly(A) to the stored mRNAs. If this notion is correct, it should be possible to demonstrate that newly synthesized polyadenylate chains are added to preformed mRNAs during the first day of germination in cotyledons. This hypothesis can be tested by germinating embryos in the presence of radioactive adenosine and phosphate, extracting the RNA from cotyledons, and determining if the radioactivity is preferentially found in the poly(A) portion of mRNA-poly(A). In the extreme case in which all the mRNA processed during the incubation period is preformed, all of the radioactivity of mRNA-poly(A) should be in poly(A). If preformed mRNA does not exist then both the mRNA and poly(A) should be proportionately labeled with phosphate, and the proportion of radioactive adenosine in each portion of the molecule should be a reflection of the length of the segment and the mole percent in adenosine in each portion.

These experiments have been performed using ^{3}H-adenosine (labeled in the 2 position) and ^{32}PO$_4 \equiv$. Further, normally germinating embryos have been compared with embryos germinated in Actinomycin D, since this antibiotic should suppress the bulk of RNA synthesis and accentuate a demonstration of newly synthesized poly(A) added to preformed mRNA. The protocol of these experiments involved a separation of mRNA-poly(A) from other RNA by chromatography on poly(U)-Sepharose, a determination of the average poly(A) chain length after enzymatic digestion of the mRNA, and an analysis of the relative amount of radioactivity in mRNA and in poly(A).

Total nucleic acid preparations from embryos germinated for 22 hrs ± Actinomycin D and pulsed with the radioactive precursors from the 12th to 22nd hr were treated with 2 *M* NaCl to precipitate high molecular weight RNA and leave DNA and low molecular weight RNA in solution. The high molecular weight RNA was loaded in high salt onto poly(U)-Sepharose columns as described by Jelinek *et al.* (1973). Three fractions were obtained from the poly(U) columns as shown in Fig. 10. The first of these washes through the column and consists primarily of ribosomal RNA as shown by its migration on SDS-polyacrylamide gels and its base composition; the synthesis

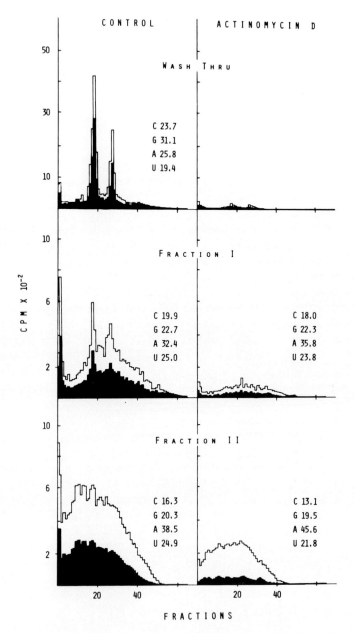

Fig. 10. Profiles of radioactivity from SDS-polyacrylamide electropherograms of fractions obtained from poly(U)-Sepharose chromatography of high molecular weight RNA from cotton embryos. Poly(U)-Sepharose columns were prepared by the method of Firtel & Lodish (1973). RNA, isolated from the cotyledons of embryos germinated in the presence of [³H] adenosine and $^{32}PO_4 \equiv$, (as described in the text) (±) Actinomycin D (20 μg/ml), was loaded on the column in 0.4 M NaCl at room temperature and the break-through fraction collected when the column was washed with additional 0.4 M NaCl. The next fraction was collected when the column was washed with 0.01 M EDTA

of this fraction is greatly suppressed in the presence of Actinomycin D. The next fraction is eluted at room temperature when the salt concentration is lowered and consists of ribosomal RNA and some heterogeneous RNA. However, upon pancreatic and Tl RNase treatment (Edmonds and Carmela, 1969) only short (10-20) nucleotide chains of poly(A) are found and, consequently, this fraction has been ignored in further consideration of poly(A) containing molecules. The last fraction is eluted from the column at 50°C in the absence of salt and appears to consist of heterogeneous high molecular weight RNA on SDS-polyacrylamide gels. This fraction of RNA has several unusual features. First, its synthesis is inhibited only 65% by Actinomycin D. Second, it contains poly(A) chains on the 3'OH end of the RNA molecules that average 100 nucleotides in length, both in normally germinating and Actinomycin D treated cotyledons. This value was ascertained from SDS-gel electrophoresis of the poly(A) moiety in 10% gels using 4 and 5s non-radioactive RNA as markers, and from electrophoretic determinations of [3H] adenosine/[3H] adenylic acid ratios after alkaline digestion of the poly(A). Third, the mole percent [3H] adenylic acid found in this fraction is very high in control cotyledons and even higher in Actinomycin D treated cotyledons. This is reflected in the gel profiles themselves when the amounts of [3H] and [32P] (shaded profile) are compared with the profiles of the other fractions. In view of these data, we have tentatively concluded that this fraction represents the mRNA-poly(A) component of the cotyledon RNA.

First of all these experiments show that poly(A) containing RNA does exist in germinating cotton embryos, and that at least some parts of the molecules are synthesized during the incubation period. Furthermore, they show that this synthesis is less sensitive to Actinomycin D inhibition than is rRNA synthesis. It would be ideal, therefore, if an inhibitor could be found which would allow only poly(A) synthesis and no other RNA synthesis so that the alleged addition of poly(A) to the putative stored mRNA could be demonstrated in a dramatic fashion. We had hoped that 3'-deoxycytidine might fulfill this role. Cordycepin completely suppresses all RNA synthesis and poly(A) polymerization in germinating cotyledons as well as reducing protein synthesis and enzyme appearance as shown in Table 2. This table also shows that 3'-deoxycytidine suppresses ribosomal RNA synthesis very effectively but is almost completely ineffective in inhibiting mRNA synthesis. Thus, we had to rely on the incomplete suppression of mRNA synthesis by Actinomycin D to serve as the means for accentuating a demonstration of the alleged phenomenon of newly synthesized poly(A) addition to performed

---

and 0.01 M Tris, pH 7.6; the last fraction was collected when the temperature of the column was raised to 50°C. The base composition of the RNA of each elution fraction was determined by the method of Lane (1963). The solid line represents the elution profile of [3H] containing material and the shaded portion that of [32P] containing material. Note the difference in the ordinate scale in the top panel as compared to the other two panels.

TABLE 2

Inhibition of the synthesis of various RNA fractions (based on incorporation of radioactivity) by the inhibitors indicated. Inhibition expressed as percent of control values. Bottom. Effect of the inhibitors on in vivo protein synthesis and on carboxypeptidase C activity.

	% OF CONTROL		
	Actinomycin D (20 μg/ml)	3'd-Cytidine (2.5 mM)	3'd-Adenosine (2.5 mM)
2 M NaCl ppt'ed RNA	9	16	~1
rRNA	5	12	~1
mRNA-poly(A)	35	95	~1
mRNA	28	95	~1
poly(A)	66	95	~1
Protein systhesis at 30 hours	130	130	20
Carboxypeptidase activity at 72 hours	95	95	<5

mRNA. It should be noted from the data in Table 2 that Actinomycin D inhibits the incorporation of radioactivity into mRNA to a greater extent than into poly(A). This coupled with the extremely high amount of ^{32}P AMP found in digests of mRNA-poly(A) from both controls and Actinomycin D treated cotyledons suggests that indeed poly(A) is put onto pre-existing RNA during germination.

Further evidence for this is given in Table 3. The first column of this table gives the predicted results for a theoretical experiment in which poly(A) is added to all newly synthesized mRNA during germination. The poly(A) chain length of 100 is known as well as the percentage of adenosine in mRNA (22% based on the base composition of pancreatic and Tl RNase sensitive portion of mRNA-poly(A)). If it is assumed that the average mRNA consists of 1000 nucleotides (Firtel and Lodish, 1973), then the proportion of [^{3}H] adenosine and ^{32}P that should be in the mRNA and in the poly(A) portions of the mRNA-poly(A) molecule can be calculated as shown in the table. However, experimentally the distribution of both isotopes between the mRNA portion and the poly(A) portions is quite different from the theoritically derived values. Far more of both isotopes are contained in the poly(A) portion. This bias in the distribution of radioactivity is even greater in the mRNA-poly(A) from Actinomycin D treated cotyledons.

TABLE 3

Analysis of the radioactivity contained in mRNA-poly(A) from control and Actinomycin D treated cotyledons. Theoritical values are those predicted if the mRNA protion of mRNA-poly(A) were newly synthesized, were 1000 nucleotides in length and had a poly(A) segment of 100 nucleotides added to it during the incubation period. Bottom. Average chain length of the mRNA calculated from the data above which is based on the radioactivity content.

	Theoretical (Assuming mRNA chain length = 1000 nucleotides)	Control	Actinomycin D Treated
Average chain length poly (A)	100	100	100
% adenosine in mRNA-poly(A)	29	39	45
% adenosine in mRNA	22	22	22
% ^{32}P of mRNA-poly(A) in poly (A)	9	20	49
% ^{3}H adenosine of mRNA-poly(A) in poly(A)	31	49	68
Calculated chain length of mRNA			
Based on poly(A) chain length, %A in mRNA-poly(A), %A in mRNA	1000	364	296
Based on poly(A) chain length, % ^{32}P of mRNA-poly(A) in poly(A)	1000	400	100
Based on poly(A) chain length, %A in mRNA, % ^{3}H adenosine of mRNA-poly(A) in poly(A)	1000	473	218

Using the experimentally derived data given in Table 3 it is possible to calculate a chain length of the mRNA portion of the mRNA-poly(A) by three different means based on radioactivity content. These calculations give a chain length in control cotyledons of about 400 nucleotides and of only 200 nucleotides in Actinomycin D treated cotyledons. This is obviously much too short a mRNA chain to code for reasonably sized polypeptides, but these values are consistent with the notion that preformed mRNA in polyadenylated during germination. That is, the amount of radioactivity in the mRNA portion does not reflect the true size of the mRNA because much of it was synthesized prior to germination, whereas the large amount of radioactivity in the poly(A) portion reflects its polymerization during germination! However, the chain length of the mRNA portion of mRNA-poly(A) isolated could conceivably have been reduced to the apparent small size by nucleases active during the isolation operations. Figure 11 shows an SDS-polyacrylamide gel electrophoretogram of mRNA-poly(A) that indicates that the apparent size is

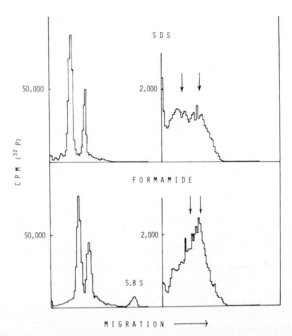

Fig. 11. Profiles of ^{32}P radioactivity from the polyacrylamide gel electrophoresis of 2 M NaCl precipitated RNA (left panels) and of mRNA-poly(A) (right panels) electrophoresed in SDS by the method of Loening (1967) (top panels) and in 99% formamide by the method of Duesberg & Vogt (1973) (bottom panels). Arrows in the right hand panels indicate migration positions of the rRNA species in the two systems.

quite large when compared to ribosomal RNA (top panels). However, mRNA-poly(A) has been shown to behave as an aggregate when not maintained under completely denaturing conditions. Consequently the size of the mRNA-poly(A) molecules was re-examined using 99% formamide, 3.5% acrylamide gels by the procedure of Deusberg and Vogt (1973). The effect of total denaturation obtained with formamide is evident in the rRNA profile (lower left panel). The large rRNA species is not as far separated from the 18s species and the 5.8s fragment released from the large rRNA species is found on the gel. Similarly the apparent size of the mRNA-poly(A) population appears to be somewhat smaller having a model size of about 19-20s. Furthermore, there is no indication of small mRNA-poly(A) molecules (6s and less). In view of these data, we would submit that the apparent short length (based on radioactivity content) of the mRNA isolated as mRNA-poly(A) and containing radioactive poly(A) (i.e. newly synthesized poly(A)) results from the fact that about 60% of the mRNA is preformed (i.e. transcribed in embryogenesis, processed during germination). The other 40% in turn would represent newly synthesized mRNA. The even smaller apparent size of the mRNA obtained from Actinomycin D treated cotyledons tends to support this conclusion. In this case about 85% of the mRNA that has had poly(A) chains added during germination is preformed; the other 15% being newly synthesized mRNA the synthesis of which is inhibited only by Actinomycin D.

VI. CONCLUSIONS

The developing cotyledons of the cotton seed provide a convenient system in which the general phenomenon of the temporal separation of transcription and translation can be studied. It is of great interest that the putative storage form of the mRNA, the RNP particles, contain some RNA transcribed from the repetitive portion of the cotton genome. Since messenger RNA is presumed to be transcribed primarily from the unique sequence DNA, we may have isolated, in a stable form, the initial transcription product from the cotton genome or an early stage in the processing of such transcripts. Such pre-mRNA molecules from most organisms are thought to be of very high molecular weight, whereas the RNA isolated from the cotton RNP population appears to be only 10-16s RNA, and is smaller even than the mRNA-poly(A) obtained from germinating cotton cotyledons. However, it is quite likely that the RNP RNA is partially degraded, when the manner of RNP particle isolation is taken into consideration. If indeed RNP RNA isolated from cotton cotyledons during the last 2 weeks of embryogenesis contains the putative stored transcripts that are to be processed and translated only after the triggering of germination, thereby bringing into existence the germination enzymes, then we should be able to show sequence homology between some fraction of this RNP RNA and the functional mRNA-poly(A) of germinating cotyledons. Such experiments are at present being attempted. Should such a relationship be confirmed, one of the mechanistic questions in understanding the regulation of gene expression in cotton seed embryogenesis and germination becomes, "What prevents the processing of that body of mRNA, destined to be 'stored', until the proper time in ontogeny?" In a general sense we know that the plant hormone abscisic acid is involved in this regulation. In view of our observations to date we feel that it will be fruitful to explore how abscisic acid interacts with the mRNA processing system.

ACKNOWLEDGMENTS

We wish to thank Antonieta Capdevila, who performed many of the cordycepin experiments, and the Upjohn Chemical Company for providing 3'-deoxycytidine. Supported in part by funds from the National Science Foundation, the United States Atomic Energy Commission and Cotton, Incorporated. One of us (V.W.) is a postdoctoral fellow of the N.I.H.

186   VIRGINIA WALBOT, BARRY HARRIS AND L. S. DURE, III

REFERENCES

Bollum, F. J. (1966). Filter paper disk techniques for assaying radioactive macro-molecules. In "Procedures in Nucleic Acid Research" (Cantoni, G. L. and Davies, D. R., eds.), pp. 296. Harper and Row, New York.

Cammarano, P., Pons, S., Romeo, A., Galdieri, M., and Gualerzi, C. (1972). Characterization of unfolded and compact ribosomal subunits from plants and their relation to those of lower and higher animals. Evidence for physicochemical heterogeneity among eucaryotic ribosomes. Biochim. Biophys. Acta 281, 571–596.

Duesberg, P. H. and Vogt, P. K. (1973). Gel electrophoresis of avian leukosis and sarcoma viral RNA in formamide. Comparison with other viral and cellular RNA species. J. Virology 12, 594–599.

Edmonds, M. and Carmela, M. G. (1969). Isolation and characterization of adenosine monophosphate-rich polynucleotides synthesized by Ehrlich ascites cells. J. Biol. Chem. 244, 1314–1324.

Firtel, R. A. and Lodish, H. F. (1973). Small nuclear precursor of messenger RNA in the cellular slime mold Dictyostelium discoideum, J. Mol. Biol. 79, 295–314.

Fisher, D. B., and Jensen, W. A. (1972). Nuclear and cytoplasmic DNA synthesis in cotton embryos. Correlated light and electron microscope autoradiographic study. Histochemie 32, 1–22.

Getz, M. J., Altenburg, L. C., and Saunders, G. F. (1972). Use of RNA labeled in vitro with iodine-125 in molecular hybridization experiments. Biochim. Biophys. Acta 287, 485–494.

Gillespie, D., and Spiegelman, S. (1965). A quantitative assay for DNA-RNA hybrids with DNA immobilized on a membrane. J. Mol. Biol. 12, 829–842.

Horowitz, B., Goldfinger, B., and Marmur, J. (1974). Effect of cordycepin (COR) and its triphosphate (COR-P) on RNA synthesis in yeast and yeast extracts. Fed. Proc. 33, 1418.

Ihle, J. N. and Dure, L. (1969). Synthesis of a protease in germinating cotton cotyledons catalyzed by mRNA synthesized during embryogenesis. Biochem. Biophys. Res. Commun. 36, 705–710.

Ihle, J. N. and Dure, L. (1970). Hormonal regulation of translation inhibition requiring RNA synthesis. Biochem. Biophys. Res. Commun. 38, 995–1001.

Ihle, J. N. and Dure, L. S. (1972a). Developmental biochemistry of cottonseed embryogenesis and germination. I. Purification and properties of a carboxypeptidase from germinating cotyledons. J. Biol. Chem. 247, 5034–5040.

Ihle, J. N. and Dure, L. S. (1972b). Developmental biochemistry of cottonseed embryogenesis and germination. II. Catalytic properties of the cotton carboxypeptidase. J. Biol. Chem. 247, 5041–5047.

Ihle, J. N. and Dure, L. S. (1972c). The temporal separation of transcription and its control in cotton embryogenesis and germination. In "Plant Growth Substances 1970" (Carr, D. J., ed.) pp. 216. Springer-Verlag, New York.

Jelinek, W., Adesnik, M., Salditt, M., Sheiness, D., Wall, R., Molloy, G., Philipson, L., and Darnell, J. E. (1973). Nuclear origin and transfer to the cytoplasm of polyadenylic acid sequences in mammalian cell RNA. J. Mol. Biol. 75, 515–532.

Lane, B. G. (1963). The separation of adenosine, guanosine, cytidine, and uridine by one-dimensional filter-paper chromatography. Biochim. Biophys. Acta 72, 110–112.

Lee, Y., Mendecki, J., and Brawerman, G. (1971). Polynucleotide segment rich in adenylic acid in the rapidly-labeled polyribosomal RNA component of mouse sarcoma 180 ascites cells. Proc. Nat. Acad. Sci. U.S. 68, 1331–1335.

Loening, U. E. (1967). The fractionation of high-molecular-weight ribonucleic acid by polyacrylamide-gel electrophoresis. *Biochem. J.* **102**, 251–257.

Marcus, A., Luginbill, B., and Felley, J. (1968). Polysome formation with tobacco mosaic virus RNA. *Proc. Nat. Acad. Sci. U.S.* **59**, 1243–1250.

Penman, S., Rosbash, M., and Penman, M. (1970). Messenger and heterogeneous nuclear RNA in HeLa cells: differential inhibition by cordycepin. *Proc. Nat. Acad. Sci. U.S.* **67**, 1878–1885.

Samarina, O. P., Lukanidin, E. M., and Georgiev, G. P. (1973). Ribonucleoprotein particles containing mRNA and pre-mRNA. *In* "Protein Synthesis in Reproductive Tissue" (Diczfalusy, ed.), pp. 130. Karolinska Institutet, Stockholm.

Walbot, V., Capdevila, A., and Dure, L. S. Action of 3'd-adenosine (cordycepin) and 3'd-cytidine on the translation of the stored mRNA of cotton cotyledons. *Biochem. Biophys. Res. Commun.* (in preparation).

Walbot, V. and Dure, L. S. Manuscript in preparation.

Wilt, F. H., Anderson, M., and Ekenberg, E. (1973). Centrifugation of nuclear ribonucleoprotein particles of sea urchin embryos in cesium sulfate. *Biochemistry* **12**, 959–966.

Chromosomal Basis for Hermaphrodism in Mice

Wesley K. Whitten

The Jackson Laboratory,
Bar Harbor, Maine 04609

I. Introduction 189
II. Observations and Results 191
    A. Hermophrodites Found among Aggregation Chimeras 191
    B. Spontaneous Hermaphrodites in BLAB/cWt Mice 192
    C. Other Spontaneous Hermaphrodites 196
    D. Morphology of Hermaphrodite Gonads 196
    E. Experiments with Hermaphrodites 197
III. Summary and Conclusions 202
    References 203

I. INTRODUCTION

Hermaphrodites are animals that contain both ovarian and testicular tissue so that each gonad may be an ovary, a testis, or an ovotestis. Germ cells may or may not be present and the degree of testicular or ovarian endocrine function may vary greatly and result in marked diversity in the secondary sex organs and characters.

Before proceeding I will define a chimera as an animal composed of cells of two or more distinct genotypes each of which is derived from a separate zygote. A mosaic is an individual containing two or more lines of cells, with different chromosomal constitutions, which are derived from a single zygote by some defective genetic process such as nondisjunction.

It may be difficult to distinguish hermaphrodites from animals with other forms of abnormal sex development. In the intersex or pseudohermaphrodite condition of female goats, homozygous for the gene polled, the phenotype may range from almost normal female to almost normal male but, in all cases, the gonads are testicular in appearance. In the recently discovered mouse counterpart, sex-reversed, (Cattanach, *et al* 1971), the adult females are phenotypically males except that the gonads appear as small testes without germ cells. However, some spermatogenesis takes place in sex-reversed, XO females and the animals behave as if a small portion of the male determining

189

region of the Y chromosome were translocated to an autosome. The gene, testicular feminization, on the X chromosome (Lyon and Hawkes, 1970) produces male mice with small abdominal testes, but no accessory structures except for a small vagina which may or may not become patent. When testicular feminization and sex-reversed are present in the same animal the reproductive tract becomes very confusing because of the mosaicism produced by random X chromosome inactivation. The masculanized reproductive tract and gonads of some bovine freemartins (female co-twins of males) in many ways resemble hermaphrodites and in fact may sometimes carry male cells in the gonads (Ohno, *et al* 1962).

I have recently observed three cases of hypospadias in male mice. In each, the lesion involved the prepuce as well as the urethera, and the penis was small and distorted. However, the remainder of the reproductive tract was normal and the epididymis contained numbers of mature spermatozoa.

Hermaphrodites have been observed in many species of mammals but the largest series was described by Hollander *et al.* (1956) in BALB/cGw mice. Braden (1957) tried to explain these as the result of immediate cleavage of the eggs followed by fertilization of both daughter blastomeres to produce a chimeric embryo. This same explanation was used later by Gartler *et al.* (1962) to account for hermaphrodism in a human chimera. However, evidence for immediate cleavage followed by simultaneous fertilization of both blastomeres has not been found under either natural or experimental conditions. My associate Dr. Peter Hoppe has developed a technique for fertilization *in vitro* (Hoppe and Pitts 1973) and has been unsuccessful in his attempts to fertilize eggs that have undergone first cleavage parthenogenetically.

In 1961, Tarkowski experimentally produced aggregation chimeras, and in 1964 reported on three newborn hermaphrodites from a series of 16 animals. Later with Mystkowska (1970) he described an adult chimeric hermaphrodite. He considered these to be models for natural hermaphrodism, but that not all spontaneous hermaphrodites developed in this manner. He stated that "the starting point for the development of a sex chromosome mosaic displaying hermaphrodite condition must be an XY egg. Non-disjunction of the Y chromosome could lead to an XO/XYY mosaic; if this is followed by a lagging of one or both chromatids an XO/XY mosaic, or an XO/XO individual will ensue." In (1969) Lyon identified a true hermaphrodite as an XO/XY mosaic.

Mintz (1971) presented evidence that hermaphrodites from her colony of BALB/c mice resulted from spontaneous fusion of two zygotes. She confirmed the finding of Dagg and myself (1961) that BALB/c embryos develop more slowly than those of other strains, but she also observed that they lose their zona pellucida at the normal time. Thus, naked morulae are present in the uterus of BALB/c mice for a prolonged period, which she argued favored spontaneous aggregation.

Dunn, Kenney and Lein (1968) found one bovine true hermaphrodite to be an XX/XY chimera and later Dunn, McEntee, and Hansell, (1970) described another as a diploid/triploid (XX/XXY) chimera. Abdel-Hameed and Shoffner (1971) found 19 intersex chickens among those selected as females soon after hatching. One was a ZZ/ZW chimera, 17 were 3A-ZZW triploids, and one was a diploid/triploid (3A-ZZZ/2A-ZW) chimera. However the gonads of the intersex chickens were deformed testes rather than ovotestes. Nevertheless these findings suggest that some form of triploidy may favour the occurrence of hermaphrodism in vertebrates, but there is no evidence to support this hypothesis for mice.

I decided to study the spontaneous hermaphrodites which occur in BALB/cWt strain of mice because the incidence is much higher than in other strains including the one used by Hollander, et al (1956). Also my colleague Dr. Leroy Stevens has made available the results of studies of the spontaneous hermaphrodites which have been found at the Jackson Laboratory, and I have acquired twelve manufactured chimeric hermaphrodites with which to make comparisons.

II. OBSERVATIONS AND RESULTS

A. Hermaphrodites Found among Aggregation Chimeras

My colleagues and I have obtained approximately 1500 mice from females into which aggregated embryos have been introduced. About 1200 of these have been overt chimeras, and from them 12 (about 1%) hermaphrodites have been identified. Four of these hermaphrodites contained some BALB/cWt cells, but relatively few chimeras were produced with this strain. This suggests that the presence of BALB/c gonads favors hermaphrodism but the numbers are too small for significance.

Ten other chimeric hermaphrodites have been described in the literature and probably a total of eight have been chimeras from within a strain. Chimeras of this type are usually made from surplus embryos and are not reported because it is impossible to determine if the process of aggregation has been successful. There is thus no way of judging the proportion which become hermaphrodites. However, the figures suggest that the absence of genetic differences between the components of chimeras may favor hermaphrodism and it is possible that surface antigens are involved in the suppression of hermaphrodism.

One of the most interesting observations on aggregation chimeras is the low incidence of overt hermaphrodism when one would expect 50% (Tarkowsky, 1964; Beatty, 1968). Mullen and Whitten (1971) and others observed a significant excess of male chimeras (Table 1). This excess approximately equals the expected number of hermaphrodites so that many of the males are cryptic hermaphrodites, or the presence of male cells prevents development of ovarian tissue. The findings of MacIntyre (1956) and of Mystkowska and Tarkowski (1970) support this conclusion. Tarkowski (1964) described 3 chimeric animals as hermaphrodites when examined histologically soon after birth and stated that one of these could have developed into an apparently normal male.

An excess of males does not occur when the chimeras are produced from embryos of different developmental vigor such as the aggregation of hybrid with inbred embryos (Mullen and Whitten 1971). In these cases some of the females may be cryptic hermaphrodites or more likely the less vigorous component will not be represented in the gonad at all. There is no evidence that more hermaphrodites develop under these circumstances, nor, when three or four embryos are aggregated to form chimeras.

If female germ cells are present in the testis of a phenotypic male and participate in spermatogenesis one would expect them to produce only female offspring. So far no male chimera has produced a significant number of only female progeny of one of the genotypes which contribute to his makeup (Mintz 1968, Mullen and Whitten, 1971). Therefore we conclude that XX cells cannot complete spermatogenesis. It also follows that some of the males that produce progeny of only one genotype are cryptic hermaphrodites, whereas those that produce two are not. Breeding records of potential XX <—> XY chimeras do not indicate any reduction in fertility, as seen in Table 2. This contrasts with the sterility seen in freemartin heifers and the recent evidence that the male co-twin may have a short productive life as studs (Dunn, Kenney, Stone, and Bendel 1968).

B. Spontaneous Hermaphrodites in BALB/cWt Mice

Dr. Meredith Runner obtained this strain from Dr. George Snell at N83 in 1956 and transferred it to Dr. Charles Dagg who left it with me. During the past few years we have obtained more than 100 hermaphrodites from this strain and some of its hybrids. My associate Mr. Dorr is particularly adept at

TABLE 1
Sex Ratios of Chimeric Mice[a]

	males	females	% males
Balanced genotypes			
B10 <—> SJL	29	11	72.5
B10CBAF$_1$ <—> SJL	25	12	67.6
(B10SJLF$_1$ x B10) <—> (BALBSJLF$_1$ x SJL)	16	8	66.7
Total	70	31	69.3
Unbalanced genotypes			
C3H <—> SJL129F$_1$	11	11	50.0
B10 <—> BALB	8	10	44.4
Total	19	21	47.5

[a] From Mullen and Whitten (1970)

TABLE 2

Breeding Performance of Chimeras

Strain combinations		Fertile males/males tested	Fertile females/females tested
C57BL/10 <——> AKR		15/15	10/10
C57BL/10 <——> BALB/c		20/20	0/0
C57BL/10 <——> C3H/He		21/21	9/9
	Total	56/56	19/19

identifying these animals visually. He has become familiar with the size and contours of the normal male and female genitalia, which vary with each strain and hybrid. Any animal which appears as a phenotypic male but has a small penis, cryptorchism, inguinal hernia, or mammary development, and any apparent female exhibiting a large urinary papilla, no vaginal opening, little mammary development, or smells like a male is inspected by laparotomy. The gonads and accessory structures are examined with the aid of a Zeiss Operation microscope at 10X magnification. Biopsy material is taken and serial sections of the gonads made. Most hermaphrodites are found when weaned at about 28 days of age. A few are detected when females are selected in proestrus for mating, or when paired animals fail to produce litters. Occasionally hermaphrodites are observed when a newborn litter is sexed and rarely when fetal gonads are examined on the 16th day of pregnancy (Fig. 1 and 2).

A striking characteristic of BALB/cWt substrain has been the low sex ratio of 38% males (Whitten and Carter, in preparation) that appears to be associated with the production of hermaphrodites. It is not directly related to the high incidence of sperm with small or large distorted heads, or the reflex ovulation which occurs soon after pairing if females have been isolated beforehand. The abnormal sperm and reflex ovulation occur in other strains of BALB/c with normal sex ratio and with only a sporadic occurrence of hermaphrodites. Particular attention was paid to reflex ovulation because Braden (1958) found that delayed mating influenced the segregation ratios of some *t* alleles, but it did not alter the sex ratio of BALB/cWt mice.

Recently ovulated eggs and early embryos from BALB/cWt mice were examined for evidence of immediate cleavage, supplementary male pronuclei, and other defects of fertilization but none have been found. Almost all recently fertilized eggs developed to blastocysts in culture so that there is no indication of preimplantation loss. A few individual blastomeres of a proportion of the embryos die but blastocysts usually develop from the remainder and the embryos probably survive. No significant amount of embryonic loss was found when females were examined on the sixteenth day of pregnancy.

From these observations it is concluded that there is insufficient pre- or post-implantation loss of male embryos to account for the abnormal sex ratio.

A low sex ratio is also observed in the offspring of SJL/Wt females mated to BALB/cWt males. They were examined for evidence of parthenogenesis possibly resulting from activation of the egg by a genetically defective sperm. If parthenogenesis occurred then a proportion of the female progeny should be pure SJL/Wt but all those examined were shown to have isozyme and hemoglobin characteristics of the BALB/c sire (Whitten 1971). Thus we can discount parthenogenesis as the cause of the excess of females. An additional indication that the abnormal sperm were not involved was that the sex ratio of the progeny from C57BL/10Wt females mated to BALB/cWt males was normal.

An association of low sex ratio and hermaphrodism has been observed in the offspring of SJL/Wt females and BALB/cWt males. The sex ratio of these hybrids was not different from that of the parental BALB/cWt and twelve hermaphrodites have been identified. However, the sex ratio of the reciprocal cross was normal and no hermaphrodites have been found. Therefore we conclude that both of these features are associated with the Y chromosome of the BALB/cWt but there must be some maternal factor in BALB/cWt and SJL/Wt eggs which permits the Y chromosome effect to be expressed.

A clue to the cause of the low sex ratio in BALB/cWt mice was obtained when females were mated to males carrying the gene tabby on the X chromosome. This was done because the sex ratio of progeny of XO females (Cattanach, 1962) is almost equal to that observed in BLAB/cWt. As a result of these matings it was apparent that the colony contained and was producing XO females.

We have examined the karyotypes of BALB/cWt hermaphrodites. Suitable testicular material was not available so we used bone marrow and found the following mosaics; 39,X/40,XY, 39,XO/41,XYY, and 39,XO/40,XY/41,XYY. This evidence, the low sex ratio, and the generation of XO females indicate that chromatid lagging or nondisjunction occurs as predicted by Tarkowski (1964). If the mitotic error takes place at the first cleavage, then cells with the following chromosomes would be observed: XO, XY, and XYY either alone or in combinations. We know that XO mice are normal females and Cattanach and Pollard (1969) have reported a large sterile mouse to be XYY, and two 39,XO/41,XYY mosaics have been reported (Lyon, 1969; Evans et al. 1969). If the error occurs at a subsequent mitosis or if chromatid lagging results in correction of some XYY cells to XY, then seven possible chromosome combinations can occur. These are set out in Table 3 with the expected phenotypes. So far we have not found any evidence for the production of sterile XYY males even though we maintain more than 60 breeding males and check their fertility.

TABLE 3

Possible Sex Chromosome Constitutions and Phenotypes
Following Y-Chromatid Lagging or
Nondisjunction in the Early Embryo

Sex chromosome constitution	Possible phenotypes of the adult			
	Female	Hermaphrodite	Male	Sterile Male
39,XO	x			
39,XO/40,XY	x	x	x	
39,XO/41,XYY	x	x	?	x
39,XO/40,XY/41,XYY	x	x	x	x
40,XY			x	
41,XYY				x
40,XY/41,XYY			x	x

If the mitotic error is a recurring phenomenon as the result of a sticky Y chromosome, then the distribution of cell types would be unpredictable. If the error takes place at the first cleavage two clones would be produced and these may or may not be distributed throughout the embryo depending both on chance and any advantage that one clone may possess. In ccllaboration with Dr. Runner we attempted to determine if it occurred at the first cleavage by separating the two daughter blastomeres and developing separate clones from them in culture. Unfortunately we could not predict the time of subsequent mitoses and there were insufficient cells to obtain metaphases at random before most of the cells differentiated to trophoblast. Perhaps a more successful approach would be to examine the first metaphases by the technique of Nesbitt and Donahue (1972) or Kaufman (1973).

We have not carried out an extensive series of chromosome studies partly because of limited facilities and partly because of difficulty in interpreting the results. It may not be possible to obtain germ cells in the hermaphrodite gonad and one can only observe the chromosomes of oocytes after maturation division commences. However in hermaphrodites we are mainly interested in the chromosomes of the cells which secrete the fetal hormones and the cells which later produce the sex steroids. The situation in an ovotestes may therefore be very complex. We are not interested in the chromosomes of stromal fibroblasts or the blood cells. The gonads we observe are the result of a *fait accompli* which may have taken place before day eight of embryonic development. Therefore we should be looking at embryonic stages and perhaps the relatively high incidence of hermaphrodism in BALB/cWt will permit these studies.

Why do we find so many hermaphrodites in BALB/cWt? Does this strain lack the mechanism which suppresses hermaphrodism or has there been an inversion or translocation involving the Y chromosome? The latter would be consistent with the high proportion of abnormal sperm. The sex ratio of BALB/cSn was normal but it dropped to its present level soon after it was transferred to Runner. Associated with this drop was an apparent increase in litter size. It is possible that a change occurred which permitted the survival of embryos which previously had died in utero. We plan to look at other BALB/c strains for such foetal loss.

So far only one spontaneous mouse chimera has been identified by genetic studies (Russell and Woodiel 1966). However the report by Mintz (1971) that the hermaphrodites of her colony of BALB/c were spontaneous chimeras suggested that there must have been many others that were not hermaphrodites. These could only be detected by a study of their chromosomes. If chimeras occurred in hybrid progeny of BALB/c females they could easily be detected particularly if they carried a suitable color gene. We have bred many such hybrids and found no chimeras. As pointed out above the hybrid hermaphrodites are the progeny of male and not female BALB/cWt mice, so they must be different from those reported by Mintz (1971).

Because of the low incidence of hermaphrodism and the difficulty of obtaining significant sex ratios from individual breeding pairs there is some risk that these characters will be lost. Therefore we have begun progeny testing of each stud male.

C. Other Spontaneous Hermaphrodites

Tarkowski (1964) reviewed the literature of hermaphrodites and has added his own findings. Over many years Dr. Leroy Stevens has collected the most extensive series from the Jackson colonies of BALB/c, A/He, 129, C57BL/6, DBA, and their derivitives. He has carefully classified this material, obtained mostly from adult animals by histological examination, and has kindly permitted me to use both his data and his slides. A summary of the data from all sources is presented in Table 4.

D. Morphology of Hermaphrodite Gonads

Tarkowski (1964) describes four types of hermaphrodites: those with (1) an ovary and a testis, (2) an ovotestis and a testis, (3) an ovotestis and an ovary, and (4) two ovotestes. He was the first to observe classes (2) and (4) in mice, and then only in newborn animals. He also remarked on the preponderence of reports of class (1) in the literature. Stevens and I have observed all four types in adult animals and again class (1) exceeds the others for the spontaneous hermaphrodites, and probably also for the chimeras. Our data also indicate a significant proportion of animals with two ovotestes which supports Tarkowski's (1964) suggestion that they may have been missed by earlier observers.

TABLE 4

Histological Classification of the Gonads of Adult
Hermaphrodite Mice

Source	Ovary and testis	Ovotestis and testis	Ovotestis and ovary	Two ovotestes
Literature (Tarkowski, 1964)	32	0	1	0
Stevens (personal communication)	38	11	6	4
Whitten				
Spontaneous hermaphrodites	16[a]	9	10	3
Chimeric hermaphrodites	5[b]	1[c]	0	1

[a] Some animals most likely in this class not included but used for experimental studies.

[b] Mintz (1968) reports six chimeric hermaphrodites, but gives data on one with a testis and a modified ovary.

[c] Mystkowska and Tarkowski (1968) describe one which had a testis and an ovotestis.

The ovarian and testicular elements of an ovotestis may be simulated at opposite poles as in Fig. 3 or adjacent as in Fig. 4. The ovotestis may be complex and contain tubules with oocytes (Fig. 5), or follicles with eggs, granulosa, and sertoli cells (Fig. 6). One may see follicles with oocytes confluent with tubules with some spermatogenesis as in Fig. 7 and 8, or tubules which are short and branched as in Fig. 9. These findings are consistent with the concept that there may be mosaicism of the gonad organizer, the germ cells, the endocrine hormones, or all three of these components. However, studies of ovotestes of adults may be misleading because an abdominal position may induce cryptorchid changes and the tubules may not be protected from an autoimmune reaction by the normal blood-testis barrier.

It is difficult to compare such variable material as the gonads of hermaphrodites. I have not seen any consistent difference between the gonads of mosaics and chimeras but the number of the latter is very small.

The sex ducts of hermaphrodites have been described by others particularly by Hollander *et al.* (1956) and by Tarkowski (1964). However I am always surprised to find animals with well developed Mullerian and Wöllfian ducts running parallel as in Fig. 10.

E. Experiments with Hermaphrodites

It has been known since the time of John Hunter that freemartin heifers are sterile. Short (1970) has argued that this is not due to testosterone or germ cells derived from the male co-twin and suggested that it may result from the action of a gonad organizer such as 'medullarin' proposed by Witschi

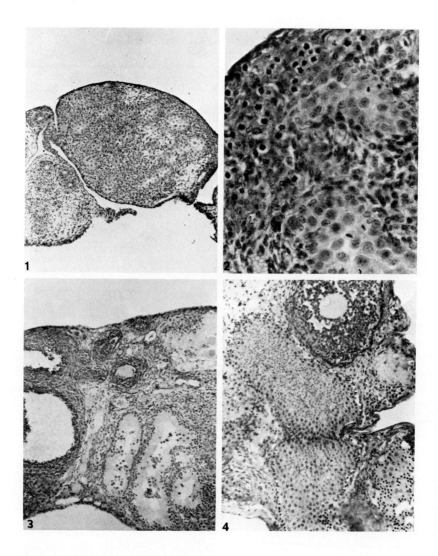

(1934). If this is the correct explanation then one would expect the ovaries of hermaphrodite mice to be sterile, but the histological appearance suggests otherwise. However, for the mouse we can test this hypothesis experimentally. We identified by laparotomy, at weaning, several completely lateral hermaphrodites. The ovaries were removed and cut in half longitudinally and grafted into the ovarian bursa of young ovariectomised females of suitable genetic background. The grafting was performed by Mr. Don Varnum using the technique described by Stevens (1958). Mr. Varnum obtains successful grafts

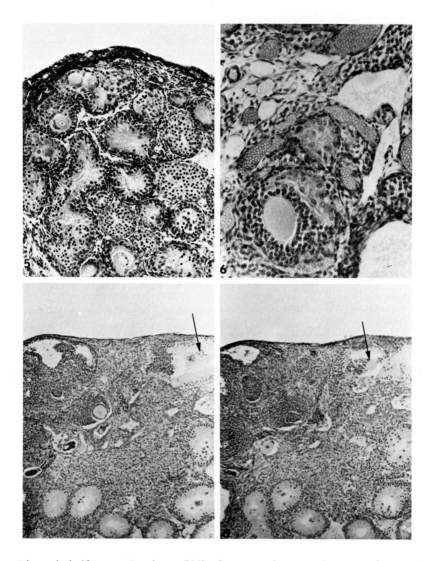

with each half ovary in about 70% of cases and repeated tests with genetic markers show negligable residual host ovarian tissue. Nevertheless we also grafted the ovaries of the hosts into four other recipients to determine the X-chromosome composition of the oocytes by mating to males carrying the tabby gene on the X chromosome. This was done because we expected the oocytes of the hermaphrodites to be 39,XO. Some of the hermaphrodites with the remaining testis were allowed to mature and were paired with suitable females to test their fertility. Eventually some were mated to the females carrying half of their original ovary to see if "selfing" were possible in a mammal. The results of these experiments are given in Table 5. They show that the ovaries and some of the testes are fertile, that the hermaphrodite oocytes are frequently 39,XO, and that "selfing" is possible in mammals. The selfing was observed in an F_1 hybrid animal but even so this suggests that residual heterozygosity may not be important for mice derived from inbred strains.

The above finding of fertile hermaphrodite gonads indicates that mice resemble marmosets, where spontaneous chimerism is frequent between unlike twins but the sexual development of the females is unimpared (Benirschke and Brownhill 1962), rather than cattle where the freemartin is sterile. The reason for the sterility of the bovine freemartin may be some endocrine phenomenon or developmental sequence characteristic of *Bovidae*.

We have seen from studies of chimeras that other things being equal a testis will outgrow an ovary during development. Why then does not the testicular component of a BALB/cWt mosaic predominate over the ovarian tissue? This could be because of the burden of an extra Y chromosome or it may be some peculiarity of the BALB/cWt testis. If it were the latter, then we sould obtain a low proportion of males in BALB/cWt <——> BALB/cWt chimeras. Making allowances for the low sex ratio, to begin with we would expect if the testis predominates, to obtain 62.8% of males, whereas, we found 28.6% in a series of 21. This difference is significant (p = 0.02), even though we have no criterion for success of aggregation, and some of the animals may in fact not be chimeras. These results contrast with those obtained with other balanced chimeras (see Table 6) and suggests that the testis of BALB/cWt does not suppress development of the ovary, but perhaps the ovary of this strain may even suppress the testis. It should be possible to examine this using the simpler experimental system of MacIntyre (1956) in which fetal ovaries and testes are grafted side by side under the renal capsule.

TABLE 5

Breeding Data From Hermaphrodites

	BALB/cWt or SJL/Wt x BALB/cWt F_1	Chimeras
Number of hermaphrodites examined	10	2
Number of ovaries grafted	10	2
Number of ovaries producing fertile grafts	9	0
Number of ovaries producing XO progeny	3	–
Number of testes examined for fertility	2	0
Number of fertile testes	2	0
Number of hermaphrodites 'selfed'	1	0

TABLE 6

*Comparison of the Sex Distribution of BALB/c <——> BALB/c
Chimeras with that of SJL <——> C57BL/O Chimeras*

	Males	Hermaphrodites	Females
		percent	
BALB/c <——> BALB/c			
Expected[a] $(0.39 + 0.61)^2$	15.2	47.6	37.2
Observed (N = 21)	28.6	4.8	66.7
SJL <——> C57BL/10			
Expected $(0.5 + 0.5)^2$	25.0	50.0	25.0
Observed (N = 40)	72.5	0	27.5

[a] Calculated from observed sex ratios.

We found one hermaphrodite in this small series and it would be interesting to know if it were a simple mosaic, a chimera between a male and a female, or between one of these and a mosaic or between two mosaics. It is possible that chimeras produced from within the BALB/cWt strain may give a workable incidence of hermaphrodism.

We are conducting another experiment which may help understand the role of germ cells in the development of the gonad. We are producing chimeras by combining normal or heterozygous embryos with those homozygous for the *W* alleles which cause anemia, defective melanocytes, and deletion of germ cells (Mintz and Russell 1957). The chimeras are made by aggregating embryos derived from the mating of suitable heterozygotes, so that only three of sixteen are the desired combinations. They can be recognized by the bold white patches, on places other than the head and belly, and a pigmented central belly spot. We have produced four of these and one appeared as a normal female which we conclude was W/W,XX<——> +/+, XX. One was a male with small testes, a section of which is seen in Fig. 11 from which it is evident that spermatogenesis is present in only some of the tubules, and we conclude the animal was probably W/W,XY <——>+/+,XY, or W/W,XY<——>W/+,XY. The third animal was a phenotypic male with a very testicular ovotestis containing very few germ cells, as seen in Fig. 12. I expect that this animal had the following constitution W/W,XY<——>+/+,XX, and that in spite of the presence of normal XX germ cells and few if any XY germ cells, the gonad developed and functioned like a testis. The remaining animal is a young male and has not been examined.

If these findings are further supported and if our interpretations are correct, then, they will confirm the suggestion by Burns (1961) that germ cells do not play a significant part in the developing gonad; and that complex

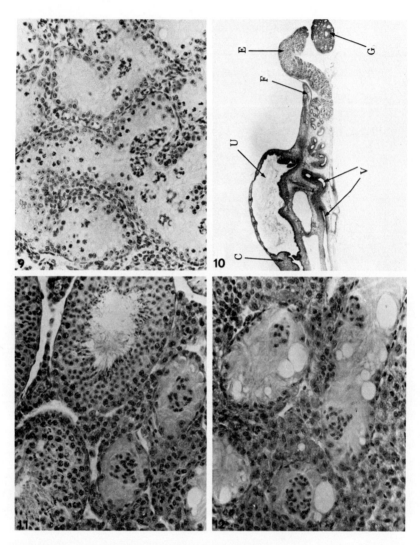

ovotestes are to be expected when the genetic sex of the endocrine, or the morphogenic, components of the gonads is mixed.

III. SUMMARY AND CONCLUSIONS

More than 100 hermaphrodites have been found in BALB/cWt and SJL/Wt x BALB/cWt F_1 hybrid mice. The karyotype, the low sex ratio, and the production of XO females indicate that a sex reversing nondisjunction of the Y chromosome occurs probably during early cleavage. This seems to be dependent on the Y chromosome of BALB/cWt and some maternal factor present in BALB/c and SJL eggs. These hermaphrodites appear similar to those which occur sporadically in other strains and to those produced by embryo aggregation. The morphology of hermaphrodite gonads particularly the ovotestes is discussed. Oogenesis and spermatogenesis may be quite normal even in adjacent portions of the same gonad. This appears to be consistent with the idea of mosaicism of both endocrine elements and germ cells. Ovarian grafts from hermaphrodites have proved fertile and have produced XO progeny. The remaining testis can also be fertile and 'selfing' has been accomplished by mating the hermaphrodite with a remaining testis to a female into which the hermaphrodite ovary had been grafted. Experiments have been described which examine the role of germ cells in gonad development and the role of the testis in inhibiting the ovary in developing hermaphrodites. The BALB/cWt testis may not be efficient in the latter regard. Chimeric BALB/cWt mice may provide the best models to study hermaphrodism. Genes causing germ cell deletion may be used to study the role of these cells in the development of gonads. Perhaps the most interesting observation is the finding of a recurring and spontaneous mitotic error affecting the Y chromosome which may be used as a model for the production of hermaphrodites and aneuploidy.

ACKNOWLEDGMENTS

I thank Dr. Leroy Stevens for much fruitful discussion, advice, and encouragement and for the free use of his material and data. I also wish to thank Mr. Stetson Carter who, supervised the BALB/cWt colony, found and succeeded in 'fixing' the low sex ratio and prepared many of the chimeras. I also want to acknowledge Mr. Darrold Dorr who found many of the hermaphrodites; Mr. Donald Varnum who successfully transplanted the ovarian tissue; and Drs. Runner and Dagg from whom I inherited the BALB/c colony.

This research was supported by NIH Research Grant HD 04083 from the National Institute of Child Health and Human Development.

REFERENCES

Abdel-Hameed, F., and Shoffner, R. N. (1971). Intersexes and sex determination in chickens. *Science* **172,** 962–963.

Beatty, R. A. (1968). Discussion. In "VIII Biennial Symposium on Animal Reproduction" (A. V. Nalbandov and D. E. Becker, eds.) (*J. Anim. Sci.* **27** *Supp. 1*), pp. 69–76.

Benirschke, K., and Brownhill, L. E. (1963). Further observations on marrow chimerism in marmosets. *Cytogenetics* **1,** 245–257.

Braden, A. W. H. (1957). Variation between strains in the incidence of various abnormalities of egg maturation and fertilization in the mouse. *J. Genet.* **55,** 476–487.

Braden, A. W. H. (1958). Influence of time of mating on the segregation ratio of alleles at the T locus in the house mouse. *Nature (London)* **181,** 786–787.

Burns, R. K. (1961). The role of hormones in the differentiation of sex. In "Sex and Internal Secretions" (W. E. Young, ed.), p. 76. Tyndall and Cox, London and Baltimore.

Cattanach, B. M. (1962). XO mice. *Genet. Res.* **3,** 487–490.

Cattanach, B. M., and Pollard, C. E. (1969). An XYY sex chromosome constitution in the mouse. *Cytogenetics* **8**, 80–96.

Cattanach, B. M., Pollard, C. E., and Hawkes, S. G. (1971). Sex-reversed mice: XX and XO males. *Cytogenetics* **10**, 318–337.

Dunn, H. O., Kenney, R. M. and Lein, D. H. (1968). XX/XY chimerism in a Bovine true hermaphrodite: an insight into understanding freemartinism. *Cytogenetics* **7**, 390–402.

Dunn, H. O., Kenney, R. M., Stone, W. H., and Bendel, S. (1968). Cytogenetic and reproductive studies of XX/XY chimeric twin bulls. *Proc. 6th Int. Cong. Anim. Reprod.*, Paris.

Dunn, H. O., McEntee, K., and Hansel, W. (1970). Diploid-triploid chimerism in a bovine true hermaphrodite. *Cytogenetics* **9**, 245–259.

Evans, E. P., Ford, C. E., and Searle, A. G. (1969). A 39,X/41,XYY mosaic mouse. *Cytogenetics* **8**, 87–96.

Gartler, S. M., Waxman, S. H., and Giblett, E. (1962). An XX/XY human hermaphrodite resulting from double fertilization. *Proc. Nat. Acad. Sci. U.S.* **4**, 332–335.

Hollander, W. F., Gowen, J. W., and Stadley, J. (1956). A study of 25 gynandromorphic mice of the Bagg albino strain. *Anat. Rec.* **124**, 223–239.

Hoppe, P. C., and Pitts, S. (1973). Fertilization *in vitro* and development of mouse ova. *Biol. Reprod.* **8**, 420–426.

Kaufman, M. H. (1973). Analysis of the first cleavage division to determine the sex-ratio and incidence of chromosome anomalies at conception in the mouse. *J. Reprod. Fert.* **35**, 67–72.

Lillie, F. R. (1917). The freemartin: a study of the action of sex hormones in the fetal life of cattle. *J. Exp. Zool.* **23**, 371–452.

Lyon, M. F. (1969). A true hermaphrodite mouse presumed to be an XO/XY mosaic. *Cytogenetics* **8**, 326–331.

Lyon, M. F., and Hawkes, S. G. (1970). X-linked gene for testicular feminization in the mouse. *Nature (London)* **227**, 1217–1219.

MacIntyre, M. N. (1956). Effects of testis on ovarian differentiation in heterosexual embryonic rat gonad grafts. *Anat. Rec.* **124**, 27–46.

Mintz, B. (1968). Hermaphroditism, sex chromosomal mosaicism and germ cell selection in allophenic mice. *J. Anim. Sci.* (Suppl. I) **27**, 51–60.

Mintz, B. (1971). Control of embryo implantation and survival. *Adv. Biosci.* **6**, 317–340.

Mintz, B., and Russell, E. S. (1957). Gene induced embryological modification of primordial germ cells in the mouse. *J. Exp. Zool.* **134**, 207–237.

Mullen, R. J., and Whitten, W. K. (1971). Relationship of genotype and degree of chimerism in coat color to sex ratios and gametogenesis in chimeric mice. *J. Exp. Zool.* **178**, 165–176.

Mystkowska, E. T., and Tarkowski, A. K. (1970). Behavior of germ cells and sexual differentiation in late embryonic and early postnatal mouse chimeras. *J. Embryol. Exp. Morph.* **23**, 395–405.

Nesbitt, M. N., and Donahue, R. P. (1972). Chromosome banding patterns in preimplantation mouse embryos. *Science* **177**, 805–806.

Ohno, S., Trujillo, J. M., Stenius, C., Christian, L. C., and Teplitz, Q. L. (1962). Possible germ cell chimeras among newborn dizygotic twin calves (*Bos taurus*). *Cytogenetics* **1**, 258–265.

Russell, L. B., and Woodiel, F. N. (1966). A spontaneous mouse chimera formed from separate fertilization of two meiotic products of oogenesis. *Cytogenetics* **5**, 106–119.

Short, R. V. (1970). The bovine freemartin: a new look at an old problem. *Phil. Trans. Roy. Soc. Lond. B.* **259**, 141–147.

Stevens, L. C. (1957). A modification of Robertson's technique of homoiotopic ovarian transplantation in mice. *Transplantation Bull.* **4,** 106–107.

Tarkowski, A. K. (1961). Mouse chimeras developed from fused eggs. *Nature (London)* **190,** 857–860.

Tarkowski, A. K. (1964). True hermaphroditism in chimaeric mice. *J. Embryol. Exp. Morphol.* **12,** 735–757.

Whitten, W. K. (1971). Parthenogenesis: does it occur spontaneously in mice? *Science* **171,** 406–407.

Whitten, W. K., and C. P. Dagg. (1961). Influence of spermatozoa on the cleavage rate of mouse eggs. *J. Exp. Zool.* **148,** 173–183.

Witschi, E. (1934). Sex differentiation in amphibia. *Biol. Rev.* **9,** 460–488.

Analysis of Determination and Differentiation in the Early Mammalian Embryo using Intra- and Interspecific Chimeras

R. L. Gardner

Department of Zoology
South Parks Road
Oxford, U.K.

I. Introduction 207
II. Differentiation of Trophoblast Versus Inner Cell Mass (ICM) 209
   A. Onset of Morphological Differentiation 209
   B. Properties of Trophoblast and ICM Tissue 210
   C. Importance of Cell Position in Determination 211
   D. Experimental Investigation of ICM Determination 212
   E. Use of Interspecific Rat:Mouse Chimeras to Overcome
      Limitations in Existing Methods of Analysis 213
III. Determination Among ICM Cells in the 3½ Day Blastocyst 213
   A. Removal of ICM Cells 214
   B. Transplantation of Isolated ICMs and Single ICM Cells to
      Host Blastocysts 219
   C. Potential Value of Rat:Mouse Chimeras in Analysis of
      Determination and Morphogenesis 223
IV. Regional Differentiation in the Trophoblast of the Blastocyst 224
   A. Morphological, Physiological, and Biochemical Features of
      Regional Differentiation 225
   B. Experimental Analysis Implicating ICM 226
   C. Hypothesis of Regional Differentiation 226
   D. Further Experiments and Observations 227
   E. Problem of the Polar Trophoblast – Blastocyst Orientation ... 228
V. Conclusions 231
   References 233

I. INTRODUCTION

It is perhaps rather sobering to reflect that each of us originated from a fertilized egg which appears relatively simple in structure and is little more than 100 μm in diameter. It serves to remind us of the amazing transformations of which developing systems are capable and helps to explain why the study of embryology has fascinated man for centuries. At first, content to observe, describe, and often speculate with abandon, he was later motivated to question and to experiment. Nevertheless, we remain more ignorant today of the development of man and other mammals than of most other animals. With few exceptions, mammalian embryos develop inside the maternal organism, and this clearly poses special problems for the experimental embryologist. However, technical advances in the culture and transplantation of embryos are rapidly reducing many of the difficulties associated with viviparity (Daniel, 1971; New, 1971; Hsu, Stevens and Rash, 1974).

A number of special problems have yet to be solved. For example, before any meaningful analysis of development can be undertaken it is essential to obtain an accurate fate map of the early embryo. This was achieved in the relatively large amphibian embryos by applying small blocks of agar impregnated with vital dyes to specific parts of their surface. By noting the subsequent deployment of the groups of dyed cells their fate could readily be established and a map constructed showing the regions of the early embryo from which the nervous system, skin, muscle, and other organs and tissues later took origin (Vogt, 1925, 1929). In certain more primitive species the exercise was still simpler, since the eggs often display intrinsic regional differences that obviate the need for dyes (e.g. Conklin, 1905; Davidson, 1968).

Unfortunately, the early mammalian embryo is rather too small for local application of vital dyes, shows no regional cytoplasmic differences in the living state, and develops its definitive organs at a stage still largely refractory to culture. Hence knowledge of the fate map of the mammalian embryo is based on examination of serial histological sections of embryos representing successive developmental stages rather than on direct tracing of cell lineages in the living embryo (Snell and Stevens, 1966).

However, in recent years, experimental techniques have been devised that enable the disaggregation, dissection, reconstitution, and combination of early embryos, and the transplantation of genetically labelled cells and tissues between them (Tarkowski, 1961; Mintz 1962; Cole and Paul, 1965; Gardner, 1971). Such manipulations have been of considerable value in assessing the accuracy of the histological fate map. They have also revealed cellular interactions important for development and differentiation that altogether elude analysis by the conventional histological approach. The aim of this paper is to discuss recent experimental investigations of early mammalian development against the descriptive backcloth provided by histology (Fig. 1). The mouse embryo will be used as a model since, though somewhat atypical in its early post-implantation morphogenesis, it has proved very amenable to experimental manipulation. Technical details will be kept to a minimum since most have been amply described elsewhere. However, the potential of certain

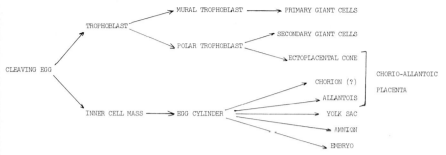

Fig. 1. Fate map of trophoblast and ICM of the 3½ day mouse blastocyst according to the histological study of Snell and Stevens (1966).

new techniques, such as the production of interspecific chimeras, will be emphasized where appropriate in relation to limitations of existing methods of analysis.

II. DIFFERENTIATION OF TROPHOBLAST VERSUS INNER CELL MASS (ICM)

Mammals exhibit several peculiar features in their very early development that serve to distinguish them for other groups of animals. For example, the embryonic genome appears to become active at a very early stage as judged by synthesis of the various classes of RNA (Graham, 1973; Church and Schultz, 1974), the action of certain lethal genes (McLaren, 1974), and the detection of paternally-coded specific gene products (Chapman, Whitten and Ruddle, 1971; Wolf and Engel, 1972; Brinster, 1973). Nevertheless, it must be remembered that although onset of genetic activity is precocious in relation to cell number, the process of cleavage is unusually slow. At least 3 days are required for its completion. Furthermore, most of the cells produced thereby make no contribution to the definitive embryo. Three quarters of the cells of the 3½ day post-coitum mouse blastocyst are trophoblast cells that are involved in establishing and maintaining an intimate relationship between the developing embryo and its mother. The remaining quarter of the 60 or so cells present at this stage form the inner cell mass (ICM) which is destined to give rise to certain extra-embryonic structures in addition to the embryo (Fig. 1; Snell and Stevens, 1966). Hence the definitive embryo probably originates from very few cells of the blastocyst (Mintz, 1970, 1971; Gardner and Lyon, 1971; Ford, Evans and Gardner, 1974).

A. Onset of Morphological Differentiation

Although subtle differences between blastomeres of the cleaving mouse egg have been detected by light and electron microscopy (Dalcq, 1957; Enders, 1971), it is not until blastulation that morphological differentiation among cells is unequivocal. The fully expanded blastocyst is at 3½ days post-coitum a hollow sphere whose surface is composed of flattened polygonal cells united by junctional complexes (Enders and Schlafke, 1965; Enders, 1971). The ICM consists of a compact group of cells totally enveloped by this outer trophoblast layer and attached to part of its inner surface. However, particularly in view of how cell morphology can vary in response to environmental changes *in vitro*, it is possibly misleading to assume that cells are differentiated on the basis of morphological criteria alone (Willmer, 1970). It is only possible to find out if determination has occurred by separating the two cell populations of the blastocyst and examining their properties and developmental potential in isolation and after transplanation. Separation of the two tissues may be achieved readily by microsurgical methods (Gardner, 1971, 1972). Cutting blastocysts into 2 pieces parallel and close to the ICM yields most of the trophoblast bordering the blastocoelic cavity (mural trophoblast) uncontaminated by ICM cells (Fig. 2). ICM tissue can be isolated by separating it mechanically from overlying (polar trophoblast) cells.

B. Properties of Trophoblast and ICM Tissue

The two living tissues appear to be entirely different when tested for their ability to accumulate blastocoelic fluid and form vesicles, to aggregate with fragments of like tissue, initiate implantation in recipient uteri, and phago-cytose small particles such as ocular melanin granules (Table 1; Gardner, 1972, 1974a). The junctional complexes between trophoblast cells may explain failure of aggregation of fragments of this tissue since clusters of ICM cells that lack these structures readily aggregate in groups or pairs. It is evident from Table 1 that it is only the trophoblast that displays properties which might be regarded as indices of differentiation. These findings support the widely held idea that trophoblast cells are precociously differentiated at the

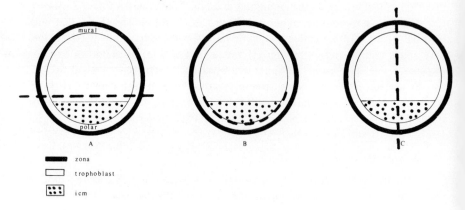

Fig. 2. Scheme of isolation of mural trophoblast (A), inner cell mass (B) and production of 'half blastocyst' (C).

TABLE 1

*Differences in Properties between ICM and Trophoblast
Tissue Isolated from 3½ Day Post-coitum Mouse Blastocysts*

Tissue	Property investigated			
	Aggregation of tissue fragments	Formation of blastocoelic fluid	Initiation of implantation in utero	Phagocytosis of melanin granules
ICM	+	−	−	−
Trophoblast	−	+	+	+

blastocyst stage. Failure of trophoblast cells, injected into the blastocoelic cavity, to colonize ICM derivatives of the host embryos also supports this view (Gardner, 1971, 1974a). Furthermore, it is tempting to conclude that the lack of specific properties in ICM cells means that they are still in an undifferentiated or developmentally naive condition. If this were indeed the case, one might anticipate that they too could differentiate as trophoblast cells under the appropriate conditions. Before discussing experiments designed to test this possibility, it is necessary to consider briefly the factors responsible for trophoblast differentiation.

C. Importance of Cell Position in Determination

The various hypotheses concerning determination of trophoblast versus ICM have been reviewed recently (Graham, 1973; Herbert and Graham, 1974; Gardner, 1974a) and will therefore not be discussed in detail. Experimental investigations of the problem attest to the developmental lability of blastomeres up to the 8-cell stage (Mintz, 1965; Tarkowski and Wroblewska, 1967; Moore, Adams and Rowson, 1968; Hillman, Sherman and Graham, 1972; Wilson, Bolton and Cuttler, 1972). Moreover, they support the idea that the relative positions of blastomeres during later cleavage is a major factor in determining their fate. Blastomeres isolated from 4- to 8-cell stage embryos made to occupy an outside position in aggregates contribute most of their progeny to the trophoblast layer of the blastocyst and to trophoblastic derivatives and yolk sac of later conceptuses (Hillman *et al.*, 1972). Conversely, isolated blastomeres or intact early cleavage embryos, placed in an inside position in aggregates, tend to contribute more descendants to the ICM and thence to the fetus or offspring (Hillman *et al.*, 1972).

If cells of cleavage stages differentiate according to position some time after the 8-cell stage as the above results imply, one would anticipate that ICM cells made to occupy an outside position would likewise form trophoblast if they really were still developmentally naive. Several different experiments have been carried out by my colleague Janet Rossant and myself in order to examine this possibility.

D. Experimental Investigation of ICM Determination

Simply by isolating an ICM it is possible to expose many of its cells to outside conditions. However, because isolated ICMs survive poorly in both culture (Coles and Paul, 1965) and *in utero* (author's unpublished observations) it is not possible to determine whether they can form trophoblast under these conditions. Janet Rossant has succeeded in injecting them into empty zonae so that they can be returned to the oviduct. They remain viable for 1 to 2 days in this environment but show no morphological evidence of trophoblast formation. Furthermore, if such ICMs are injected into host blastocysts differing in genotype for glucose phosphate isomerase (Gpi-1), after recovery from the oviduct chimeric post-implantation embryos can be formed. No contribution of cells of isolated ICMs treated in this way to trophoblastic derivatives of the host embryos has been found in 5 chimeras analyzed so far (Rossant, unpublished observations).

The second experimental approach has been to aggregate isolated ICMs homozygous for one electrophoretic variant of Gpi-1 with naked 8- or 16-cell morulae homozygous for the other isozyme. Aggregation occurs readily and leads within 24 to 36 hours to the formation of blastocysts of normal overall size which contain approximately twice the usual number of ICM cells. Following implantation *in utero* these composite embryos can develop into morphologically normal conceptuses in which the donor isozyme is detectable in the definitive embryo and membranes but again not in trophoblastic derivatives (Rossant, unpublished observations). The ICM cells are thus apparently unable to form trophoblast cells when placed on the outside of later cleavage stages but nevertheless find their appropriate location in the host embryo.

Very recently it has been established by autoradiography that [^{3}H]-thymidine-labelled teratocarcinoma cells can be incorporated into the trophoblast monolayer of the blastocyst after injection into the blastocoelic cavity (Evans and Gardner, unpublished observations). Nevertheless, whole isolated ICMs or individual ICM cells injected similarly have never been found to contribute cells to trophoblastic tissues of host embryos later in pregnancy, though they regularly colonized ICM derivatives (Gardner, 1971, 1974b; Gardner and Lyon, 1971).

Although the above experiments argue strongly against the ability of ICM cells to form trophoblast, unequivocal proof that they cannot do so is more difficult to obtain at present than might be supposed for the following reason. The majority of trophoblast cells of the blastocyst are mural trophoblast cells that surround the blastocoelic cavity. These cells cease dividing at the late blastocyst stage and transform into primary trophoblastic giant cells (Dickson, 1966) probably by continued endoreduplication of their DNA (Sherman, McLaren and Walker, 1972). Therefore, if ICM cells contributed to the mural trophoblast following aggregation or injection they would be unlikely to be detected by the genetic markers employed (Gardner, 1974a; Gardner and Johnson, 1974).

It is only if they were to colonize the polar trophoblast overlying the ICM that their proliferation is likely to be sustained. Although it has not proved possible to inject ICM cells into the optimal site for colonizing the polar trophoblast, isolated ICMs can be inserted into vesicles of mural trophoblast (Gardner, 1971, 1974b). As discussed later these 'reconstituted' blastocysts exhibit normal trophoblast proliferation. No ICM contribution to the proliferated trophoblast can be detected following implantation when the two recombined tissues differ in Gpi-1 genotype (Gardner, Papaioannou, and Barton, 1973).

E. Use of Interspecific Rat:Mouse Chimeras to Overcome Limitations in Existing Methods of Analysis

There is, therefore, compelling evidence that ICM cells cannot differentiate as polar trophoblast cells, though differentiation as mural trophoblast cells cannot yet be ruled out. This latter possibility can only be excluded by use of a marker that would allow detection of single cells. Prior labelling of the donor ICM cells by [^{3}H]-thymidine provides one such marker, though the extreme sensitivity of these cells to the label would preclude unambiguous interpretation of the results (Snow, 1973a, 1973b). An alternative single cell marker is afforded by using rat and mouse blastocysts as donor and host embryos. Individual cells of either species origin can thus be identified unequivocally using an indirect immunofluorescent technique to visualize the distribution of species antigens (Gardner and Johnson, 1973, 1974; Johnson and Gardner, 1974).

Chimeric embryos formed by transplanting rat ICMs into mouse blastocysts show wide distribution of rat cells by immune-fluorescence analysis in post-implantation stages. Once again, donor cells are conspicuously absent from all parts of the trophoblast of these embryos (Gardner and Johnson, 1973, 1974). Perhaps the most critical experiment is the combination of ICMs isolated from blastocysts of one species with morulae of the other. This has yet to be done.

III. DETERMINATION AMONG ICM CELLS IN THE 3½ DAY BLASTOCYST

As mentioned earlier the ICM gives rise both to the definitive embryo and to extra-embryonic structures including the amnion, allantois, and yolk sac. Is it, therefore, composed of a developmentally heterogeneous population of cells, or is the fate of ICM cells still unspecified as regards the ICM derivatives to which their progeny can contribute? Again, several different types of experiment have been devised to answer this question.

A. Removal of ICM Cells

If the cells of the ICM were already committed to different developmental pathways, removal or destruction of some of them would be expected to result in deficiencies or defects in the embryo or its membranes later in development. Lin (1969) developed a micro-suction device with which he could remove part of the ICM. It is not clear to what extent the ICM was reduced in these experiments, since cells were completely disrupted during removal and no blastocysts were prepared for cell counts post-operatively. Only those blastocysts that had expanded and appeared morphologically normal after culture were transferred to recipient uteri. Eight out of twenty eight transferred embryos developed to term. The offspring were normal, though one was small. However, these results are not conclusive because damage to the ICM was variable and anomalous embryos are likely to have been resorbed.

Reduction of the ICM can also be achieved by cutting blastocysts in half perpendicular to the surface of the ICM (Fig. 2). This differs from Lin's technique in that both ICM and trophoblast tissue are reduced to an equivalent extent. 'Half-blastocysts' are in fact likely to contain less than half the number of cells of intact blastocysts because cells in the path of the micro-scalpel will be destroyed. Nevertheless, the majority recavitate after a brief period in culture (Fig. 3).

Forty six 'half-blastocysts' were transferred bilaterally to 5 recipients on the third day of pseudopregnancy, regardless of their post-operative morphology. One recipient carrying 8 embryos died shortly after operation; the remaining 4 all had implantation sites when killed and examined 7 to 8 days later. Twenty six decidual swellings were found altogether, so that the implantation rate compared favorably with that of intact blastocysts and trophoblastic fragments (Gardner, 1972, 1974b). Furthermore, 12 decidua contained fetuses or embryos. Six fetuses were within 24 hours of the stage of development attained by transferred intact blastocysts. A further 4 fetuses, one of which showed degeneration of endocardial cells, were retarded by more than 24 hours. All fetuses seemed to be morphologically normal and to contain their full complement of organs and tissues in the correct relationship. The two remaining embryos were pre-somite stages. One was equivalent to a normal 7½ day embryo, and the other possibly anomalous. The latter was difficult to interpret because histological processing and the plane of sectioning were unsatisfactory. It appeared to be abnormal morphologically rather than deficient in constituent parts.

The remaining 14 decidua did not contain definitive embryos. In fact, 5 altogether lacked cells of donor embryo origin. The sixth was typical of a

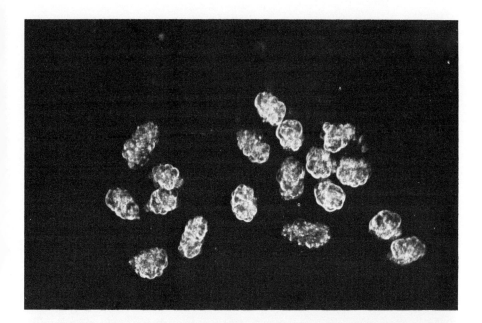

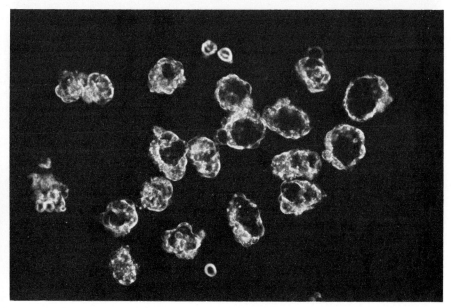

Fig. 3. Eighteen 'half blastocysts' (A) before, and (B) after culture for approximately 1½ hrs. All but 4 show some cavitation after culture. Note that the top left fragment in (B) is a single 'half blastocyst' though it resembles a pair of such fragments in contact.

decidual swelling formed by the transplantation of mural trophoblast. It contained only 2 trophoblastic giant cells (Gardner, 1972). Abundant giant cells were apparent in each of a further 4 decidua, but because extensive degenerative changes had occurred, it was not possible to establish whether ICM derivatives had also developed.

The last 4 implantations that had developed from 'half-blastocysts' that failed to form embryos all contained ICM derivatives in addition to abundant giant cells. Ectoplacental trophoblast was also present in 3 of them. The first had a small oval of Reichert's membrane material with presumptive distal endoderm cells embedded in it as the only tissue of ICM origin in a mass of giant trophoblast cells (Fig. 4). The second contained a vesicle composed of a thick Reichert's membrane, again lined with distal endoderm cells and enclosing clumps of cells of uncertain identity. The ectoplacental cone was well-developed and was situated appropriately on the mesometrial side of the vesicle (Fig. 5). Reichert's membrane was interrupted in the ectoplacental region as in a normal conceptus. The third contained a similar but larger vesicle completely surrounded by Reichert's membrane (Fig. 6). The ecto-

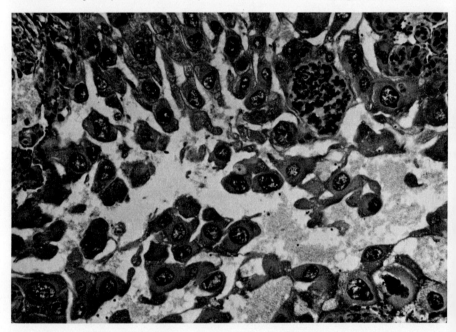

Fig. 4. Implantation chamber of 'half blastocyst' 8 days after transfer to the uterus. Note the small oval matrix of Reichert's membrane material with cells embedded in it, towards the top right-hand corner of the figure. Trophoblastic giant cells are abundant, but no ectoplacental tissue remains. The mesometrial surface of the chamber is at the top of this figure, as also in Figs. 5-7.

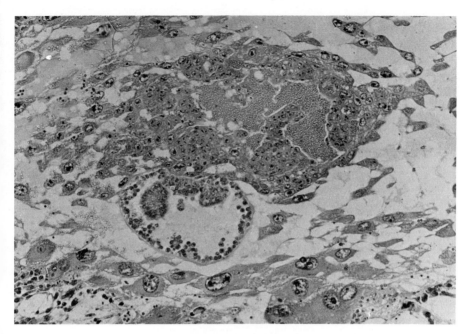

Fig. 5. Implantation chamber of 'half blastocyst' 7 days after transfer. Note the small vesicle formed by a thick Reichert's membrane is discontinuous mesometrially in the region of the ectoplacental cone. Giant cells, which extend beyond the limits of the figure, are abundant.

placental cone was again situated mesometrially, but separated from the vesicle by several layers of giant cells. The whole of the inner surface of the vesicle was covered with presumptive distal endoderm cells as were internal tongues and islands of this extra-cellular material. In all 3 of the above implants the presumptive ICM derivatives thus closely resembled murine yolk sac teratocarcinoma embedded in neoplastic hyalin with which distal endoderm and its secretion, Reichert's membrane, are almost certainly homologous (Pierce, Midgley, Sri Ram and Feldman, 1962; Midgley and Pierce, 1963).

The last implantation site in this group was most intriguing: both embryo and amnion were missing. Nevertheless, it contained a well-developed yolk sac with typical blood islands surrounded by normal ectoplacental and giant trophoblast (Fig. 7). Furthermore, formation of a chorio-allantoic placenta seemed to have proceeded normally with union between the chorion and base of the ectoplacenta. Allantoic tissue was also evident in the placental region (Fig. 7).

Halving the number of ICM cells in the 3½ day blastocyst does not result in specific defects later in development but in a continuous spectrum of

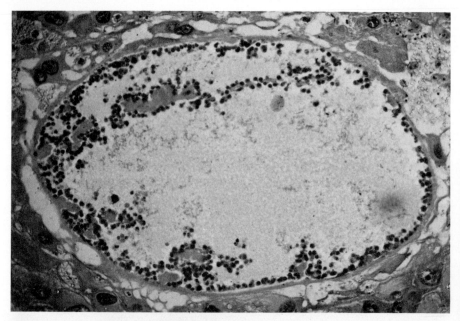

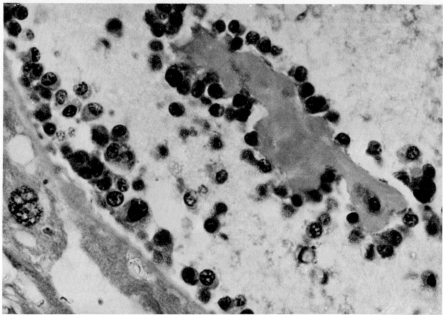

Fig. 6. Implantation chamber of a 'half blastocyst' 8 days after transfer. Reichert's membrane is thick and continuous mesometrially (A). An ectoplacental cone was situated above the part of the chamber depicted in this figure, several layers of giant cells being interposed between it and the Reichert's membrane-bound vesicle. Tongues and islands of Reichert's membrane material surrounded by presumptive distal endoderm cells extend into the vesicle (B).

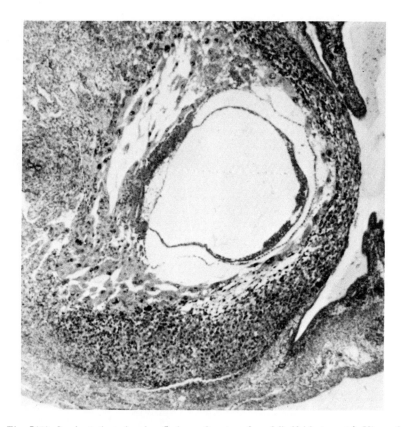

Fig. 7(A). Implantation chamber 7 days after transfer of 'half blastocysts'. View of decidual swelling at low magnification. The chamber is filled with a large yolk sac, but the latter lacks any vestige of embryo and amnion. Note that orientation of this conceptus is atypical, the placenta lying laterally rather than mesometrially.

developmental forms ranging from fetuses through retarded embryos to those in which ICM derivatives are minimal or absent altogether. This pattern of development is very similar to that encountered following transfer of intact preimplantation stages to ectopic sites (e.g. Billington, Graham and McLaren, 1968), though implants containing morphologically normal embryos are considerably more frequent in the former than the latter case.

B. Transplantation of Isolated ICMs and Single ICM Cells to Host Blastocysts

Experimental transplantation of second ICMs to the blastocoelic cavity of 3½ day post-coitum blastocysts further attests to the regulative capacity of this tissue. Donor ICMs typically aggregate with those of the host blastocysts to yield chimeric embryos and normal-sized chimeric offspring (Gardner,

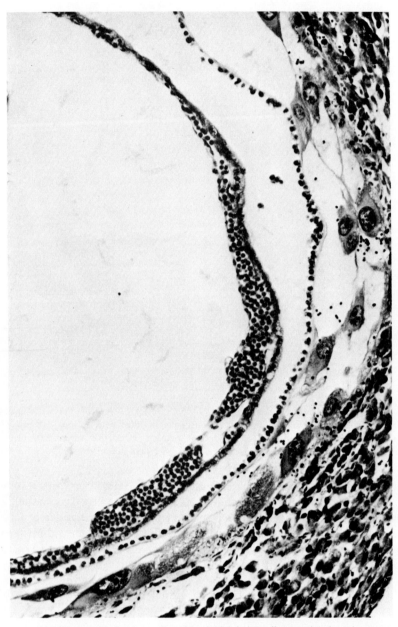

Fig. 7(B). Detail of (A), to show blood islands in yolk sac splanchnopleure.

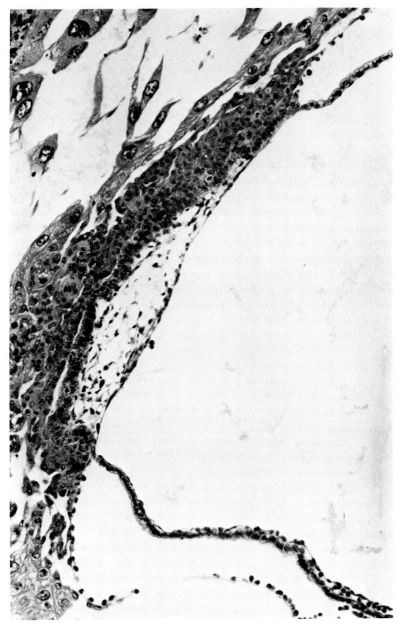

Fig. 7(C). Detail of placental region showing what appears to be allantoic tissue.

1971). However, a more informative approach to the question of determination among ICM cells was made possible by the discovery that a high proportion of overtly chimeric mice could be produced by injecting just one ICM cell into host blastocysts. This technique of inducing chimerism not only minimizes the problem of size regulation imposed by aggregation of cleavage stages or transplanting whole ICMs (Buehr and McLaren, 1974), but also enables the clonal history of a single cell to be followed *in vivo*.

Approximately one quarter of the offspring developing from blastocysts receiving a single donor cell were chimeric (Gardner and Lyon, 1971; Gardner, 1974b). Since, on the basis of cell morphology and dye exclusion tests, only viable cells were selected for transplantation, it seemed likely that the majority of injected cells must have colonized the extra-embryonic tissues of host embryos. Blastocysts of Gpi-1a/Gpi-1a genotype that had received single Gpi-1b/Gpi-1b ICM cells were transferred to recipient uteri in order to test this hypothesis. Implanted embryos were recovered 6 days later when ICM derivatives are usually sufficiently well developed to enable embryonic and extra-embryonic fractions to be analyzed individually for Gpi while still separable from trophoblastic tissues. Eighty embryos were analyzed altogether (Table 2). In no case was the trophoblast colonized by progeny of the injected cell, in accord with data presented earlier regarding determination of this tissue.

TABLE 2

Distribution of Chimerism in Postimplantation Embryos Developed from Gpi-1a/Gpi-1a Blastocysts Injected with a Single Gpi-1b/Gpi-1b ICM Cell*

	Chimeras	
Distribution of chimerism	No. of definite	No. of questionable
Embryonic fraction only	3	1
Extra-embryonic fraction only	4	2
Embryonic + extra-embryonic fraction	4	3
Trophoblastic fraction	0	0

*Embryos were recovered for analysis 6 days after transfer to the uteri of 2½ day post-coitum pseudopregnant female mice. A total of 80 embryos were analyzed, 63 of which showed no evidence of chimerism in any fraction. The amount of tissue in a fraction was only sufficient for one or at most 2 electrophoresis runs. Hence the uncertainty regarding chimerism in 6 embryos. (Gardner, Papioannou and Barton, unpublished data).

Unexpectedly, the proportion of definite chimeras was considerably lower (14%) rather than higher than in offspring obtained in the earlier experiments. In some cases the injected cell had colonized only the extra-embryonic parts of the conceptus, in others only the definitive embryonic region. However, most interestingly, in a similar number of cases progeny of the cell colonized both ICM-derived regions of the same embryo (Table 2). Clearly, at least some ICM cells are not committed to embryonic versus extra-embryonic differentiation at 3½ days post-coitum. Both these regions consist of 3 germ layers at the stage chosen for analysis so that determination of ICM cells for different germ layers is not excluded by these experimental results.

The possibility that germ layer commitment had occurred was examined by repeating the foregoing experiment, but analyzing the results considerably later in development when, although trophoblastic and ICM derivatives cannot be separated in the placental region, extra-embryonic membranes and fetal organs can be analyzed individually. Preliminary results are summarized in Table 3. The percentage of chimeras is higher (25%), suggesting that some were missed by carrying out the electrophoretic analysis earlier in development. The need to classify some of the latter as questionable chimeras (Table 2) also supports this conclusion. However, of greater interest is the observation that the progeny of the single injected cell can be widely disseminated in chimeric conceptuses. Furthermore, they are evident in organs derived from all 3 germ layers of the fetus. These experiments are currently being extended to check whether the 3 cases of yolk sac and placental colonization are due to a sub-population of determined cells or are simply the result of chance distribution of the progeny of pluripotent ICM cells. These preliminary data indicate that ICM cells are probably not restricted in potency at 3½ days post-coitum, except as regards formation of trophoblast.

C. Potential Value of Rat:Mouse Chimeras in Analysis of Determination and Morphogenesis

It is to be hoped that by extending this type of experimental analysis to progressively more advanced stages both the nature and time of determination among cells of the ICM may be accurately defined. Evidence of determination would be provided by finding reproducible, restricted patterns of colonization by individual cells rather than the very broad spectrum of donor cell contributions apparent in most of the above experiments. Precision is limited by use of currently available mouse markers because they are not very sensitive and because tissues must be dissected apart and disrupted prior to analysis. Many organs and membranes are composed of intimate mixtures of tissues of different germ layer origin that cannot be separated readily. These difficulties can be circumvented by using rat:mouse chimeras so that all cells of either species origin may be identified in serially sectioned embryos in

TABLE 3

Distribution of the Progeny of Single Gpi-1^b/Gpi-1^b ICM Cells Injected into 3½ Day Post-coitum Gpi-1^a/Gpi-1^a Blastocysts in 15½ - 17½ Day Post-coitum Chimeric Conceptuses*

	Organs and tissues analyzed	Number of conceptus					
		1	2	3	4	5	6
Extra embry-onic	Placenta	–	–	–	+ (1/5)	+ (1/4)	+ (2/3)
	Umbilical cord	+ (1/8)	–	–	–	–	–
	Yolk sac	–	–	–	+ (2/3)	+ (1/8)	+ (2/3)
	Amnion	+ (1/2)	–	–	–	–	–
Fetal	Brain & Spinal cord	+ (1/3)	+ (1/3)	–	–	–	–
	Right kidney	+ (1/1)	+ (1/1)	–	–	–	–
	Left kidney	+ (1/1)	+ (1/3)	+ (1/8)	–	–	–
	Right adrenal	+ (1/4)	n.a.	–	–	–	–
	Left adrenal	+ (1/2)	n.a.	–	–	–	–
	Right gonad	+ (1/4)	+ (1/3)	–	–	–	–
	Left gonad	+ (2/3)	–	–	–	–	–
	Right lung	+ (1/3)	+ (1/4)	–	–	–	–
	Left lung	+ (1/1)	+ (1/5)	–	–	–	–
	Heart	+ (2/3)	+ (1/4)	+ (1/5)	–	–	–
	Liver	+ (2/3)	+ (3/4)	+ (1/5)	–	–	–
	Carcass	+ (1/1)	+ (1/1)	–	–	–	–

* Twenty-four conceptuses were analyzed altogether, 18 of which appeared to be non-chimeric.

† n.a. means tissue not analyzed. (Gardner, Papaionnou and Barton, unpublished data).

Numbers in brackets are visual estimates of the ratio of donor to host isozyme in chimeric tissues.

which normal cellular relationships are maintained (Gardner and Johnson, 1973). It has been demonstrated very recently that single injected rat ICM cells can be incorporated into the ICM of host mouse blastocysts and continue to divide (Gardner and Johnson, 1974). This provides us with a very powerful method of investigating cellular determination in the early rodent embryo.

IV. REGIONAL DIFFERENTIATION IN THE TROPHOBLAST OF THE BLASTOCYST

So far the trophoblast has proved more amenable than the ICM to examination of determination and differentiation within as opposed to between tissues of the early embryo. This is largely because both morphological and physiological criteria are available by which regional differentiation may be established. Furthermore, progress has been made in understanding how this differentiation may be controlled. I do not intend to discuss the experiments in detail because this has been done elsewhere (Gardner, 1971, 1972, 1974a, 1974b; Gardner et al., 1973). The aim is rather to summarize the principal findings, outline the conclusions that have been reached, and draw attention to certain problems that have yet to be resolved.

A. Morphological, Physiological and Biochemical Features of Regional Differentiation

All trophoblast cells of the 3½ day blastocyst are morphologically alike. They also behave similarly as regards immobilization and phagocytosis of ocular melanin granules. Many granules adhere to the outer surface of all trophoblast cells following brief culture in a suspension of these particles. Some granules are phagocytosed within 2 to 3 hours, but show little enlargement once in the cytoplasm of the trophoblast cells (Gardner, 1974a).

Two distinct populations of trophoblast cells are found in blastocysts recovered from the uterus at 4½ days post-coitum. The mural trophoblast cells bordering the blastocoelic cavity are larger than the polar cells overlying the ICM, and the former exhibit enhanced uptake and intracellular digestion of melanin granules compared with the latter (Gardner, 1974a). Enlargement of the mural trophoblast cells begins at the abembryonic pole of the blastocysts at about midnight on day 4 of development and is completed approximately 12 hours later (Dickson, 1966). Examination of progressively more advanced embryos sectioned in utero reveals that the mural cells continue to increase in size without dividing, thereby forming primary giant cells, while polar cells remain mitotically active and presumably diploid. A mass of trophoblast cells, the ectoplacental cone, is thus formed over the ICM derivatives. Later, cells in the periphery of the ectoplacental cone also enlarge to form numerous secondary giant cells identical in appearance to the primary ones of the mural trophoblast (Duval, 1892; Snell and Stevens, 1966). Studies on the mode of formation of these remarkable cells are based on the secondary ones because they are much more numerous than primary, though the identical morphology of the two generations makes it very likely that conclusions apply to the latter as well. The DNA content increases in parallel with size (Barlow and Sherman, 1972), apparently by endoreduplication of the entire genome rather than by cell fusion (Sherman et al., 1972; Chapman, Ansell, and McLaren, 1972; Gearhart and Mintz, 1972; Gardner et al., 1973).

DNA levels higher than 4 times the haploid value have been detected in nuclei of advanced blastocysts cultured in vitro and also those recovered from the uterus (Barlow & Sherman, 1972; Barlow, Owen & Graham, 1972), though the techniques employed did not enable determination of the part of the blastocysts whence they came. However, since a direct correlation between nuclear size and DNA content has been established for trophoblast cells it appears very likely that they belong to the mural trophoblast (Barlow and Sherman, 1972).

Thus, in summary, initial regional differentiation of murine trophoblast appears to be completed in a period of less than 24 hours during early implantation. It involves loss of the capacity to divide accompanied by continued DNA synthesis and enhanced phagocytic activity in mural trophoblast cells versus maintenance of diploid mitotic activity in the polar trophoblast cells overlying the ICM. The fact that the fate of trophoblast cells differs according to whether or not they are in contact with the ICM suggests that this tissue may be involved in the differentiation.

B. Experimental Analysis Implicating ICM

Vesicles of mural trophoblast isolated from 3½ day blastocysts by the procedure described earlier (Fig. 2) usually form only a few trophoblastic giant cells following uterine implantation, in accord with their fate in the intact blastocyst (Gardner, 1972). Lack of sustained mitotic activity is not simply due to reduction in number of trophoblast cells because, as noted earlier, half blastocysts can produce extensive trophoblasts in a high proportion of cases. By inserting ICM tissue into such trophoblastic vesicles before returning them to the uterus, their proliferation is ensured (Gardner, 1971). Furthermore, by arranging that the two recombined tissues differed in Gpi genotype, it was found that the proliferated trophoblast was derived exclusively from the mural trophoblast (Gardner et al., 1973). This result demonstrates that at least some mural trophoblast cells are able to alter their fate and become polar trophoblast cells at 3½ days post-coitum. Also, it suggests that the ICM is involved in this change in a way that is perhaps best described as an inductive interaction (Speman, 1938; Gardner, 1974a). This raises the perennial question of the nature of the inductive stimulus (Saxen and Wartiovaara, 1974). All that can be said at present is that the interaction is not strictly species-specific because rat ICMs that develop separately from host mouse ICMs after injection into mouse blastocysts can stimulate host mural trophoblast to form a second ectoplacental cone (Gardner and Johnson, 1974).

C. Hypothesis of Regional Differentiation

We have proposed that all trophoblast cells lose their intrinsic capacity for cell division at the late blastocyst stage while retaining the ability to replicate their DNA. Continued division beyond this stage requires local inductive support from the ICM and thus leads to formation of the ectoplacental cone in the polar region. However, as the cone enlarges its more peripheral cells are removed from the sphere of influence of the ICM and therefore cease dividing. Many of these peripheral cells continue to replicate their DNA and thus transform into secondary giant cells (Gardner, 1972, 1974a; Gardner et al., 1973).

D. Further Experiments and Observations

Other experimental results and observations accord with the above hypothesis. Snow (1973a, 1973b) has discovered that continuous culture of cleaving eggs from before the 16 cell stage in the presence of low levels of [³H]-thymidine totally suppresses development of the ICM. Trophoblastic vesicles produced in this way show no increase in cell number during outgrowth *in vitro*, while control blastocysts typically do (Ansell and Snow, 1974). These workers also noted that the rate of outgrowth was faster in trophoblastic vesicles than blastocysts. A possible *in vivo* counterpart of this phenomenon is indicated by the observation that implanted trophoblastic vesicles rapidly form a loose network of trophoblast cells while blastocysts retain a vesicular trophoblast wall (Gardner, 1972, Fig. 9).

Mitoses in the trophoblast of serially sectioned post-implantation embryos examined between 6 and 9 days post-coitum are almost exclusively limited to the ectoplacental region. Mitoses in the later mural trophoblast surrounding Reichert's membrane have been seen only exceptionally, though cells judged to be diploid on the basis of nuclear diameter are not uncommon (author's unpublished observations). Hence it would seem that it is the loss of mitotic activity that is primary and that DNA synthesis may continue in fairly rapid cycles or possibly be resumed after a quiescent period. The ectoplacental cone itself may be resolved histologically into a basal solid core of trophoblast cells and a more loosely organized peripheral zone in which the cells are permeated by maternal blood sinusoids (Gardner, 1974a, Fig. 4). All core cells appear to be diploid on the basis of nuclear diameter, early secondary giant cells being confined to the outer zone. It would be most useful if quantitative studies of the DNA content of trophoblast nuclei in different regions of successive post-implantation stages could be made in such a way as to maintain normal cellular relations.

Further evidence favoring an inductive role for the ICM in trophoblast differentiation comes from the study of embryos developing abnormally *in utero* and in ectopic sites (Gardner, 1972, 1974b; Johnson, 1972). Such embryos can be assigned to one of two categories, those in which abundant trophoblast is present, and those containing only a few giant cells. ICM derivatives are almost invariably in evidence in the former and always absent in the latter category. The few exceptions to this generalization may be apparent rather than real because ICM derivatives may be overlooked when small (Fig. 4), particularly in uterine implantation sites that are resorbing and in ectopic grafts in which morphogenesis is often aberrant. The second category of embryos containing a limited number of giant cells and no ICM tissues have only recently been described in ectopic grafts (Johnson, 1972), probably because only macroscopic 'takes' were examined histologically in most earlier studies.

Although no evidence was found to suggest that the later trophoblast received an ICM contribution in the blastocyst reconstitution experiments, a trophoblast contribution to presumptive ICM derivatives was observed consistently (Gardner *et al.*, 1973). The extra-embryonic ectoderm of rat egg-cylinders that developed separately in mouse blastocysts was in all 3 cases composed entirely of mouse cells (Gardner and Johnson, 1974). It thus seems very probable that this tissue does not originate from the ICM as the histological fate map indicates (Snell and Stevens, 1966), but from the polar trophoblast. The extra-embryonic ectoderm forms part of the chorion which we attempted to isolate with the embryonic rather than the trophoblastic fraction in the reconstitution experiments (Gardner *et al.*, 1973). The chorion fuses with the base of the ectoplacenta a little later in development, so its ectodermal layer may be the source of one or more of the distinctive trophoblast cell populations found in the mature placenta (Duval, 1892). This and other aspects of trophoblast differentiation in relation to interaction with the ICM are currently being investigated.

E. *Problems of the Polar Trophoblast – Blastocyst Orientation*

Available experimental data and observations thus present a consistent picture regarding the regional differentiation of trophoblast. However, it must be remembered that the experiments implicating the ICM were based on an investigation of mural rather than polar trophoblast. Therefore, it could be argued that in normal development polar trophoblast can proliferate quite independently of the ICM. This would seem unlikely, particularly since trophoblastic vesicles produced in an entirely different way behave similarly when development of the ICM is totally suppressed (Snow, 1973a, 1973b). However, it is conceivable that the isotope could interfere with polar trophoblast as well as ICM formation.

The obvious way to test this alternative hypothesis is to study the post-implantation development of blastocysts in which all ICM cells have been destroyed so as to leave the overlying polar trophoblast cells intact. Unfortunately, this experiment has not proved technically feasible so far. However, the following consideration suggested a possible way round this difficulty. When the mouse blastocyst implants in the uterus it invariably does so with the ICM directed towards the mesometrium, although prior to implantation the position of the ICM is variable. The late David Kirby and his colleagues (Kirby, Potts, and Wilson, 1967) advanced the attractive hypothesis that orientation occurs by migration of the ICM round the inner surface of the trophoblast in response to some undefined gradient in the uterus. Mural trophoblast is invariably isolated prior to the time of blastocyst orientation in our experiments. Therefore, if the hypothesis is true, it means that the polar trophoblast at the time of microsurgery is not necessarily, or indeed usually, the final polar trophoblast, and hence mural and polar cells are unlikely to be different at this stage.

The way to test this hypothesis is to mark individual trophoblast cells that bear a specific spatial relationship to the ICM, return the marked blastocysts to the uterus, and check whether the ICM has shifted its position after orientation has taken place. A suitable cell marker would be one that is conspicuous, inert, and resistant to conventional histological processing. Melanin granules isolated from the eyes of pigmented mice seemed to meet these requirements, though attempts to inject them into trophoblast cells were unsuccessful. The chance observation that such granules are readily phagocytosed by trophoblast cells provided a means of overcoming this particular problem.

Melanin granules were injected locally under the zona pellucida of 3½ day blastocysts by the technique illustrated in Fig. 8, so as to avoid damage to the trophoblast and contraction of the blastocyst. They could in this way be trapped between the zona and outer surface of the trophoblast so that they were phagocytosed by one or a few cells during a brief culture interval. They provided a conspicuous, local intracellular marker that persists for at least 48 hours (Fig. 9).

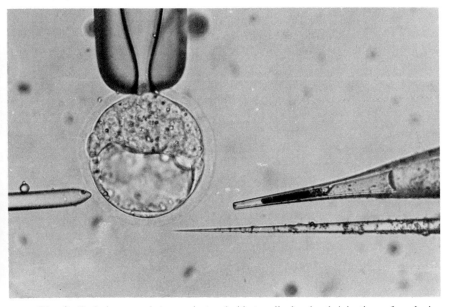

Fig. 8. Technique used to mark trophoblast cells by local injection of melanin granules under the zona pellucida. The blastocyst is oriented for abembryonic polar injection in this figure, and is held by suction. A slit is made in the zona with the sharp needle on the right, the left blunt needle being used to steady the blastocyst. The blastocyst is then rotated through 90° around the embryonic-abembryonic axis so that the slit in the zona is vertical. The pipette on the right containing a dense suspension of melanin granules is then inserted in the slit.

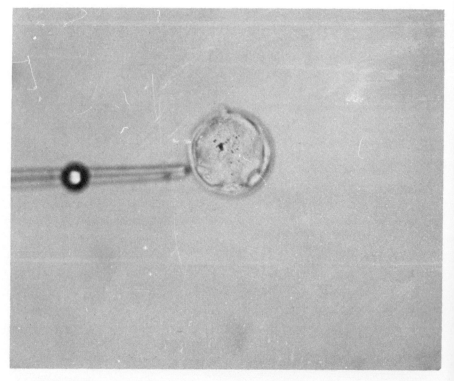

Fig. 9. A blastocyst marked at the abembryonic pole. The local dense area is injected melanin granules lying over one trophoblast cell. These granules would be phagocyted by the cell during a period in culture. The more widely distributed paler dots are other trophoblastic inclusions.

Subzonal injections were made over the ICM, at the abembryonic pole, or laterally half way between the two poles of the blastocyst. Both poles of the same blastocyst were marked in a few cases.

The blastocysts were re-examined after culture to check that the position of the ICM had not shifted during this interval and were then transplanted to the uteri of mice on the third day of pseudopregnancy. Some uterine horns were processed for histology and others flushed between 6 p.m. the following day and 10 a.m. the day after. Fifteen blastocysts were sectioned serially *in utero* and the ICMs of all but one were correctly oriented at this stage. Marker melanin granules were evident in trophoblast cells of every one. Distortion of blastocyst shape due to fixation precluded precise determination of the position of the ICM relative to the marked trophoblast cell or cells, even after serial reconstruction of the embryos. Nevertheless, gross changes in ICM position could be ruled out in every case. A further 26 blastocysts were flushed out of the recipient uteri at the same time interval after transplantation. Most were fully expanded at recovery, though zona loss had occurred and giant cell transformation was well advanced in every case. The marked cell or cells could be identified readily in 19 blastocysts. Their position relative to the ICM was noted approximately in the few contracted blastocysts, and these were then cultured until they had re-expanded. The relationship could be checked accurately in the expanded and cultured re-expanded blastocysts by observing the blastocysts in the microscope while rotating them with a blunt glass needle held in a micromanipulator. Once again, no shift in ICM position relative to the labelled trophoblast cells could be discerned. Therefore, the hypothesis of Kirby *et al.* (1967) appears to be incorrect; other explanations must be sought for the mechanism of blastocyst orientation, and the problem of the nature of the polar trophoblast remains to be established.

A possible way of obtaining pure polar trophoblast emerged during the above experiments. When melanin granule injections were done over the ICM region, the marked polar trophoblast sometimes herniated through the slit in the zona leaving the ICM behind. Two blastocoelic cavities were thus formed, demonstrating that polar trophoblast cells can also pump fluid (Gamow and Daniel, 1970; Tuft and Boving, 1970). Attempts are now being made to encourage sufficient herniation to permit isolation and transfer of pure trophoblast that includes the polar region. It is to be hoped that the present issue may be resolved in this way.

V. CONCLUSIONS

It is seldom possible in experimental embryology to answer a question decisively by means of a single experimental approach. Usually, a particular problem has to be attacked from several angles using as many different experimental procedures. If the weight of evidence is still not decisively in favor of one conclusion rather than another, it is often necessary to employ intuition to tip the scales. The work discussed in this paper is no exception to this generalization.

Available data support the conclusion that blastomeres of the cleaving mouse egg are totipotent up to the 8-cell stage. Approximately 3 cell cycles later, at about the 64-cell stage blastocyst, this is no longer the case. Two distinct tissue compartments then exist, the cells of which are committed to mutually exclusive paths of development in the sense that they can no longer be interconverted by experimental manipulation. Hence determination of cells for trophoblastic versus ICM differentiation must occur between these 2 stages at a time that has yet to be precisely defined. Cleavage stage embryos must be cultured in the presence of [^{3}H]-thymidine from before the 16-cell stage in order to suppress totally formation of the ICM (Snow, 1973a, 1973b). This is of interest because the fate of a blastomere evidently depends on whether it occupies an inside or outside position, and it is between the 8- and 16-cell stage that cells can first become wholly enclosed (Graham, 1971). The inside cells divide faster than outside cells from the time that they first occur (Barlow et al., 1972). Thus it seems likely that critical changes in cells of the cleaving egg take place around the 16-cell stage, though some cells may still be labile in the very early blastocyst (Stern and Wilson, 1972).

Although cells are segregated into two distinct compartments by 3½ days post-coitum, the fate of cells within each compartment is probably as yet undetermined. The trophoblast is composed of approximately 45 cells at this stage in the strains of mice with which we work. Conclusive evidence has been obtained that mural trophoblast cells can be converted into polar trophoblast cells (Gardner et al., 1973) and the reverse is most probably also true. The immediate fate of a trophoblast cell is concerned with whether or not it continues to divide beyond the blastocyst stage. This depends on whether it is in contact with the cells of the ICM. Since, apparently, the ICM does not move relative to the overlying trophoblast, regional differentiation of the trophoblast in normal development will thus be determined by the site of attachment of the developing ICM.

The ICM contains about 15 cells at 3½ days post-coitum. The fate of most, if not all, of these cells appears to be unrestricted except in as much as they are unable to form trophoblast. Morphological differentiation of cells within this tissue occurs by 4½ days post-coitum with delamination of the primitive endoderm along its blastocoelic surface. Observations on the development of isolated mouse ICMs and on rat ICMs injected into mouse blastocysts suggest that cell position is important in determination of endoderm as well as trophoblast versus ICM (Rossant, unpublished data; Gardner and Johnson, 1974).

A variety of techniques are now available which may help not only to solve some of the problems persisting in the present work, but also to enable direct analysis of morphogenesis and determination in the hitherto inaccessible post-implantation stages of development. Interspecific chimeras between the rat and mouse are particularly promising in terms of the resolution they provide for clonal analysis. However, crossing the species barrier raises immunological and other problems that must be investigated before these unique organisms can be accepted as valid models for the analysis of normal development. These problems have been discussed elsewhere (Gardner and Johnson, 1974).

ACKNOWLEDGMENTS

I wish to thank Dr. C. F. Graham and Miss J. Rossant for their comments, Mrs. S. C. Barton, Dr. V. E. Papaioannou and Miss Rossant for allowing me to quote unpublished results, and Miss C. Merriman and Mrs. P. Williams for able secretarial assistance. The author's own work discussed in this paper was supported by the Ford Foundation, the Medical Research Council, and the World Health Organization.

REFERENCES

Ansell, J. D. and Snow, M. H. L. (1974). The development of trophoblast *in vitro* from blastocysts containing varying amounts of inner cell mass. *J. Embryol. Exp Morph.* in press.

Barlow, P. W. and Sherman, M. I. (1972). The biochemistry of differentiation of mouse trophoblast: studies in polyploidy. *J. Embryol. Exp. Morph.* **27**, 447–465.

Barlow, P. W., Owen, D., and Graham, C. F. (1972). DNA synthesis in the preimplantation mouse embryo. *J. Embryol. Exp. Morph.* **27**, 431–445.

Billington, W. D., Graham, C. F., and McLaren, A. (1968). Extra-uterine development of mouse blastocysts cultured *in vitro* from early cleavage stages. *J. Embryol Exp. Morph.* **20**, 391–400.

Brinster, R. L. (1973). Parental glucose phosphate isomerase activity in 3-day mouse embryos. *Biochem. Genet.* **9**, 187–191.

Buehr, M. and McLaren, A. (1974). Size regulation in chimaeric mouse embryos. *J. Embryol. Exp. Morph.* **31**, 229–234.

Chapman, V. M., Ansell, J. D. and McLaren, A. (1972). Trophoblast giant cell differentiation in the mouse : expression of glucose phosphate isomerase (GPI-1) electophoretic variants in transferred and chimaeric embryos. *Develop. Biol.* **29**, 48–54.

Chapman, V. M., Whitten, W. K. and Ruddle, F. H. (1971). Expression of paternal glucose phosphate isomerase - 1 (Gpi-1) in preimplantation stages of mouse embryos. *Develop. Biol.* **26**, 153–158.

Church, R. B. and Schultz, G. A. (1974). Differential gene activity in the pre- and postimplantation mammalian embryo. *In* "Current Topics in Developmental Biology" (A. A. Moscona and A. Monroy, eds.), pp. 179–202. Academic Press, New York.

Cole, R. J. and Paul, J. (1965). Properties of cultured preimplantation mouse and rabbit embryos, and cell strains derived from them. *In* "Preimplantation Stages of Pregnancy" (G. E. W. Wolstenholme and M. O'Connor, eds.), pp. 82–112. Churchill, London.

Conklin, E. G. (1905). The organization and cell lineage of the ascidian egg. *J. Acad. natn. Sci. Philad.* **13**, 5–119.

Dalcq, A. M. (1957). *Introduction to General Embryology*. Oxford University Press.

Daniel, J. C. Jr. (1971). *Methods in Mammalian Embryology*. Freeman, San Francisco.

Davidson, E. H. (1968). *Gene Activity in Early Development*. Academic Press, New York.

Dickson, A. D. (1966). The form of the mouse blastocyst. *J. Anat.* **100**, 335–348.

Duval, M. (1892). *Le Placenta des Rongeurs*. Extrait du Journal de l'Anatomie et de la Physiologie Années 1889 – 1892 (F. Alcan, ed.), Paris, Ancienne Librarie Gemmer Baillière.

Enders, A. C. (1971). The fine structure of the blastocyst. *In* "The Biology of the Blastocyst" (R. J. Blandau, ed.), pp. 71–94. University of Chicago Press, Chicago.

Enders, A. C. and Schlafke, S. J. (1965). The fine structure of the blastocyst : some comparative studies. *In* "Preimplantation Stages of Pregnancy" (G. E. W. Wolstenholme and M. O'Connor, eds.), pp. 29–54, Churchill, London.

Ford, C. E., Evans, E. P. and Gardner, R. L. (1974). Marker chromosome analysis of two mouse chimaeras. (submitted for publication).

Gamow, E. and Daniel, J. C., Jr. (1970). Fluid transport in the rabbit blastocyst. *Wilhelm Roux' Arch.* **164**, 261–278.

Gardner, R. L. (1971). Manipulations on the blastocyst. *In* "Advances in the Biosciences, No. 6. Schering Symposium on Intrinsic and Extrinsic Factors in Early Mammalian Development" (G. Raspé, ed.), pp. 279–296. Pergamon, Oxford.

Gardner, R. L. (1972). An investigation of inner cell mass and trophoblast tissue following their isolation from the mouse blastocyst. *J. Embryol. Exp. Morph.* **28**, 279–312.

Gardner, R. L. (1974a). Origin and properties of trophoblast. *In* "The Immunobiology of trophoblast" in press.

Gardner, R. L. (1974b). Microsurgical approaches to the study of early mammalian development. *In* "The Seventh Harold C. Mack Symposium: Birth Defects and Fetal Development, Endocrine and Metabolic Disorders" (K. S. Moghissi, ed.), pp. 212–233. Thomas, Springfield.

Gardner, R. L. and Johnson, M. H. (1973). Investigation of early mammalian development using interspecific chimaeras between rat and mouse. *Nature New Biol.* **246**, 86–89.

Gardner, R. L. and Johnson, M. H. (1974). Investigation of cellular interaction and deployment in the early mammalian embryo using interspecific chimaeras between the rat and mouse. *In* "Ciba Foundation Symposium on Pattern Formation" in press.

Gardner, R. L. and Lyon, M. F. (1971). X-chromosome inactivation studied by injection of a single cell into the mouse blastocyst. *Nature (London)* **231**, 385–386.

Gardner, R. L., Papaioannou, V. E. and Barton, S. C. (1973). Origin of the ectoplacental cone and secondary giant cells in mouse blastocysts reconstituted from isolated trophoblast and inner cell mass. *J. Embryol. Exp. Morph.* **30**, 561–572.

Gearhart, J. D. and Mintz, B. (1972). Glucosephosphate isomerase subunit-reassociation tests for maternal-fetal and fetal-fetal cell fusion in the mouse placenta. *Develop. Biol.* **29**, 55–64.

Graham, C. F. (1971). The design of the mouse blastocyst. *In* "Control Mechanisms of Growth and Differentiation: Symposia of the Society for Experimental Biology. No. 25" (D. D. Davies and M. Balls, eds.), pp. 371–378. Cambridge University Press.

Graham, C. F. (1973). The necessary conditions for gene expression during early mammalian development. *In* "Genetic Mechanisms of Development, 31st Symposium of The Society for Developmental Biology" (F. H. Ruddle, ed.), pp. 201–224, Academic Press, New York.

Herbert, M. C. and Graham, C. F. (1974). Cell determination and biochemical differentiation of the early mammalian embryo. *In* "Current Topics in Developmental Biology" (A. A. Moscona and A. Monroy, eds.), pp. 151–178, Academic Press, New York.

Hillman, N., Sherman, M. I. and Graham, C. F. (1972). The effect of spatial arrangement on cell determination during mouse development. *J. Embryol. Exp. Morph.* **28**, 263–278.

Hsu, Y. C., Stevens, L. C. and Rash, J. E. (1974). Development *in vitro* of mouse embryos from the 2-cell stage to the early somite stage. *J. Embryol. Exp. Morph.* **31**, 235–245.

Johnson, M. H. (1972). Relationship between inner cell mass derivatives and trophoblast proliferation in ectopic pregnancy. *J. Embryol. Exp. Morph.* **28**, 306–312 (Appendix).

Johnson, M. H. and Gardner, R. L. (1974). Analysis of rat:mouse chimaeras by immunofluorescence : a preliminary report. *In* "Proceedings of the 1st International Congress of the Immunology of Obstetrics and Gynaecology, Padua, 1973. Excerpta Medica in press.

Kirby, D. R. S., Potts, D. M. and Wilson, I. B. (1967). On the orientation of the implanting blastocyst. *J. Embryol. Exp. Morph.* **17**, 527–532.

Lin, T. P. (1969). Microsurgery of inner cell mass of mouse blastocysts. *Nature, (London)* **222**, 480–481.

McLaren, A. (1974) Embryogenesis. *In* "Physiologic and Genetic Aspects of Reproduction" (A. Hollaender and E. Cantinno, eds.) in press.

Midgley, A. R. and Pierce, G. B. (1963). Immuno-histochemical analysis of basement membranes of the mouse. *Am. J. Path.* **43**, 929–943.

Mintz, B. (1962). Experimental recombination of cells in the developing mouse egg: normal and lethal mutant genotypes. *Am. Zool.* **2**, 145a.

Mintz, B. (1965). Experimental genetic mosaicism in the mouse. *In* "Preimplantation Stages of Pregnancy." (G. E. W. Wolstenholme and M. O'Connor, eds.), pp. 194–207. Churchill, London.

Mintz, B. (1970). Gene expression in allophenic mice. *In* "Control Mechanisms in the Expression of Cellular Phenotypes; Symposia of the International Society for Cell Biology, Vol. 9," (H. A. Padykula, ed.), pp. 15–42. Academic Press, New York.

Mintz, B. (1971). Clonal basis of mammalian differentiation. *In* "Control Mechanisms of Growth and Differentiation : Symposia of the Society for Experimental Biology, No. 25, (D. D. Davies and M. Balls, eds.), pp. 345–370. Cambridge University Press.

Moore, N. W., Adams, C. E. and Rowson, L. E. A. (1968). Developmental potential of single blastomeres of the rabbit egg. *J. Reprod. Fert.* **17**, 527–537.

New, D. A. T. (1971). Culture of fetuses *in vitro*. *In* "Advances in the Biosciences, No. 6. Schering Symposium on Intrinsic and Extrinsic Factors in Early Mammalian Development" (G. Raspé, ed.), pp. 367–378. Pergamon, Oxford.

Pierce, G. B., Midgley, A. R., Sri Ram, J. and Feldman, J. D. (1962). Parietal yolk sac carcinoma : clue to the histogenesis of Reichert's membrane of the mouse embryo. *Am. J. Path.* **41**, 549–566.

Saxen, L. and Wartiovaara, J. (1974). Embryonic induction. *In* "Developmental Biology of Plants and Animals" (P. F. Wareing and C. F. Graham, eds.), Blackwell, Oxford in press.

Sherman, M. I., McLaren, A. and Walker, P. M. B. (1972). Mechanism of accumulation of DNA in giant cells of mouse trophoblast. *Nature New Biol.* **238**, 175–176.

Snell, G. D. and Stevens, L. C. (1966). Early embryology. *In* "Biology of the Laboratory Mouse" (E. L. Green, ed.), pp. 205–245. McGraw-Hill, New York.

Snow, M. H. L. (1973a). The differential effect of [^{3}H]-thymidine upon two populations of cells in pre-implantation mouse embryos. *In* "The Cell Cycle in Development and Differentiation" (M. Balls and F. S. Billett, eds.), pp. 311–324. Cambridge University Press.

Snow, M. H. L. (1973b). Abnormal development of pre-implantation mouse embryos grown *in vitro* with [^{3}H] thymidine. *J. Embryol. Exp. Morph.* **29**, 601–615.

Spemann, H. (1938). *Embryonic Development and Induction.* Yale University Press.

Stern, M. S. and Wilson, I. B. (1972). Experimental studies on the organization of the pre-implantation mouse embryo. 1. Fusion of asynchronously cleaving eggs. *J. Embryol. Exp. Morph.* **28**, 247–254.

Tarkowski, A. K. (1961). Mouse chimaeras developed from fused eggs. *Nature (London)* **190**, 857–860.

Tarkowski, A. K. and Wroblewska, J. (1967). Development of blastomeres of mouse eggs isolated at the 4- and 8-cell stage. *J. Embryol. Exp. Morph.* **18**, 155–180.

Tuft, P. H. and Boving, B. G. (1970). The forces involved in water uptake by the rabbit blastocyst. *J. Exp. Zool.* **174**, 165–172.

Vogt, W. (1925). Gestaltungsanalyse am amphibienkeim mit örtlicher vitagfärbung. Vorwort über wege und ziele. I. Methodik u. wirkungsweise der ortlichen vitalfärbung mit agar als farbträger. *Wilhelm Roux' Arch.* **106**, 542–610.

Vogt, W. (1929). Gestaltungsanalyse am amphibienkeim mit ortlicher vitalfarbung. II. Teil : gastrulation und mesodermbildung bei urodelen und anuren. *Wilhelm Roux' Arch.* **120**, 385–706.

Willmer, E. N. (1970). *Cytology and Evolution :* Academic Press, New York.

Wilson, I. B., Bolton, E. and Cuttler, R. H. (1972). Preimplantation differentiation in the mouse egg as revealed by micro-injection of vital markers. *J. Embryol. Exp. Morph.* **27**, 467–479.

Wolf, U. and Engel, W. (1972). Gene activation during early development of mammals. *Humangenetik* **15**, 99–118.

IV. Hormonal Controls in Reproduction

Estrogen Feed-back and Gonadotrophin Secretion[1]

Béla Flerkó

Department of Anatomy,
University Medical School,
Pécs, Hungary

I. Introduction 239
II. Results 241
  A. Localization of LRF-containing Neurons in the Hypothalamus . 241
  B. Role of the Preoptic Area in Regulating Adenohypophyseal
     Gonadotropin Secretion 244
  C. Localization and Role of Estrogen-Sensitive Neurons in
     the Hypothalamus of Normal and Androgen-Sterilized Rats . . . 244
  D. Effects of Intraventricular Norepinephrine, and Dopamine
     in Inducing Ovulation in Normal and Androgen-Sterilized Rats . 248
  References 250

I. INTRODUCTION

Current concepts invoke a dual brain mechanism controlling cyclic gonadotrophin secretion (Barraclough and Gorski, 1961; Flerkó, 1962; Halász, 1969). The first of these, termed a tonic mechanism by Barraclough and Gorski (1961), stimulates the continuous "tonic" discharge of gonadotrophic hormones in sufficient quantity to maintain follicular growth and estrogen secretion, but cannot initiate the ovulatory surge of gonadotrophins. Barraclough and Gorski (1961) localized this mechanism in the ventromedial-arcuate nucleus. For the anatomical localization of the tonic mechanism, I use the hypophysiotrophic area (HTA) expression, a term coined by Halász *et al.* (1962) for the half moon-shaped area of the hypothalamus (Fig. 1). This area is unique because it is the only part of the body in which the normal structure and function of implanted anterior pituitary tissue is maintained. These findings of Halász *et al.* (1962) suggested that the neural elements which synthesize the hypophysiotrophic or releasing factors are in the HTA and the releasing factors which reach the cells of the intrahypothalamic

---

[1] Research carried out recently in the Pécs Anatomy Department and mentioned in this paper was partly supported by the Population Grant M72.85.

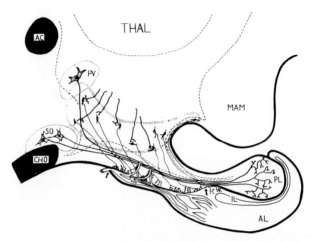

Fig. 1. Diagram of a midsagittal section of the hypothalamus, showing arrangement of the coarse fibered supraoptico-neurohypophyseal tract and the fine fibered tubero-infundibular tract. The former originates from the large cells of the supraoptic nucleus (SO) and is joined by similar fibers arising from the paraventricular nucleus (PV). Its fibers can be traced in Golgi pictures clearly and exclusively to the posterior lobe (PL). This tract is crossed by the fine axons of the tubero-infundibular tract arising from small nerve cells that are situated in the halfmoon-shaped "hypophysiotrophic area" (surrounded by hatched zone) immediately beneath the walls of the 3rd ventricle. The fine axons of this tract terminate exclusively in the surface zone (*zona palisadica*) of the median eminence and of the most proximal part of the pituitary stalk. (Rostral and caudal ends of this zone are indicated by arrows.) Nerve endings of the axons belonging to the tubero-infundibular tract are situated in the immediate vicinity of the hypophysial portal vessels or on the capillary loops of the portal vessels penetrating the median eminence. Modified after Szentagothai (1964). Abbreviations: AC = anterior commissure; AL = anterior lobe of the pituitary gland; CHO = optic chiasm, IL = intermediate lobe of the pituitary gland; MAM = mamillary body; PL = posterior lobe of the pituitary gland; THAL = thalamus.

pituitary graft by diffusion stimulate the hormone synthesis and release from the cells of the intrahypothalamic pituitary graft.

At the time of Halász' experiments with the intrahypothalamic pituitary grafts, Szentágothai located the origin, in the brain, of the fibers that terminate on or in the immediate vicinity of the capillary loops of the hypophysial portal system penetrating the median eminence. He succeeded in Golgi-staining of these nerve fibers (Fig. 1 and 2c) and termed this system of fibers tubero-infundibular tract (TIT) or parvicellular neurosecretory system (Szentágothai, 1962; 1964). According to Szentágothai's finding, the fibers of the TIT originate from the small cells of the arcuate and anterior periventricular nucleus, and from those of the retro-suprachiasmatic area (Fig. 1). The fibers enter the pituitary stalk and terminate on or in the vicinity of the capillary loops of the median eminence. The fine calibered fibers of the TIT can easily be distinguished from the thicker fibers of the paraventriculo-supraoptico-hypophysial tract terminating in the posterior lobe of the pituitary gland. The functional significance of the TIT seemed to be obvious: they are the structural basis of the neuro-humoral link between the hypothalamus and the anterior pituitary established through the portal circulation. Thus, Szentágothai concluded in 1962 that the TIT contains the neurons which produce the various releasing hormones. It is obvious from Fig. 1 that the HTA (surrounded by hatched zone) corresponds to the distribution of the nerve cells from which the tubero-infundibular tract to the median eminence arises. It has later been shown by Halász (1969) and his co-workers that the HTA, in spite of physical interruption of all nervous pathways coming to it, is capable of maintaining a fairly normal structure and function of the thyroid, adrenal, and testis, but not of the ovary. Female rats with HTA isolated neurally from the rest of the brain have shown polyfollicular ovaries and permanent vaginal cornification similar to rats with preoptic-anterior hypothalamic lesions or to rats exposed to neonatal androgen action or to continuous illumination. It appears from the foregoing that the HTA acts directly on the adenohypophysis by its hypophysiotrophic or releasing factors carried by the hypophysial portal circulation to the anterior lobe of the pituitary gland.

II. RESULTS

In order to localize the neurons which synthesize the various releasing factors Mess and Martini (1968) put electrolytic lesions in various parts of the HTA of the rat and measured the content of the various releasing factors in the hypothalamus. Their findings have suggested that nerve cells synthesizing the follicle-stimulating hormone-releasing factor (FRF) would be situated in the antero-dorsal region of the HTA and those synthesizing the luteinizing hormone-releasing factor (LRF) would be located mainly in the antero- and postero-ventral part of the HTA. On the other hand, the whole HTA would be involved in the production of thyrotrophin-releasing factor. No appreciable changes in corticotrophin-releasing factor (CRF) content were observed with any of the lesions. Several pieces of evidence support the view that CRF production would be restricted to the medio-ventral part of the HTA (Porter et al., 1970; Csernus et al., 1973).

A. Localization of LRF-containing Neurons in the Hypothalamus

A joint effort for the exact anatomical localization of the LRF-containing neurons in the rat hypothalamus has recently been attempted by the research efforts of members of the Pécs Anatomy Department and the New Orleans Endocrine and Polypeptide Laboratories. Using specific antibodies generated in rabbits by Arimura and co-workers (1973) against synthetic LRF, we succeeded in showing a system of nerve fibers and terminals which contain immunoreactive LRF (Sétáló et al., 1974). Fig. 2a is a montage photomicrograph showing a paramedian-sagittal section through the medial-basal part of the hypothalamus of an adult rat investigated with the indirect peroxidase-labeled antibody method (Nakane and Pierce, 1967; Mazurkiewicz and Nakane, 1972). A strong bundle of nerve fibers positive to anti-LRF serum is seen in the palisade layer of the anterior and posterior part of the median eminence. The arrangement of the fibers is mainly parallel with the surface of the median eminence. However, fibers running perpendicularly to the surface are also seen especially in the middle part of the neural stalk. Some of these fibers running perpendicularly can be traced back to the arcuate nucleus, although no LRF-positive cell body can be seen either in the arcuate or in other nuclei of the hypothalamus and adjoining brain parts. Terminations of the LRF-positive nerve fibers on or in the immediate vicinity of the capillary loops penetrating the median eminence are clearly visible on photomicrographs with a higher magnification (Fig. 2b). A comparison of the LRF-positive neural elements (Fig. 2b) and a photomicrograph from the original Golgi material of Szentágothai (1962), demonstrating the nerve endings of the tubero-infundibular tract in the palisade layer of the median eminence (Fig. 2c), shows that the two structures are identical. Of course, only a small minority (approximately 10 percent) of the nerve terminals of the tubero-infundibular tract shows a positive reaction, while all other endings of the palisade layer are unequivocally negative (Fig. 2b).

The arrangement of the LRF-positive fiber system resembles a funnel distorted by transverse compression, so that it has a longer sagittal axis. In frontal sections of the hypothalamus, various parts of the funnel become visible (Sétáló et al., 1974). In the more rostral part of the median eminence, the positive fibers accumulate in a flat superficial layer receiving numerous fibers from the retrochiasmatic area. Slightly posteriorly, where the stalk begins to emerge from the median eminence, the positive fibers gather into two major bundles at both sides of the origin of the stalk. Many positive fibers join these bundles from above and from the lateral side to emerge among the cells of the arcuate nucleus. On the other hand, several fibers appear to detach themselves from the two lateral bundles and to be distributed in the lateral part of the stalk. Some endings, descending towards the palisade layer, begin gradually to disappear or to terminate on or in the vicinity of the capillary loops. More caudally, where the pituitary stalk becomes clearly established, the two LRF-positive bundles are localized on both edges of the stalk and next to the sulcus separating the stalk from the base of the hypothalamus. Terminations in the superficial zone of the stalk are most abundant in the lateral parts, but with a higher magnification, the terminals can be found all over the palisade layer and around the capillary

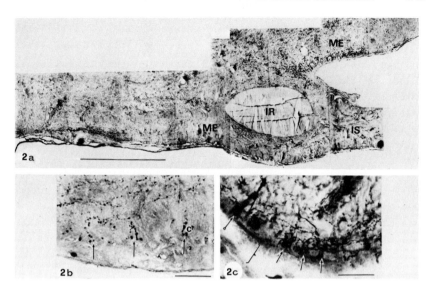

Fig. 2a. Montage photomicrograph of a paramedian-sagittal section through the medial-basal part of the hypothalamus. LRF-containing nerve fibers and their terminals are indicated by the reaction product of diaminobenzidine. Indirect peroxidase-labeled antibody method. Abbreviations: IR = infundibular recess; IS = infundibular stalk; ME = median eminence. Scale: 500 μm. 2b. Terminations of the LRF-containing nerve fibers (arrows) on or in the immediate vicinity of the capillary loops penetrating the median eminence. Indirect peroxidase-labeled antibody method. C = capillary. 2c. Golgi picture of the same site showing the silverdichromate stained nerve terminals of the *zona palisadica* (arrows). Ringed arrow points to glial end-feet. The cleft indicated by bracket was brought about by the detachment of the pars tuberalis from the median eminence surface. (By the courtesy of J. Szentágothai, 1962.)

loops of the median eminence. In the posterior part of the median eminence the positive fibers are found in a flat superficial layer. Most of these fibers run toward the posterior margin of the stalk, and only a few terminations can be seen in the palisade layer of the stalk. These findings indicate that the pathway of the LRF-positive nerve fibers coincides completely with the course of the nerve fibers belonging to the tubero-infundibular tract described by Szentágothai (1962). It appears, therefore, that the LRF-positive fiber system arises from small nerve cells that are situated in the HTA, immediately beneath the walls of the 3rd ventricle, as suggested by Szentágothai and Halász (1964). Recently, Palkovits et al. (1974) determined the LRF-content of 16 separate hypothalamic nuclei and the median eminence in the rat by radioimmunoassay. High concentrations of LRF were measured in the median eminence and in the arcuate nucleus. Other nuclei of the hypothalamus contained little or no LRF, which coincides fairly well with our findings concerning the localization of the nerve fibers containing LRF.

B. *Role of the Preoptic Area in Regulating Adenohypophyseal Gonadotrophin Secretion*

The second or higher level of neural control for gonadotrophin secretion includes all the brain structures outside of the HTA, structures that can modulate, i.e. enhance or inhibit, the activity of the FRF- and LRF-producing neurons. Such neurons appear to be concentrated in the preoptic and anterior hypothalamic area as well as in the limbic system, especially in the amygdala, septal complex, hippo-campus, and in the midbrain reticular formation (Flerkó, 1973).

A center for the neural mechanism that triggers the burst of LH-release responsible for ovulation has been localized by Everett (1961) in the preoptic area and termed the preoptic LH-trigger or LH-control mechanism. The preovulatory estrogen surge facilitates the activity of these neurons which, in turn, stimulate the FRF- and LRF-producing neurons to release into the hypophysial portal vessels an amount of FRF and LRF sufficient to discharge an ovulatory dose of gonadotrophin from the anterior pituitary. Since the pioneering studies of Sawyer *et al.* (1949) it has been known that a positive neuro-hormonal estrogen feedback is involved in the ovulatory release of gonadotrophic hormones. Recently, Ferin *et al.* (1969a, b) gave strong evidence for this assumption, having shown that antibodies to estradiol will inhibit ovulation in both normally cycling and pregnant mare serum gonadotrophin-treated immature rats.

The first evidence for the role of a negative neuro-hormonal estrogen feedback in the control of gonadotrophin secretion came from studies carried out in the Pécs Anatomy Department (Flerkó, 1956; 1957). Subsequent studies by Flerkó and Szentágothai, (1957) indicated that there were estrogen-sensitive neurons in the preoptic-anterior hypothalamic area. The postulate that this part of the brain contains estrogen sensitive neurons was later supported by the finding that individual nerve cells in this part of the brain, as well as in the medial-basal part of the hypothalamus, accumulate estradiol (Michael, 1962; Attramadal, 1964; Pfaff, 1968; Stumpf, 1970), and that the preoptic-anterior hypothalamic area, similarly to the medial-basal part of the hypothalamus, takes up and retains estradiol in the same way as the peripheral estrogen-responsive organs (Eisenfeld and Axelrod, 1965; Kato and Villee, 1967; Flerkó *et al.*, 1969).

C. *Localization and Role of Estrogen-Sensitive Neurons in the Hypothalamus of Normal and Androgen-Sterilized Rats*

Our earlier experimental findings indicated that the estrogen-sensitive neurons in the preoptic-anterior hypothalamic area were instrumental in the negative estrogen feedback through which FSH secretion was inhibited by a slight, physiological, elevation of the estrogen level in the blood (Flerkó,

1956; 1957). On the other hand, absence of compensatory ovarian hyper-trophy following hemispaying in rats with anterior hypothalamic lesions has supported the idea that these anterior hypothalamic estrogen-sensitive neurons play a role also in that process by which a decrease in blood estrogen level enhances FSH release (D'Angelo and Kravatz, 1960; Flerkó and Bárdos, 1961). On the basis of these and similar experimental findings, an FSH control mechanism in the anterior hypothalamus has been postulated by us and simultaneously, on the basis of completely different experimental findings, by Donovan and van der Werff ten Bosch (1959a, b). This assumption has recently been supported by Kalra et al. (1971) who showed that FSH release was stimulated only when the stimulating electrodes were located in the anterior hypothalamic area. Stimulation in the preoptic area or septal complex, which was very effective in evoking LH-release, failed to alter FSH-levels. Studies with the deafferentation technique by Halász and Gorski (1967) as well as by Köves and Halász (1969) have shown that no compensa-tory ovarian hypertrophy occurred in rats bearing a frontal cut that separated the anterior hypothalamic area from the HTA. Also Kalra et al. (1970) observed that separation of the anterior hypothalamic area from the HTA inhibited the postcastration rise of FSH in immature parabiotic rats.

It has been suggested by these and other experimental findings that the so-called "cycle-mechanism" of hypothalamic control for gonadotrophin secretion involves FSH- and LH-control mechanisms, both being composed of estrogen-sensitive neurons. According to the varying estrogen levels in the blood, the cycle-mechanism does influence the activity of the tonic mech-anism, i.e. the activity of the FRF- and LRF-producing neurons and, in this way, it maintains ovulation and the cyclic release of gonadotrophic hormones.

On the basis of this hypothesis, it is easy to explain why rats with electrolytic lesions or a frontal cut made by the Halász-knife in the anterior hypothalamus will not ovulate. When electrolytic lesions or a frontal cut in the anterior hypothalamus separates the cycle-mechanism from the tonic mechanism, i.e. from the HTA, the rat will not ovulate in the absence of a preoptic trigger to stimulate the FRF- and LRF-producing neurons to ovulatory action. Furthermore, in the absence of the subsequent negative sex steroid feedback, which in the intact animal operates through the cycle-mechanism, a moderate, but continuous outflow of FSH and LH will be maintained by the activity of the tonic mechanism, i.e., by the spontaneous activity of the FRF- and LRF-producing neurons. This continuous outflow of FRF-FSH and LRF-LH will stimulate follicular growth and estrogen secretion, which, in turn, depresses LRF- and LH-release, apparently by a continuous negative estrogen feedback on the LRF-producing neurons in the HTA.

The polyfollicular ovary syndrome, induced by a single injection of testosterone given to newborn rats (Barraclough, 1961), is completely similar to that induced by anterior hypothalamic lesions. The identity of the two experimental syndromes have suggested that identical brain structures might suffer damage in both syndromes. Indeed, Barraclough and Gorski (1961) have shown that the preoptic LH-trigger was deranged in androgen-sterilized rats. Looking for the mechanism by which early postnatal androgen action may elicit this effect, we have found that the estradiol-binding capacity of the peripheral target tissues, such as anterior pituitary and uterus as well as anterior and middle hypothalamus, was significantly reduced in androgen-sterilized rats as compared to control females without early postnatal androgen treatment (Flerkó et al., 1969; 1971). This finding has been confirmed by McGuire and Lisk (1969), Tuohimaa et al. (1969), Vértes and King (1969), and by McEven and Pfaff (1970).

The finding that androgen-sterilized rats had significantly lower anterior and middle hypothalamic radioactivity levels than did the control animals raised the possibility that perinatal androgen action might interfere with the synthesis of estrogen-binding proteins and, hence, with normal uptake and/or retention of estradiol by the hypothalamic estrogen-sensitive neurons. In this way, these neurons might become unresponsive or less responsive to estradiol and functionally inactive in mediating estrogen feedback on the FRF- and LRF-producing neurons, and through these neurons on pituitary cells producing gonadotrophic hormones.

It was also considered possible that constant light might interfere with the synthesis of the estradiol binding proteins in the estrogen-responsive hypothalamic neurons and, in this way, block the ovulatory gonadotrophin release. To test this hypothesis, we investigated recently the estradiol binding capacity of the hypothalamus, parietal cortex, anterior pituitary, and uterus of rats exposed to constant light (Illei-Donhoffer et al., 1974). In the continuously illuminated rats, the amount of radioactivity present in the tissues investigated was significantly lower than that present in the controls kept under normal diurnal lighting conditions (Table 1). However, the estradiol-binding capacity of the pituitary and uterus of continuously illuminated rats was not reduced in spite of appearance, i.e., in spite of the significant difference in connections with pituitaries and uteri. This is clearly indicated by the fact that increase of weight of the pituitary and uterus of continuously illuminated rats, induced by the continuous estrogen action, matched the reduction of the amount of radioactivity present in the pituitary and uterus of these animals (Table 2). This indicates that the estrogen receptors in the brain of the continuously illuminated rats were not available in the same amount as in the intact female rat.

Appropriately timed injections of anti-estrogens such as MER-25 (Shirley et al., 1968), chlomiphene (Labhsetwar, 1970a), I.C.I. 46474 (Labhsetwar, 1970b), or implantation of I.C.I. 46474 into the hypothalamus and anterior

TABLE 1

Radioactivity 2 hr After Injection of 50 μCi Tritiated Estradiol into Spayed Control and into Spayed Rats Exposed to Continuous Illumination

Cpm/mg wet tissues Exp.	No. of rats	Ant. hypoth.[1]	Middle and post. hypoth.[1]	Cortex[1]	Ant. pituit.[1]	Uterus[1]
Controls	21	37.5±1.3[2]	27.1±1.0[2]	6.3±0.4[3]	508.1±34.8[4]	364.7±29.2[2]
Rats exposed to constant light	25	28.3±1.0[2]	20.5±0.6[2]	4.7±0.2[3]	424.8±14.1[4]	115.9±8.3[2]

[1] Mean ± SEM.
[2] $p < 0.001$.
[3] $p < 0.01$.
[4] $p < 0.05$.

pituitary (Brainbridge and Labhsetwar, 1971) have been shown to prevent ovulation. However, LRF overcomes the ovulation-blocking effect of anti-estrogens (Labhsetwar, 1970a, b). This suggests that interference with ovulation stems from the central effects exerted at hypothalamic and probably also at the pituitary levels. Apparently the anti-estrogens, attached to the hypothalamic and pituitary estrogen receptors, eliminated the positive estrogen feedback that is indispensable to the induction of the ovulatory surge of gonadotrophin release. This assumption is supported by the findings of several investigators (Michael, 1962; Attramadal, 1964; Eisenfeld and Axelrod, 1966; Eisenfeld, 1967; Kato and Villee, 1967; Pfaff, 1968; Flerkó et al., 1969; Stumpf, 1970) who reported preferential accumulation of radioactive estradiol

TABLE 2

Mean Body, Ovary, Anterior Pituitary, and Uterus Weight of Control and Rats Exposed to Permanent Illumination. Percentile Increase of Weight and Decrease of Radioactivity in the Anterior Pituitary and Uterus of Rats Exposed to Continuous Illumination.

	Body weight, g[1]	Ovaries, mg[1]	Anterior pituitary			Uterus		
			weight, mg[1]	increase of weight	decrease of radioactivity	weight, mg[1]	increase of weight	decrease of radioactivity
Controls	307±8.6	58.5±4.4[2]	11.3±0.5	—	—	233±27.4[3]	—	—
Rats exposed to constant light	322±4.9	41.9±1.9[2]	12.6±0.4	12 %	16 %	384±31.3[3]	60 %	70 %

[1] SEM.
[2] $p < 0.001$.
[3] $p < 0.01$.

in the anterior and middle hypothalamus and in the anterior pituitary after systemic administration. However, it is possible to displace this estradiol by appropriately timed injections of anti-estrogens, clomiphene (Roy *et al.*, 1964: Eisenfeld and Axelrod, 1966; Kato *et al.*, 1968; Maurer and Woolley, 1971) or U-11100A (Eisenfeld and Axelrod, 1966). These findings show that, in the rat, when estrogen receptors of the hypothalamus and anterior pituitary are blocked, the pro-estrous estrogen surge cannot initiate the ovulatory gonadotrophin-release. In our androgen- or light-sterilized rats an insufficient number of estrogen-receptors are available for the ovulatory gonadotrophin surge-inducing action of estradiol because of a hitherto unknown disturbance of the replacement of the estradiol-binding proteins, brought about by the action of neonatal androgen or by constant light. Thus, the absence of the specific trapping mechanism of certain hypothalamic neurons may account for the loss of the positive estrogen feedback. This, in turn would result in the loss of ovulation and cyclic gonadotrophin release in rats exposed to constant light or to neonatal androgen action.

D. Effects of Intraventricular Norepinephrine, and Dopamine in Inducing Ovulation in Normal and Androgen-Sterilized Rats

The common feature of experimental procedures inducing anovulatory sterility associated with a polyfollicular ovary appears to be that they destroy or functionally inactivate the estrogen receptor-containing preoptic and anterior hypothalamic neurons, which trigger the ovulatory activity of the FRF- and LRF-producing neurons. It appears very likely that the neuro-humoral transmitter substance between the preoptic-anterior hypothalamic neurons and the FRF- and LRF-producing neurons would be norepinephrine (NE). Therefore, it was to be expected that ovulation could be induced by this substance in experimental anovulatory rats. We tested this by injecting norepinephrine into the third ventricle of rats made anovulatory either by anterior hypothalamic lesions or by exposure to neonatal androgen or to continuous illumination. For further experimentation, we employed only rats which showed persistent vaginal cornification at least for a month prior to experimentation (Tima and Flerkó, 1974).

As is shown in Table 3, intraventricular injection of 100 µg of NE, but not that of dopamine, induced ovulation in more than 50 percent of rats made anovulatory by electrolytic lesions in the preoptic-anterior hypothalamic area, which lesions were reported to reduce NE concentration of the medial-basal hypothalamus (Endröczi and Tallián, 1973). Ovulation occurred also two days after intraventricular injection of 50 µg of NE in more than 50 percent of rats made anovulatory by exposure to continuous illumination (Table 4). The discrepancy between the number of rats having tubal ova and those bearing fresh corpora lutea can be explained by assuming that, 48 hr

TABLE 3

Occurrence of Tubal Ova and Fresh Corpora Lutea Two Days after Intraventricular Injection of Norepinephrine (NE) into Rats Made Anovulatory by Anterior Hypothalamic Lesions.

Treatment	No. of rats	Tubal ova		Fresh corp. lutea	
100 μg NE	21	12/21	57 %	18/21	86 %
50 μg NE	23	9/23	39 %	11/23	48 %
100 μg dopamine	14	1/14	7 %	1/14	7 %
100 μl saline	11	0/11	0 %	0/11	0 %

after intraventricular injection of NE, in some animals ova had already passed through the Fallopian tubes before inspection under the dissecting microscope. We have assumed that in the present experiments NE injected into the third ventricle induced ovulation by stimulating the FRF- and LRF-producing neurons to release an ovulatory dose of these releasing factors into the hypophysial portal circulation. By assuming this, we can explain why NE induced ovulation in a lower percentage of the androgen-sterilized rats (Table 5). Studies of Barraclough (1967) and Kurcz *et al.* (1969) have shown that besides inactivating the preoptic gonadotrophin trigger, neonatal androgen, when given in adequate doses at the appropriate time, damages also the gonadotrophin-releasing factor-producing neurons to such an extent that the capacity of these neurons to synthesize and/or release FRF and LRF is reduced as compared to intact animals. The reduced functional capacity of the gonadotrophin-releasing factor-producing neurons in androgen-sterilized rats may account for the difference between the ovulatory action of NE observed in androgen-sterilized and hypothalamus-lesioned or light-sterilized rats. Results of these experiments lend further support to the concept that NE is the synaptic transmitter which releases FRF and LRF from the gonadotrophin-releasing factor-producing neurons.

On the other hand, the above experimental findings have clearly shown that the hypothalamo-hypophysial system of the anovulatory persistent estrous rat contains FRF, LRF, and gonadotrophins in an amount sufficient

TABLE 4

Occurrence of Tubal Ova and Fresh Corpora Lutea Two Days after Intraventricular Injection of Norepinephrine (NE) into Rats Made Anovulatory by Exposure to Continuous Illumination.

Treatment	No. of rats	Tubal ova		Fresh corp. lutea	
50 μg NE	17	9/17	53 %	12/17	71 %
50 μl saline	15	0/15	0 %	0/15	0 %

TABLE 5

Occurrence of Tubal Ova and Fresh Corpora Lutea Two Days after Intraventricular Injection of Norepinephrine (NE) into Rats Made Anovulatory by Neonatal Testosterone Administration.

Treatment	No. of rats	Tubal ova	Fresh corp. lutea
100 μg NE	20	6/20 30 %	11/20 55 %
100 μl saline	15	2/15 13 %	3/15 20 %

to induce ovulation when released. However, in rats made anovulatory by the experimental procedures mentioned before, the preoptic-anterior hypothalamic cycle-mechanism, the neural trigger of the FRF- and LRF-producing neurons for ovulatory action, is absent. In rats with anterior hypothalamic lesions, the estrogen-receptor-containing neurons of the cycle mechanism are physically separated from the FRF- and LRF-producing neurons. The action of the neonatal androgen or constant light is to reduce the estrogen-receptor content of the neurons contributing to the cycle-mechanism. In this way, the estrogen-responsive preoptic and anterior hypothalamic neurons, instrumental in the neuro-hormonal estrogen feedback, become functionally inactive. In the absence of the neuro-hormonal estrogen feedback the spontaneous tonic activity of the FRF- and LRF-producing neurons can maintain only the male-type, i.e., a non-cyclic gonadotrophin release without ovulation. This suggests that in all forms of the anovulatory persistent estrus, the absence (physical or functional) of the preoptic-anterior hypothalamic estrogen-sensitive trigger mechanism may account for the absence of cyclic gonadotrophin release and ovulation.

REFERENCES

Arimura, A., Sato, H., Kumasaka, T., Worobec, R. B., Debeljuk, L., Dunn, J. and Schally, A. V. (1973). Production of antiserum to LH-releasing hormone (LH-RH) associated with gonadal atrophy in rabbits: development of radioimmunoassays for LH-RH. *Endocrinology* 93, 1092–1103.

Attramadal, A. (1964). The uptake and intracellular localization of oestradiol-17β-6,7-H^3 in the anterior pituitary and the hypothalamus of the rat. *Acta path. microbiol. scand.* 61, 151–152.

Barraclough, C. A. (1961). Production of anovulatory, sterile rats by single injections of testosterone propionate. *Endocrinology* 68, 62–67.

Barraclough, C. A. (1967). Modifications in reproductive function after exposure to hormones during the prenatal and early postnatal period. *In* "Neuroendocrinology" (L. Martini and W. F. Ganong, eds.), vol. II, pp. 61–99. Academic Press, New York.

Barraclough, C. A. and Gorski, R. A. (1961). Evidence that the hypothalamus is responsible for androgen-induced sterility in the female rat. *Endocrinology* 68, 68–79.

Brainbridge, J. G. and Labhsetwar, A. P. (1971). The role of oestrogens in spontaneous ovulation: location of site of action of positive feedback of oestrogen by intracranial implantation of the anti-oestrogen I.C.I. 46474. *J. Endocrin.* **50,** 321–327.

Csernus, V., Lengvari, I. and Halasz, B. (1973). Data on the localization of CRF-producing neural elements. *In* "Abstracts of the Seventh Conference of European Comparative Endocrinologists" (J. Szentagothai and F. Hajos, eds.), p. 126. Akademiai Kiado, Budapest.

D'Angelo, S. A. and Kravatz, A. S. (1960). Gonadotrophic hormone function in persistent estrous rats with hypothalamic lesions. *Proc. Soc. Exp. Biol. Med., N.Y.* **104,** 130–133.

Donovan, B. T. and van der Werff ten Bosch, J. J. (1959a). The hypothalamus and sexual maturation in the rat. *J. Physiol., Lond.,* **147,** 78–92.

Donovan, B. T. and van der Werff ten Bosch, J. J. (1959b). The relationship of the hypothalamus to oestrus in the ferret. *J. Physiol. Lond.* **147,** 93–108.

Eisenfeld, A. J. (1967). Computer analysis of the distribution of ^{3}H-estradiol. *Biochim. Biophys. Acta* **136,** 498–507.

Eisenfeld, A. J. and Axelrod, J. (1965). Selectivity of oestrogen distribution in tissues. *J. Pharmacol. Exp. Theor.* **150,** 469–475.

Eisenfeld, A. J. and Axelrod, J. (1966). Effect of steroid hormones, ovariectomy, oestrogen pretreatment, sex and immaturity on the distribution of ^{3}H-estradiol. *Endocrinology* **79,** 38–42.

Endröczi, E. and Tallian, F. (1973). Catecholaminergic control of pituitary function. *In* "Hormones Metabolism and Stress" (S. Németh, ed.), pp. 15–24. Publishing House of the Slovak Academy of Sciences, Bratislava.

Everett, J. W. (1961). The preoptic region of the brain and its relation to ovulation. *In* "Control of Ovulation" (C. A. Villee, ed.), pp. 101–112. Pergamon Press, New York.

Ferin, M., Zimmering, P. E. and Van de Wiele, R. L. (1969). Effects of antibodies to estradiol-17β on PMS-induced ovulation in immature rats. *Endocrinology* **84,** 898–900.

Ferin, M., Tempone, A., Zimmering, P. E. and Van de Wiele, R. L. (1969). Effects of antibodies to 17β-estradiol and progesterone on the estrous cycle of the rat. *Endocrinology* **85,** 1070–1078.

Flerkó, B. (1956). Die Rolle hypothalamischer Strukturen bei der Hemmungswirkung des erhöhten Östrogenblutspiegels auf die Gonadotrophinsekretion. *Acta physiol. Acad. Sci. hung.* **9 (Suppl.),** 17–18.

Flerkó, B. (1957). Le rôle des structures hypothalamiques dans l'action inhibitrice de la folliculine sur la sécrétion de l'hormone folliculo-stimulante. *Arch. Anat. micr. et Morph. exp.* **46,** 159–172.

Flerkó, B. (1962). Hypothalamic control of hypophyseal gonadotrophic function. *In* "Hypothalamic Control of the Anterior Pituitary" (J. Szentágothai, B. Flerkó, B. Mess and B. Halász, eds.), pp. 192–246, Akadémiai Kiadó, Budapest.

Flerkó, B. (1972). Hypothalamic mediation of neuroendocrine regulation of hypophysial gonadotrophic functions. *In* "MTP International Review of Science, Physiology Section, Series 1, Reproductive Physiology" (R. O. Greep, ed.), pp. 1–32. Butterworths, London.

Flerkó, B. and Bárdos, V. (1961). Absence of compensatory ovarian hypertrophy in rats with anterior hypothalamic lesions. *Acta endocrin., Kbh.* **36,** 180–184.

Flerkó, B., Illei-Donhoffer, A. and Mess, B. (1971). Oestradiol-binding capacity in neural and non-neural target tissues of neonatally androgenized female rats. *Acta biol. Acad. Sci. hung.* **22,** 125–130.

Flerkó, B., Mess, B. and Illei-Donhoffer, A. (1969). On the mechanism of androgen sterilization. *Neuroendocrinology* **4,** 164–169.

Flerkó, B. and Szentágothai, J. (1957). Oestrogen sensitive nervous structures in the hypothalamus. *Acta endocrin., Kbh.* **26**, 121–127.

Halász, B. (1969). The endocrine effects of isolation of the hypothalamus from the rest of the brain. *In* "Frontiers in Neuroendocrinology" (W. F. Ganong and L. Martini, eds.), pp. 307–342, Oxford University Press, New York.

Halász, B. and Gorski, R. A. (1967). Gonadotropic hormone secretion in female rats after partial or total interruption of neural afferents to the medial basal hypothalamus. *Endocrinology* **80**, 608–622.

Halász, B., Pupp, L. and Uhlarik, S. (1962). Hypophysiotrophic area in the hypothalamus. *J. Endocr.* **25**, 147–154.

Illei-Donhoffer, Flerkó, B. and Mess, B. (1974). Reduction of estradiol-binding capacity of neural target tissues in light-sterilized rats. *Neuroendocrinology* **14**, 187–194.

Kalra, S. P., Ajika, K., Krulich, L., Fawcett, C. P., Quijada, M. and McCann, S. M. (1971). Effects of hypothalamic and preoptic electrochemical stimulation on gonadotrophin and prolactin release in pro-oestrous rats. *Endocrinology* **88**, 1150–1158.

Kalra, S. P., Velasco, M. E. and Sawyer, C. H. (1970). Influences of hypothalamic deafferentation on pituitary FSH release and oestrogen feedback in immature parabiotic rats. *Neuroendocrinology* **6**, 228–235.

Kato, J., Kobayashi, T. and Villee, C. A. (1968). Effect of clomiphene on the uptake of estradiol by the anterior hypothalamus and hypophysis. *Endocrinology* **82**, 1049–1052.

Kato, J. and Villee, C. A. (1967). Preferential uptake of estradiol by the anterior hypothalamus of the rat. *Endocrinology* **80**, 567–575.

Köves, K. and Halász, B. (1969). Data on the location of the neural structures indispensable for the occurrence of ovarian compensatory hypertrophy. *Neuroendocrinology* **4**, 1–11.

Kurcz, M., Maderspach, K. and Horn, G. (1969). Damaging effect of neonatal androgen treatment on the production of gonadotrophic hormones. *Acta. biol. Acad. Sci. hung.* **20**, 303–310.

Labhsetwar, A. P. (1970a). The role of oestrogens in spontaneous ovulation: evidence for positive estrogen feedback in the 4-day oestrous cycle. *J. Endocrin.* **47**, 481–493.

Labhsetwar, A. P. (1970b). The role of oestrogen in spontaneous ovulation as revealed by the use of oestrogen-antagonist I.C.I. 46474. *Nature (London)* **224**, 80–81.

Maurer, R. and Woolley, D. (1971). Distribution of ³H-estradiol in clomiphene-treated and neonatally androgenized rats. *Endocrinology* **88**, 1281–1287.

Mazurkiewicz, J. E. and Nakane, P. K. (1972). Light and electron microscopic localization of antigens in tissues embedded in polyethylene glycol with a peroxidase-labeled antibody method. *J. Histochem, Cytochem.* **20**, 969–974.

McEwen, B. S. and Pfaff, D. W. (1970). Factors influencing sex hormone uptake by rat brain regions. I. Effects of neonatal treatment, hypophysectomy, and competing steroid on estradiol uptake. *Brain Res.* **21**, 1–6.

McGuire, J. L. and Lisk, R. D. (1969). Oestrogen receptors in androgen or estrogen sterilized female rats. *Nature (London)* **221**, 1068–1069.

Mess, B. and Martini, L. (1968). The central nervous system and the secretion of anterior pituitary trophic hormones. *In* "Recent Advances in Endocrinology" (V. H. T. James, ed.), pp. 1–49. Churchill, London.

Michael, R. P. (1962). Estrogen-sensitive neurons and sexual behavior in female cats. *Science* **136**, 322–323.

Nakane, P. K. and Pierce, G. B., Jr. (1967). Enzyme-labeled antibodies for the light and electron microscopic localization of tissue antigens. *J. Cell Biol.* **33**, 307–318.

Palkovits, M., Arimura, A., Brownstein, M., Schally, A. V., Saavedra, J. M. (1974). Luteinizing hormone releasing hormone (LH-RH) content of the hypothalamic nuclei in rat. *Endocrinology,* in press.

Pfaff, D. W. (1968). Autoradiographic localization of radioactivity in rat brain after injection of tritiated sex hormones. *Science* **161**, 1355–1356.

Porter, J. C., Mical, R. S., Tippit, P. R. and Drane, J. W. (1970). Effect of selective surgical interruption of the anterior pituitary's blood supply on ACTH release. *Endocrinology* **86**, 590–599.

Roy, S., Mahesh, V. B. and Greenblatt, R. B. (1964). Effect of clomiphene on the physiology of reproduction in the rat. III. Inhibition of uptake of radioactive oestradiol by the uterus and pituitary gland of immature rat. *Acta endocrin. Kbh.* **47**, 669–675.

Sawyer, C. H., Everett, J. W. and Markee, J. E. (1949). A neural factor in the mechanism by which estrogen induced the release of luteinizing hormone. *Endocrinology* **44**, 218–233.

Sétáló, G., Vigh, S., Schally, A. V., Arimura, A. and Flerkó, B. (1974). LH-RH-containing neural elements in the rat hypothalamus. *Endocrinology,* in press.

Shirley, B., Wolinsky, J. and Schwartz, N. B. (1968). Effects of a single injection of an estrogen antagonist on the estrous cycle of the rat. *Endocrinology* **82**, 959–968.

Stumpf, W. E. (1970). Estrogen-neurons and estrogen-neuron systems in the periventricular brain. *Amer. J. Anat.* **129**, 207–217.

Szentágothai, J. (1962). Anatomical considerations. *In* "Hypothalamic Control of the Anterior Pituitary" (J. Szentágothai, B. Flerkó, B. Mess and B. Halász, eds.), pp. 19–105, Akadémiai Kiadó, Budapest.

Szentágothai, J. (1964). The parvicellular neurosecretory system. *Prog. Brain Res.* **5**, 135–146.

Szentágothai, J. and Halász, B. (1964). Regulation des endokrinen Systems über Hypothalamus. *Nova Acta Leopoldina* **28**, 227–248.

Tima, L. and Flerkó, B. (1974). Ovulation induced by norepinephrine in rats made anovulatory by various experimental procedures. *Neuroendocrinology,* in press.

Tuohimaa, P., Johansson, R. and Niemi, M. (1969). Oestradiol binding in the hypothalamus and uterus of the androgenized rats. *Scand. J. clin. Lab. Invest. 23 suppl.* **108**, 42.

Vértes, M. and King, R. J. B. (1969). The influence of androgen on oestradiol binding by rat hypothalamus. *J. Endocrin.* **45**, xxi-xxiii.

Ontogeny of Development of the Hypothalamic Regulation of Gonadotropin Secretion: Effects of Perinatal Sex Steroid Exposure

Charles A. Barraclough[1] and Judith L. Turgeon

Department of Physiology
University of Maryland Medical School
Baltimore, Maryland

I. Introduction 255
II. Materials and Methods 259
III. Results 259
    A. Luteinizing Hormone Concentrations in Plasma of 6-20 day old Female Rats; Pituitary Response to Luteinizing Hormone-Releasing Hormone (LH-RH) 259
    B. Pituitary Responsiveness to LH-RH in Adult Male and Female Rats 260
    C. Effects of Testosterone and Estradiol on Pituitary Responsiveness to LH-RH in Adult Castrated Rats 262
    D. Response of Androgen-Sterilized Rats to LH-RH 263
    E. Changes which occur in Peripheral Plasma Progesterone Concentration during the Estrous Cycle of the Rat 266
    F. Effects of Medical Preoptic Electrochemical Stimulation on LH Release in Androgen-Sterilized Rats 267
IV. Conclusions 269
V. References 272

I. INTRODUCTION

While the genetic sex of mammals is determined at the time of fertilization, the phenotype following sex chromosomal union in animals can be altered by numerous factors. In rodents, male or female patterns of gonadotropin secretion seem not to depend upon the genetic sex of the animal but rather upon the exposure of the hypothalamo-hypophyseal system to sex steroids (or the lack thereof) secreted by the perinatal gonad. As early as 1936, Pfeiffer observed that the exchange of gonads between male and female rats resulted in the adult males having patterns of gonadotropin secretion which resembled the female. Similarly, castrated females with prepubertally

255

implanted testes had male gonadotropin secretion patterns. Since neither the sex steroids nor the gonadotropic hormones had been identified in these early years, we reinvestigated this phenomenon, originally in mice (Barraclough and Leathem, 1954), and subsequently in rats (Barraclough, 1961). In these original studies we were particularly interested in defining the specific age during which the administration of a single injection of testosterone propionate (TP) would permanently alter the reproductive endocrinology of the experimental animal. Rats were injected with 1-1.25 mg TP at 2, 5, 10, or 20 days of age and examined at various ages into adulthood. At autopsy (100 days of age) the ovaries of all rats which received androgen at 2 or 5 days of age contained numerous large vesicular follicles and had persistently cornified vaginal smears but corpora lutea were absent. Androgen administration at 10 days of age had a less drastic effect on ovarian structure with only 4 of 10 animals failing to ovulate. All rats treated with androgen at 20 days of age were normal at autopsy. As a consequence of these observations we proposed that a period of steroid sensitivity exists in the female rat to approximately the tenth day of age during which administration of androgen will result in subsequent sterility (Barraclough, 1961). The next series of studies were designed to establish the minimum dosage of TP required to produce sterility. As is shown in Table 1, as little as 10 μg, administered at 5 days of age, produced sterility in 70% of the treated animals (Gorski and Barraclough, 1963). In later studies, it was determined that early androgen exposure detrimentally affected function of the CNS preoptic area, the hypothalamus, the hypophysis, and even the ovary itself. Recently, Hayashi and Gorski (1974) have observed that 48 hrs of exposure of the basal hypothalamus to crystalline TP was required for the induction of sterility by 100 days of age, and even exposure for 72 hrs was only capable of inducing sterility in 43% of the animals by 50 days of age.

TABLE 1

Effects of Various Doses of Testosterone Propionate
on Induction of Sterility in the Female Rat

Amount injected on day 5 of age (μg)	Number of rats injected	Number of rats sterile	Percent sterility*
1250	1000	998	99.8
10	136	96	70.6
5	25	11	44.0
1	10	3	30.0

*These figures represent a compilation of data obtained in several studies (Barraclough, 1961; Gorski and Barraclough, 1963).

With the development of radioimmunoassay procedures, several laboratories have examined the changes in plasma concentrations of luteinizing
hormone (LH), follicle stimulating hormone (FSH), and testosterone which
occur with early postnatal development. Miyachi *et al.* (1973) observed in
male rats that serum LH decreased from high concentrations at 5 days of age
to a nadir at 18 days of age (Fig. 1). At 19 days, it increased sharply from
65.0 ± 2.4 to 103.5 ± 12.3 ng/ml and then decreased to an average value of
65 ng/ml through day 35. Serum FSH decreased from day 5 to a nadir at day
16 and then increased to its highest values at 18 days of age. Thereafter it did
not differ significantly from successive values to 30 days of age. Of importance in these studies was the observation that testosterone was measurable in
serum even at 5 days of age and reached its peak at 369 ng/ml at 18 days or
one day prior to the LH peak at 19 days of age. The significance of timing of
these peaks remains obscure. Thus, the perinatal testes secrete sufficient
testosterone at a critical developmental period to imprint permanently upon
the hypothalamus the type of regulatory role it will exert in adulthood (tonic

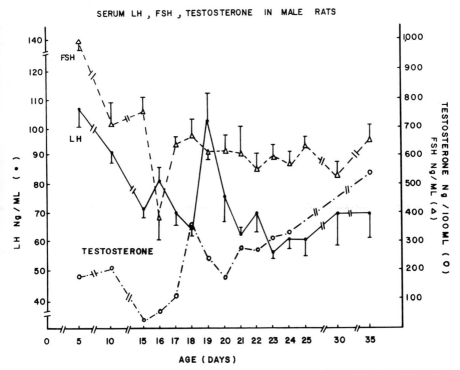

Fig. 1. Serum LH, FSH and testosterone levels in male rats from 5-35 days. (Miyachi
et al., 1973).

vs. cyclic). The early development of a negative feedback relationship between testes and pituitary-hypothalamus also has been demonstrated by Goldman *et al.* (1971) who found that castration of neonatal male rats caused elevations in plasma concentrations of both LH and FSH.

The patterns and concentrations of the sex steroids secreted by the perinatal ovary (if any) are unknown. However, Goldman *et al.* (1971) and Döhler and Wüttke (1974) have observed higher LH concentrations in females than in males at 2 days of age. In female rats, they observed further that LH decreased after the early postnatal elevation but showed peaks at 10, 12, 14, 17, and 19 days of age. Since estrogen secretion by the prepubertal ovary presumably is absent until after this period, such peaks in LH may be similar to the pulsatile discharges of LH observed after castration of adult female rats and upon removal of the negative feedback action of estrogen (Gay and Sheth, 1972).

Consequently, the current literature suggests that feedback actions of testosterone on gonadotropin secretion are established, at least in male rodents, very early in life and that pituitary responsiveness to luteinizing hormone-releasing hormone (LH-RH) is not altered even though plasma gonadotropin concentrations are changing throughout the first 20 days of prepubertal development (Miyachi *et al.*, 1973). Rather, the reproductive changes which occur in the male during the early postnatal developmental period seem more related to testicular development and its secretions rather than to alterations in function of the hypothalamo-pituitary unit.

The current series of studies were designed to investigate further the alterations produced in the hypothalamo-pituitary gonadal axis of prepubertal male and female rats at critical postnatal developmental ages (5, 15, 20 days) and to ascertain how early exposure to sterilizing doses of testosterone or estradiol-17β initially altered responsiveness of one component of this axis, the pituitary, to the action of luteinizing hormone-releasing hormone. Our objective was to establish whether estradiol-17β or testosterone could exert an effect at the pituitary level even prior to the suspected time of beginning of ovarian estrogen secretion, i.e., prior to puberty. We further wished to examine the longevity of action of a sex steroid on hypothalamo-pituitary function following its injection at 5 days of age. To analyse these data critically we found it necessary to understand more completely normal female reproductive neuroendocrinology, and consequently have examined more fully the preopticotuberal components of the hypophysiotropic system and their relationships to the discharge of pituitary LH. We also have examined in greater detail the positive and negative feedback actions of estrogen and progesterone in facilitating the ovulatory discharge of LH. Finally, we have compared the normal reproductive endocrine events which occur in the adult female rat to those obtained after the early sex steroid exposure which induces permanent sterility in adult female animals.

II. MATERIALS AND METHODS

The Sprague-Dawley rats employed in the prepubertal studies were born in our colony and at 5 days of age were injected either with sesame oil vehicle, with 1.25 mg TP or with 5 μg estradiol benzoate (EB) s.c. in oil. Animals were divided so that each litter received only androgen or estrogen. Sufficient non-steroid treated rats were retained in each litter to serve as controls for each of the experimental groups. Synthetic LH-RH (Beckman) was injected at 6, 15 or 20 days of age (250 ng i.p.) and blood was collected 30 min later by decapitation of 6-day old rats or from the abdominal aorta of etherized 15 or 20 day old animals. LH was measured by the radioimmuno-assay (RIA) procedure according to the method of Niswender et al. (1968) and the LH values were expressed in terms of a laboratory standard which is 0.17 x NIH-LH-Sl standard.

III. RESULTS

A. *Luteinizing Hormone Concentrations in Plasma of 6-20 day old Female Rats; Pituitary Response to Luteinizing Hormone-Releasing Hormone (LH-RH)*

As is shown in Fig. 2, plasma LH was high in females at 6 days of age but had decreased in concentration by 20 days of age. In contrast, male plasma LH values remained relatively constant throughout the ages studies. In response to 250 ng LH-RH it was observed that the control female pituitary was exquisitely more responsive than the male pituitary and this responsiveness was retained throughout development into adulthood. The pituitaries of male animals injected wtih 5 μg EB or 1.25 mg TP at 5 days of age and tested for responsiveness to LRF (LH-RH) 24 hrs later showed no ill effects of such steroid treatment. When challenged with LRF at 15 or 20 days of age, the responses of pituitaries of such steroid treated rats were similar to those obtained in the normal control. When the pituitary response of the androgen-sterilized female rat was tested to LH-RH at 24 hrs after treatment (6 days of age), LH release was markedly suppressed. Furthermore, the decreased sensitivity of the female gland persisted throughout the later prepubertal ages into adulthood whereas the male gland at 15 or 20 days of age responded in a manner equivalent to that of the normal male rat.

These data suggest that even at 6 days of life receptors for LH-RH exist in the pituitary gland which are capable of resulting in the discharge of LH. Furthermore, early exposure to sterilizing dosages of androgen or estrogen profoundly affects the responsiveness of this gland to LH-RH even beyond the time of initial exposure of the pituitary gonadotrophs to the sex steroids. The reason for the reduced responses obtained at 15 or 20 days of age awaits further elucidation. Makino (1973) has demonstrated *in vitro* that LH-RH

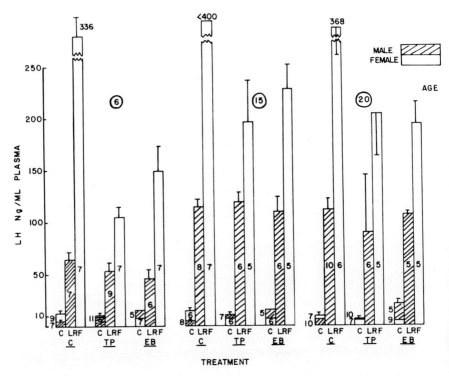

Fig. 2. Plasma LH concentrations in normal (C) and experimental 6, 15 and 20 day old male and female rats. Experimental male and female rats received either 1.25 mg testosterone propionate (TP) or 5 μg estradiol benzoate (EB) s.c. in oil at 5 days of age. Both control and experimental animals were injected i.p. with 250 ng LH-RH and blood was collected 30 min later and assayed for LH.

markedly stimulates the adenyl cyclase system of the pituitary concomitant with the release of LH. Similarly, addition of $N^6,O^{2'}$-dibutyrl cyclic AMP increased release of LH into the medium. Perhaps, early steroid treatment affects the ability of this system to respond to a pulse of LH-RH through an action on the gonadotroph cell membrane receptor.

B. Pituitary Responsiveness to LH-RH in Adult Male and Female Rats

To elucidate further the mechanism of action of the sex steroids in regulating adenohypophyseal LH secretion, we have examined the responses obtained in adult normal male and proestrous female rats following a pulse of LH-RH and have compared them to patterns and concentrations of LH released into plasma following similar LH-RH injections into androgen-sterilized persistent estrous rats. As is illustrated in Fig. 3, pulse injections of varying dosages of LH-RH into the Nembutal-blocked proestrous rat results in

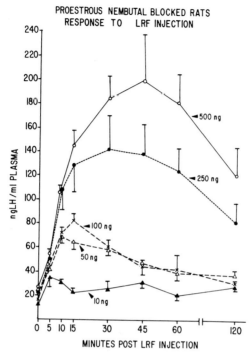

PROESTROUS NEMBUTAL BLOCKED RATS
RESPONSE TO LRF INJECTION

Fig. 3. Plasma LH concentrations in sequential bleedings from an indwelling jugular cannula following i.v. injection of LH-RH on the afternoon of proestrus in Nembutal-blocked rats. Pulse injections of LH-RH varied form 10 to 500 ng.

a reasonable dose response relationship between amount of LH-RH injected and peak concentrations of LH released into plasma. Such responses have been obtained by many other laboratories. However, in these studies, external jugular cannulae were inserted to the level of the right atrium and sequential blood samples were obtained throughout the afternoon of proestrus. LH was radioimmunoassayed as described previously. From the data obtained by sequential bleedings it can be observed that peak LH concentrations were obtained 15 min after LH-RH injection. Furthermore, as the dosage of LH-RH was increased the disappearance time of LH decreased resulting in "bell-shaped" plasma LH patterns with the higher 250 and 500 ng LH-RH dosages. Seemingly, secretion and disappearance rates of LH between 15-60 min after injection are equal. However, since the animals were anesthetized with Nembutal, this could be an effect of lowered blood pressure, decreased degradation of circulating LH, or other nonspecific phenomena. It also should be noted in Fig. 3 that a peak LH concentration of 198 ng/ml was obtained following injection of 500 ng of LH-RH. In comparison to these data are those obtained in unanesthetized male rats with indwelling external jugular cannulae (Fig. 4). Peak LH concentrations were also obtained within 15 min after injection. However, only with 1000 ng was a semblance of the continued secretion of LH observed. At other dosage levels, the disappearance of LH from plasma corresponded to the t 1/2 for LH in circulation. Of particular importance was the peak concentrations of LH obtained after LH-RH injection. With 500 ng only 38 ng/ml peak LH was obtained as compared to 198 ng/ml observed in the female rat.

C. Effects of Testosterone and Estradiol on Pituitary Responsiveness to LH-RH in Adult Castrated Rats

Since the intact male pituitary in these studies was subjected to the secretions of androgens whereas the proestrous female pituitary was under the influence of ovarian estrogens, the differential responsiveness between males and females most likely was the result of the sex steroid in circulation at the time of experimentation. Indeed, Schally *et al.* (1973) have shown, in male rats, that testosterone propionate suppresses the release of LH and FSH after LH-RH, and small dosages of estrogen given to female rats increases this gland's responsiveness. McEwen *et al.* (1970) have suggested that receptors to estradiol-17β and testosterone exist in both male and female pituitary glands

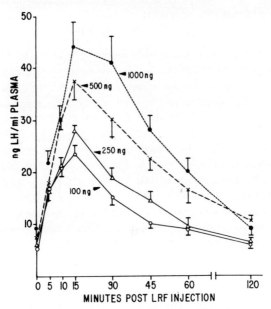

Fig. 4. Plasma LH concentrations in sequential bleedings following i.v. injection of LH-RH in unanesthetized normal male rats with an indwelling jugular cannula. Pulse injection of LH-RH varied from 100 to 1000 ng.

and consequently it is possible that "maleness" or "femaleness" of the pituitary is not dictated by the genetic sex of the animal but rather by the gonadal steroids released into circulation. To test this possibility, adult male and female Sprague-Dawley rats were castrated, and 7 days later groups of 7-9 animals were injected with either 10 μg EB/100 gm body wt. or with 500 μg TP s.c. in oil. Twenty-four hrs later, external jugular cannulae were inserted under ether anesthesia in the morning and 2-3 hrs later a single injection of 250 ng of synthetic LH-RH was injected i.v. As is shown in Fig. 5, both castrated male and female pituitaries, subjected 24 hrs earlier to EB, responded in almost identical manner with peak concentrations being achieved within 15 min with the same magnitude of LH being released into plasma. However, in males, disappearance of LH from plasma was more rapid than in females. Similarly, male and female pituitaries dominated by testosterone showed equivalent suppressed responses to LH-RH.

D. Response of Androgen-Sterilized Rats to LH-RH

Harris (1955) originally hypothesized that during development, the anterior pituitary remains plastic or pluripotential in its functional capacity and that its activity in the male or female animal depends on the stimulus the gland receives from the hypothalamus. However, the current data suggest that the quantity and temporal patterns of LH released in response to a given LH-RH stimulus from the hypothalamus is directly dependent upon the sex steroid bound to the gonadotropic cells.

With this basic information available, we next tested the *in vivo* response of pituitaries of adult rats previously subjected to 1.25 mg TP at 5 days of age. The injection of 250 ng LH-RH to Nembutalized-persistent estrous rats resulted in peak LH values of 85 ng/ml which were intermediate between the 145 ng/ml obtained in the normal proestrus and the 38 ng/ml obtained in Nembutal-anesthetized male rats. This observation is not surprising since the polyfollicular ovaries of the sterile rat are secreting estrogen which can affect pituitary sensitivity to LH-RH. However, even though LH was released, ovulation did not occur in these intact animals. We have previously reported that peak LH concentrations of 44 ng/ml are required in normal proestrous rats to induce complete ovulation (Turgeon and Barraclough, 1973). To permit a more meaningful comparison between the sterile rat's pituitary response and that of adult normal male or female glands, 8 sterile rats were castrated and 7 days later were injected with EB or TP as described previously. In Fig. 6 it can be observed that although estrogen-dominated glands released greater amounts of LH as compared to TP dominated pituitaries, both responses were far less than the nonsterilized male or female castrates subjected to identical steroid treatment. This difference was particularly obvious in the sterile-TP treated group (compare Figs. 5-6). Consequently, the depressed response to LH-RH observed during the prepubertal periods in sterilized rats continues into adulthood and even estrogen therapy

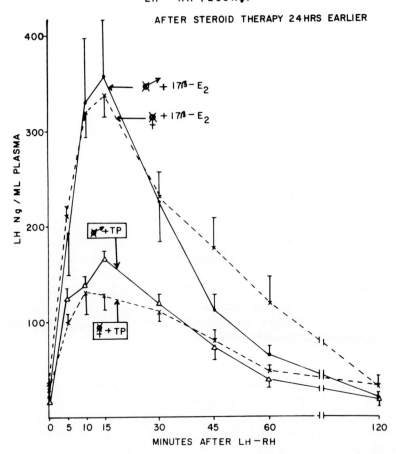

Fig. 5. Changes in plasma LH patterns and concentrations in male and female rats castrated 7 days earlier and injected s.c. with 10 μg/100 gm body wt. estradiol-17β or 500 μg testosterone propionate. Twenty-four hours after steroid treatment 250 ng LH-RH were injected i.v. through an indwelling external jugular cannula.

fails to overcome this partial pituitary refractoriness. The importance of these observations is twofold: (1) even provided that the hypothalamus releases an "ovulatory quota" of LH-RH, pituitary responsiveness is depressed as a consequence of early androgen treatment and thus the quantity of LH released into plasma could be less than required for ovulation; (2) while the concentration of LH released from the sterile rat pituitary may be sufficient to induce ovulation in proestrous rats, ovarian responsiveness to LH also seems markedly reduced to such a degree that even endogenous peak values of 85 ng/ml failed to induce ovulation.

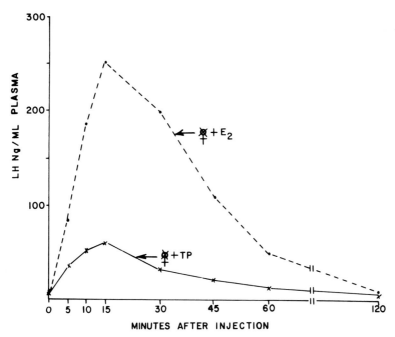

CASTRATED ASR RESPONSE TO LH-RH(250Ng)

AFTER STEROID THERAPY 24HRS EARLIER

Fig. 6. Changes in plasma LH patterns and concentrations in response to a 250 ng LH-RH injection in androgen-sterilized female rats castrated and subjected to the same steroid treatment as described in Fig. 5.

While such data provide information on one component of the hypothalamo-hypophyseal-gonadal axis, the pituitary, the major events occurring within this system to result in an ovulatory LH discharge are far more complex than pituitary LH-RH sensitivity alone. A rather simplified version of changes which occur within the rat estrous cycle recently has been presented by Barraclough (1973). Some of the prime events which occur in the 24-32 hrs preceding the ovulatory release of LH have been investigated by Mann and Barraclough (1973) and Turgeon and Barraclough (1973). The importance of estrogen in promoting the proestrous surge of LH has been emphasized by the studies of Ferin et al. (1969). By use of specific antibodies produced to estradiol-17β (E_2) they demonstrated that injection of such antisera delayed ovulation in 33 of 35 rats if injected into 4 day cycling rats at 1700 hrs or earlier on diestrous-2. All animals treated at 1000 or 1500 hrs on proestrus ovulated at the expected time. Similarly, administration of such nonsteroidal estrogen antagonists as MER-25 (Shirley et al., 1968), or CN-55,945-27 (Callantine et al., 1966) to adult cyclic rats effectively prevented the spontaneous release of LH from occurring. In the rhesus monkey Karsch *et al.* (1973) have emphasized that estrogen is the only steroid required to precipitate the ovulatory LH surge and that progesterone does not play a significant role in this physiological event. In the rat, Everett (1948) observed that the injection of estrogen at mid-diestrum regularly induced ovulation 24 hrs earlier than expected. However, this same investigator reported treatment of 5-day cyclic rats with progesterone on diestrus-3 also advanced ovulation by 24 hrs. Seemingly, in the rodent, both estrogen and progesterone are essential for the spontaneous release of LH.

E. Changes which occur in Peripheral Plasma Progesterone Concentration during the Estrous Cycle of the Rat

Recently we have investigated the changes which occur in peripheral plasma concentrations of progesterone throughout the 4-day estrous cycle of rats and the contributions to this pool from the adrenal gland and the ovary. Ovariectomy and/or adrenalectomy was performed on diestrus-2 at 0100 hr and peripheral concentrations and ovarian secretion rates of progesterone (P) were measured 24 hrs later. Ovariectomy alone reduced peripheral plasma P by only 59% in 24 hrs whereas adrenalectomy resulted in a 75% decrease without altering ovarian secretion rates of this steroid. Removal of both adrenals and ovaries resulted in an almost complete disappearance of P from the peripheral plasma pool. Shown in Fig. 7 are the temporal changes which exist in peripheral plasma concentrations and ovarian secretion rates of P during the estrous cycle (Mann and Barraclough, 1973). From these studies it is apparent that a diurnal rhythm exists; plasma levels of P are at nadir between 1000 and 1400 hrs and then rise to a peak during the early morning hrs (0100-0500) on each day of the cycle. Considered together with data published on plasma E_2 concentrations (Shaikh, 1971) and the timing of the LH surge in our rat colony (Turgeon and Barraclough, 1973) we have constructed a composite diagram of the plasma steroid hormonal concentration changes which occur prior to LH release (Fig. 8). The studies of Mann and Barraclough (1973) suggest, in the normal sequence of events, that it is the ratio of E_2/P which is the essential factor in such neuroendocrine conditioning rather than one steroid acting alone.

The physiological significance of these steroid events resides in the fact that by altering thresholds of excitability within the preoptico-tuberal system, hypothalamic activation of terminal median eminence structures to release LH-RH is permitted. Simultaneously, E_2 alters pituitary sensitivity to the LH-RH released into the portal circulation and consequently a maximal output of LH at the time of the ovulatory surge is achieved. The degree to which quantums of LH-RH are released in response to extrinsic activation of medial preoptic area (MPOA) have been investigated by Turgeon and Barraclough (1973). In Nembutal-blocked proestrous rats, a bipolar concentric

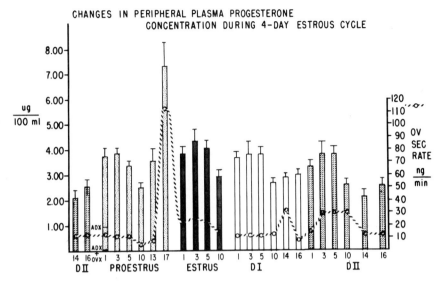

Fig. 7. Peripheral plasma concentrations (μg/100 ml) and ovarian secretion rates (ng/min) of progesterone during the 4-day estrous cycle of rats. At 0100 hr proestrus, a comparison is made of intact, adrenalectomized (adx), and adrenalectomized-ovariectomized (ovx) animals (solid portion of bar). Bar graphs indicate peripheral plasma concentrations, and ovarian secretion rates are designated by the striped line. Standard errors are represented by vertical lines (Mann and Barraclough, 1973).

electrode was stereotaxically oriented into the MPOA and various current intensities were delivered over a 60 sec period. As is shown in Fig. 9, as the current was increased form 20-100 μA/60 sec progressive increases in plasma LH concentrations were obtained. Such increases in current were found to result in increased amounts of MPOA tissue being stimulated and reinforce Everett's (1964) original hypothesis that a "point-to-point" relationship exists between the preoptic neurons activated and terminal median eminence structures releasing LH-RH. As more MPOA neurons are activated, a greater number of LH-RH cells are stimulated to release their hormone thus causing increased amounts of LH to be discharged by the adenohypophysis into circulation.

F. Effects of Medical Preoptic Electrochemical Stimulation on LH Release in Androgen-Sterilized Rats

In the final series of studies we examined the ability of the MPOA of adult rats sterilized with 1.25 mg TP to respond to electrochemical stimulation. Shown in Fig. 10 are the results obtained. Normal Nembutal-blocked proestrous rats released peak LH concentrations of 120-130 ng/ml LH following a 60 μA/60 sec stimulus. Identical stimulation of 10 androgen-sterilized rats (ASR) resulted in one-half releasing similar concentrations as the

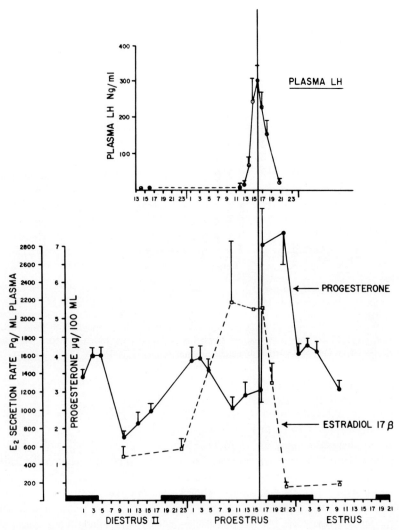

Fig. 8. Composite diagram depicting peripheral plasma LH concentrations (Turgeon and Barraclough, 1973), peripheral plasma progesterone concentrations (Mann and Barraclough, 1973) and ovarian venous plasma estradiol-17β concentrations (Shaikh, 1971) during diestrus-II, proestrus and estrus of the rat.

normal and in one-half releasing only minimal LH concentrations into plasma. All control rats ovulated whereas none of the sterile animals had tubal ova at autopsy 20 hrs later. In 1964 we suggested (Barraclough *et al.*, 1964) that perhaps the absence of corpora lutea in the ASR ovaries resulted in reduced plasma concentrations of progesterone (P), a steroid thought to be essential for the gonadotropin ovulatory surge to occur. However, detailed studies of replacement therapy with P revealed that restoration of normal plasma

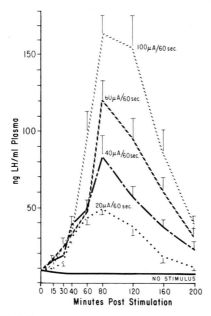

Fig. 9. Plasma LH following unilateral MPOA electrochemical stimulation in pro-estrous, Nembutal-blocked rats. Each group represents sequential collections from 7-9 stimulated rats. Vertical bars represent the SE (Turgeon and Barraclough, 1973).

concentrations of P alone was not conducive to LH release even though vaginal cyclicity could be artificially maintained. Since the endpoint of LH release was ovulation in these early studies, we have re-examined the effects produced by P on plasma LH concentrations if given alone or used in conjunction with MPOA stimulation. Adult androgen-sterilized rats were injected with P (2 mg s.c. in oil) and at the time of return to a vaginal proestrous smear (2-3 days later), these animals were anesthetized with Nembutal, an external jugular cannula was inserted, and the MPOA was stimulated electrochemically with 60 μA/60 sec. As is shown in Fig. 11 copious amounts of LH, in excess of 250 ng/ml, were released from the pituitaries of the androgen sterilized-progesterone-primed rat which at peak concentrations was approximately 2X that released from the normal pro-estrous rat subjected to identical MPOA stimulation. Furthermore, all andro-gen sterilized-progesterone primed-MPOA stimulated rats ovulated.

IV. CONCLUSION

From the present studies it can be observed that exposure of prepubertal rats to androgen not only produces anovulatory sterility but that the sites of action of early androgen treatment are multiple. Follicular responsiveness to endogenous LH release induced either by exogenous or endogenous LH-RH is

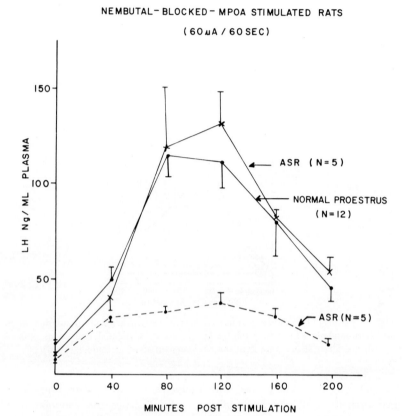

Fig. 10. Plasma LH patterns and concentrations in normal Nembutal-blocked proestrous rats in which the medial preoptic area (MPOA) was electrochemically stimulated with 60 μA/60 sec. Also illustrated are the responses obtained by identical MPOA stimulation of 10 androgen sterilized rats (ASR). Note 50% released LH in concentrations similar to the normal whereas 50% released significant but physiologically unimportant amounts of LH. All normal rats but none of the ASR ovulated.

suppressed and only when peak LH concentrations in plasma exceed 200 ng/ml does ovulation occur.

While such ovarian refractoriness may be due to a reduction in LH receptors, the patterns and concentrations of endogenous FSH in the androgen sterilized rat have not been investigated in detail. Since FSH and prolactin both are released together with LH at the time of the ovulatory gonadotropin surge, perhaps the preovulatory follicular swelling and other events require the presence of sufficient FSH which, acting in conjunction with LH, brings about ovum expulsion from the follicle. Furthermore, the effects of LH-RH on release of LH in normal rats depends primarily on whether or not the gland is

MPOA — STIMULATED, NEMBUTAL — BLOCKED RATS

60 μA / 60 SEC

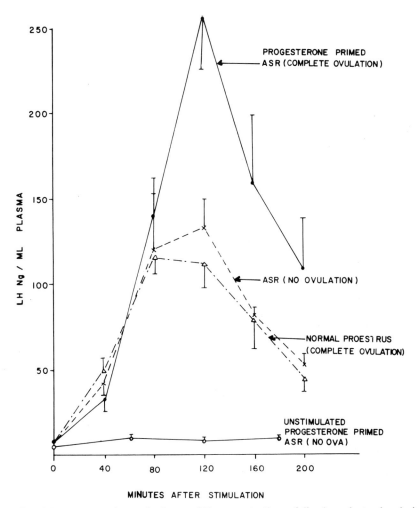

Fig. 11. A comparison of plasma LH concentrations following electrochemical stimulation in normal proestrous rats, ASR and progesterone-primed ASR. All rats were Nembutal-blocked. When the MPOA of progesterone-primed (2 mg s.c.) rats who returned to a vaginal proestrous smear (2-3 days after injection) were electrochemically stimulated, plasma LH concentrations were considerably in excess of the control. All such ASR ovulated. Treatment of ASR with progesterone alone (no stimulation, no Nembutal) did not alter basal LH concentrations.

dominated by testosterone or estradiol rather than the genetic sex of the animal. Debeljuk *et al.* (1973) have observed an increase in castrate male pituitary sensitivity to LH-RH after estrogen treatment, but not after estradiol treatment of intact male rats. Rippel *et al.* (1972) have examined pituitary responsiveness to LH-RH throughout the rat estrous cycle. They observed that the serum LH elevation following LH-RH was greater at proestrus than at metestrus or diestrus day 1, and least at diestrus day 2.

However, even after steroid replacement in the androgen sterilized rat, pituitary responsiveness to LRF is reduced suggesting that increased concentrations of LRF would be required to induce an ovulatory release of LH under any circumstance. Finally, the most damaging malfunction produced by early androgen treatment is the inability of the preoptic area to initiate those spontaneous events essential for the release of LH-RH into portal blood. This loss of the spontaneous neural trigger alone would account for the anovulatory syndrome even if other components of the hypothalamo-hypophyseal-gonadal system were capable of responding normally.

ACKNOWLEDGMENT

These studies were supported by a grant from the USPHS, HD-02138.

REFERENCES

Barraclough, C. A. (1961). Production of anovulatory, sterile rats by single injections of testosterone propionate. *Endocrinology* **68**, 62–67.

Barraclough, C. A. (1973). Sex steroid regulation of reproductive neuroendocrine processes. *In* "Handbook of Physiology" (R. O. Greep, ed.), Section 7, Vol. II, pp. 29–56. American Physiological Soc., Washington, D. C.

Barraclough, C. A. and Leathem, J. (1954). Infertility induced in mice by a single injection of testosterone propionate. *Proc. Soc. Exp. Biol. Med.* **85**, 673–674.

Barraclough, C. A., Yrarrazaval, S. and Hatton, R. (1964). A possible hypothalamic site of action of progesterone in the facilitation of ovulation in the rat. *Endocrinology* **75**, 838–845.

Callantine, M. R., Humphrey, R. R., Lee, S. L., Windson, B. L., Schottin, N. H. and O'Brien, O. P. (1966). Action of an estrogen antagonist on reproductive mechanisms in the rat. *Endocrinology* **79**, 153–167.

Debeljuk, L., Arimura, A. and Schally, A. V. (1973). Effect of estradiol on the response to LH-RH in male rats at different times after castration. *Proc. Soc. Exp. Biol. Med.* **143**, 1164–1167.

Döhler, K. D. and Wüttke, W. (1974). Serum LH, FSH, prolactin and progesterone from birth to puberty in female and male rats. *Endocrinology* **94**, 1003–1008.

Everett, J. W. (1948). Progesterone and estrogen in the experimental control of ovulation time and other features of the estrous cycle in the rat. *Endocrinology* **43**, 389-405.

Everett, J. W. (1964). Preoptic stimulative lesions and ovulation in the rat: 'Thresholds' and LH-release time in late diestrus and proestrus. *In* "Major Problems in Neuroendocrinology" (E. Bajusz and G. Tasmin, eds.), pp. 346–366. Williams and Wilkens Co., Baltimore, Md.

Ferin, M., Tempone, A., Zimmering, P. E. and Vande Wiele, R. L. (1969). Effects of antibodies to estradiol-17β and progesterone on the estrous cycle of the rat. *Endocrinology* **85**, 1070–1078.

Gay, V. L. and Sheth, N. A. (1972). Evidence for a periodic release of LH in castrated male and female rats. *Endocrinology* **90**, 158–168.

Goldman, B. D., Grazia, Y. R., Kamberi, I. A. and Porter, J. C. (1971). Serum gonadotropin concentrations in intact and castrated neonatal rats. *Endocrinology* **88**, 771–776.

Gorski, R. A. and Barraclough C. A. (1963). Effects of low dosages of androgen on the differentiation of hypothalamic regulatory control of ovulation in the rat. *Endocrinology* **73**, 210–216.

Harris, G. W. (1955). Neural control of the pituitary gland. Edward Arnold (Publishers) LTD. London.

Hayashi, S. and Gorski, R. A. (1974). Critical exposure time for androgenization by intracranial crystals of testosterone propionate in neonatal female rats. *Endocrinology* **94**, 1161–1167.

Karsch, F. J., Dierschke, D. J., Weick, R. F., Yamaji, T., Hotchkiss, J. and Knobil, E. (1973). Positive and negative feedback control by estrogen of luteinizing hormone secretion in the rhesus monkey. *Endocrinology* **92**, 799–804.

Makino, T. (1973). Study of the intracellular mechanism of LH release in the anterior pituitary. *Amer. J. Obstet. Gyn.* **115**, 606–614.

Mann, D. R. and Barraclough, C. A. (1973). Changes in peripheral plasma progesterone during the rat 4-day estrous cycle: An adrenal diurnal rhythm. *Proc. Soc. Exp. Biol. Med.* **142**, 1226–1229.

McEwen, B. S., Pfaff, D. W. and Zigmond, R. E. (1970). Factors influencing sex hormone uptake by rat brain regions. III. Effects of competing steroids on testosterone uptake. *Brain Res.* **21**, 29–38.

Miyachi, Y., Nieschlag, E. and Lipsett, M. B. (1973). The secretion of gonadotropins and testosterone by the neonatal male rat. *Endocrinology* **92**, 1–5.

Niswender, G. D., Midgley, A. R., Monroe, S. E. and Reichert, L. E. (1968). Radioimmunoassay for rat luteinizing hormone with antiovine LH serum and ovine LH-[131]I. *Proc. Soc. Exp. Biol. Med.* **128**, 395–404.

Pfeiffer, C. (1936). Sexual differences of the hypophyses and their determination by the gonads. *Am. J. Anat.* **58**, 195–226.

Rippel, R. H., Johnson, E. S. and White, W. F. (1972). Ovulation and serum luteinizing hormone in the cycling rat following administration of gonadotropin releasing hormone. *Proc. Soc. Exp. Biol. Med.* **143**, 55–58.

Schally, A. V., Kastin, A. J., Arimura, A., Coy, D., Coy, E., Debeljuk, L. and Redding, T. W. (1973). Basic and clinical studies with luteinizing hormone-releasing hormone (LH-RH) and its analogues. *J. Reprod. Fert. Suppl.* **20**, 119–136.

Shaikh, A. A. (1971). Estrone and estradiol levels in the ovarian venous blood from rats during the estrous cycle and pregnancy. *Biol. Reprod.* **5**, 297–307.

Shirley, B., Wolinsky, J. and Schwartz, N. B. (1968). Effects of a single injection of an estrogen antagonist on the estrous cycle of the rat. *Endocrinology* **82**, 959–968.

Turgeon, J. and Barraclough, C. A. (1973). Temporal patterns of LH release following graded preoptic electrochemical stimulation in proestrous rats. *Endocrinology* **92**, 755–761.

V. Implantation

The Relationships Between the Early Mouse Embryo and its Environment

Michael I. Sherman and David S. Salomon

Department of Cell Biology
Roche Institute of Molecular Biology
Nutley, New Jersey 07110

I. Introduction 277
  A. Maternal-Fetal Interactions 277
  B. Development of Mouse Embryos *In Vitro* 279
II. *In Vitro* Blastocyst Implantation Studies 280
  A. Previous Studies 280
  B. Morphological Studies of Blastocyst Implantation on
    Uterine Monolayers 281
  C. Effects of Substratum on Blastocyst Attachment 286
  D. Effect of Dextran-Norite Treated Serum on Blastocyst
    Attachment 290
  E. Factors Controlling Blastocyst Attachment in Culture 293
    1. Time in Culture 293
    2. Serum Concentration 294
    3. Uterine Fluid 294
  F. Assessment of the System 296
III. Biochemistry of Trophoblast Differentiation *In Vivo* and *In Vitro* . 297
  A. Previous Studies 297
  B. Expression of Trophoblast Δ^5, 3β-Hydroxysteroid Dehydrogenase
    Activity *In Vivo* and *In Vitro* 299
  C. Significance 304
IV. Conclusion 305
  References 305

I. INTRODUCTION

A. Maternal-Fetal Interactions

The mouse embryo undergoes its entire development in the reproductive tract. However, the degree of intimacy of the embryo with the mother varies throughout gestation. During propulsion through the oviduct and upon entry into the uterus the embryo is surrounded by the zona pellucida and is suspended in the fluids of the reproductive tract. Shortly after hatching from the zona, attachment of the blastocyst to the uterine epithelium occurs. This, in turn, is followed by implantation, an invasive event in the mouse, during which trophoblast cells burrow through the epithelial layer of the uterus and anchor the embryo to the uterine wall. Proliferation of the trophoblast leads to the establishment of these cells as the fetal portion of the placenta. Initially, the embryo proper is physically prevented from direct contact with maternal tissue by the trophoblast layer. The yolk sac and amnion subsequently develop, and they further isolate the embryo. However, after the eleventh day of gestation (the day of observation of the copulation plug is considered the first day) the giant trophoblast cells at the abembryonic pole of the mouse placenta begin to break down, so that eventually, surviving trophoblast cells are restricted to the intermediate layers of the placental disc, and the yolk sac assumes direct contact with the uterine fluid.

Fetal-maternal interactions are constantly occurring throughout development as a result of the close contact between mother and conceptus. Glass (1970) has demonstrated the selective passage of macromolecules from the mother to the egg before ovulation; after fertilization, there is further transfer of maternal proteins from the oviducal fluid to the cleaving embryo as it passes down the oviduct. The blastocoelic fluids of rodent and rabbit blastocysts contain proteins representative of those present in the uterine fluid (Brambell, 1958; Sugiwara and Hafez, 1965). Even enzymes present in the uterine fluid appear to be taken up by the blastocyst, presumably into the blastocoelic fluid (Sherman, 1974b). Undoubtedly, small molecules in the reproductive tract penetrate the preimplantation embryo.

Just prior to implantation, the blastocyst hatches from the zona pellucida. While mechanical processes might adequately explain the rupture of the zona (Cole and Paul, 1965; Cole, 1967), it has been suggested that a maternally derived zona lysin facilitates hatching (McLaren, 1970a; Mintz, 1971; Pinsker et al., 1974). This maternal proteinase may also play a role in implantation (Mintz, 1971; Pinsker et al., 1974). Implantation is further regulated maternally in that ovariectomy at the appropriate time in gestation effectively blocks this event (Chambon, 1949), leading to "implantation delay." This phenomenon can be reversed by steroid hormone therapy (Psychoyos and Alloiteau, 1962; Nutting and Meyer, 1963). The requirement for a maternal source of steroid hormones in the maintenance of pregnancy can be demonstrated throughout most of the gestation period in the mouse (Robson, 1938; Hall and Newton, 1947).

Since mouse trophoblast cells are bathed in maternal blood after implantation, and since this layer of cells completely envelops the embryo, it is reasonable to assume that the passage of nutrients from mother to fetus before the establishment of the umbilical cord is via the trophoblast layer. Besides low molecular weight compounds (Brunschweig, 1938), at least one enzyme, a non-specific esterase, is selectively taken up from the serum and passes from the trophoblast to the yolk sac. However, the enzyme is prevented from penetration of the embryo proper (Sherman and Chew, 1972). In the latter third of pregnancy, maternal immunoglobulins are taken up by the fetus and passive immunity is established (Anderson, 1959; Koch *et al.*, 1967; Brambell, 1970). It has been proposed that the yolk sac is the route of penetration of these immunoglobulins (Brambell, 1970).

B. Development of Mouse Embryos In Vitro

It is possible that while some material may pass from mother to embryo in order to satisfy nutrient requirements, other maternal molecules or macromolecules may function as inducers or triggers of differentiation of various embryonic or extraembryonic cell types. If this were so, it would be expected that such differentiation should not occur outside the reproductive tract. It has been known for many years that blastocysts, or even two-cell embryos, can survive when explanted to sites as diverse as the anterior chamber of the eye, the testis, or the brain of an adult host (see Kirby, 1965 for a review). In most of these studies, giant trophoblast cells were the only egg-derived cells that survived after several days of incubation in the ectopic site. Judging from their size, morphological characteristics, and invasiveness, these trophoblast cells were well differentiated. Recent studies on ectopic blastocyst implants by Billington *et al.* (1968) indicate that morphologically normal early egg cylinder embryos can also develop in some cases.

A series of improvements in culture conditions for preimplantation embryos (see Brinster, 1969) has culminated in a report by Whitten (1971) that one-cell embryos develop to blastocysts at high frequency in his medium. Hatching from the zona also occurs. Furthermore, these blastocysts can give rise to normal young when transferred to foster mothers. These results suggest that preimplantation embryos are capable of developing to the blastocyst stage in the absence of the maternal environment. It should, however, be noted that blastocysts derived from cleavage stage embryos *in vitro* may contain less than half the number of cells (Graham, 1971), and show poorer postblastocyst development in culture (Menke and McLaren, 1970; Hsu *et al.*, 1974; Sherman, 1974c), than do blastocysts removed from the uterus.

An appropriate medium for postblastocyst development was first reported by Mintz (1964). Cole and Paul (1965) and Gwatkin (1966a,b) subsequently developed other media which supported postblastocyst development. In these studies, and in early investigations in this laboratory (Sherman, 1972a; Barlow and Sherman, 1972), trophoblast cells grew out along the culture dish, while the inner cell mass (ICM; the inner cells of the blastocyst from which the embryo proper and extraembryonic membranes are derived) failed to develop in all but a small number of cases (Cole and Paul, 1965). Improvements in culture conditions were subsequently reported in this laboratory (Sherman, 1972b, 1974c; Bell and Sherman, 1973) and in others (Hsu, 1971, 1973; Hsu *et al.*, 1974; Spindle and Pedersen, 1973). Under our present culture conditions, ICM cells proliferate in a high percentage of cases and large vesicles commonly appear. These vesicles resemble yolk sac, both morphologically and biochemically (Bell and Sherman, 1973; Sherman, 1974b,c). The most recent culture conditions utilized by Hsu (1973; Hsu *et al.*, 1974) include a collagen substratum, human cord serum, and very frequent medium changes. The result is the appearance, albeit at low frequency, of morphologically normal early somite stage embryos, like those seen *in vivo* on the eighth or ninth day of gestation. Although these embryos fail to develop further, New and Daniel (1969) have demonstrated that 8.5 day rat embryos at the egg cylinder stage can develop *in vitro* to the anterior limb bud stage. Embryos explanted at later gestation age develop still further (New, 1971).

It is clear from these observations that at least at the gross morphological level, substantial differentiation of both embryonic and extraembryonic tissues can occur *in vitro*. Therefore, the rodent embryo in its development does not appear to rely to any great degree upon maternal inducers or triggers. On the other hand, with a single exception (Jenkinson and Wilson, 1973) to be discussed later, postblastocyst development has not been found to occur in the absence of serum, usually fetal, which may contain many of the same factors present in the uterine fluid and maternal blood which bathes the trophoblast layer. Furthermore, while the studies described above have demonstrated development *in vitro* through various stages, and while in some cases, these stages overlap, complete embryonic development in culture has not yet been feasible. (Of course, the desirability of such an achievement may be dubious.)

Although a large variety of components has been used in early embryo culture studies, there have been very few reports on the effect of steroid hormones on development *in vitro* (e.g., Whitten, 1957; Daniel, 1964; Glenister, 1965; Kirkpatrick, 1971; Grant, 1973), and these have involved nonphysiological levels of the hormones under study. We found it curious that a role for steroid and polypeptide hormones has not been established *in vitro* while they are crucial for implantation and maintenance of pregnancy *in vivo* and might in some way trigger developmental events. We have therefore initiated a set of experimental approaches to study this problem. In the next section, we shall discuss the effect of removal of steroid hormones on "implantation" *in vitro*, and in the subsequent section we shall consider whether the production of progesterone by trophoblast cells may be under maternal control.

II. *IN VITRO* BLASTOCYST IMPLANTATION STUDIES

A. Previous Studies

Most of the postblastocyst culture studies described above involved attachment of blastocysts to a plastic or glass culture dish or to an acellular collagen substratum. However, as early as 1961, Glenister observed that rabbit blastocysts could attach to, and implant into, isolated strips of endometrial tissue in the presence of a complicated variety of biological components, such as chick embryo extracts, rabbit serum, and albumin. More recently, Grant

(1973) demonstrated that mouse blastocysts could implant when trapped within intact uterine horns in culture. In these studies, however, continuous inspection of the implantation process was not possible, and the implantation frequency in each experiment could only be determined by sectioning and histological analysis. Cole and Paul (1965) utilized live or UV-irradiated HeLa or L cells as feeder layers on which to culture mouse blastocysts. They noted that blastocysts could attach to this type of substratum. We have devised a similar system wherein blastocysts attach to, and subsequently "implant" into, primary monolayer cultures of uterine cells. In this way, the entire process can be monitored visually, and the exact times of implantation of large numbers of blastocysts can be noted, allowing quantitative studies to be done. The technical aspects of this procedure are described elsewhere (Salomon and Sherman, 1974).

B. Morphological Studies of Blastocyst Implantation on Uterine Monolayers

Blastocysts utilized in these studies were taken either directly from the uterus on the 4th day of gestation (Fig. 1A), from two-cell (2nd day) embryos grown to blastocysts in the medium of Whitten and Biggers (1968) for 3 days (Fig. 1B), or from delayed blastocysts removed from the uterus 7 days after ovariectomy of the mother (Fig. 1C). In all three cases the blastocysts generally expand shortly after being placed in culture. Delayed blastocysts are characteristically more oval in shape. In previous studies (Sherman and Barlow, 1972) we found that on average fourth day blastocysts contained 64 cells, while delayed blastocysts contained between 40 and 200. Since two-cell embryos develop about a half day more slowly when cultured in vitro than in vivo (Bowman and McLaren, 1970; Graham, 1971), it would be expected that blastocysts derived from two-cell embryos after three days in culture would also consist of about 64 cells.

When blastocysts from any of these sources attach to a plastic culture dish, the site of attachment appears to be somewhat random, although the lateral or abembryonic trophoblast cells are most often initially involved. Shortly after attachment, the trophoblast cells above the ICM inevitably separate and migrate along the culture dish, leaving the ICM as a ball of cells attached only to the monolayer of cells below it (Fig. 2). The mechanism by which the trophoblast cells separate is unknown. Curiously, just prior to implantation, separation of trophectoderm (precursor trophoblast) cells by proteolytic enzymes is extremely difficult. This separation, of course, is a departure from the morphological nature of blastocysts implanting in vivo.

As the trophoblast cells grow out along the culture dish, they become enlarged and often highly vacuolated, although those cells which were apparently at the embryonic pole often show a lesser degree of vacuolation (Fig. 2). At later stages in the culture, most trophoblast cells are no longer vacuolated; instead, the cytoplasm adjacent to the nucleus of these cells becomes densely granulated (e.g., Fig. 4C). The nature of neither vacuoles nor

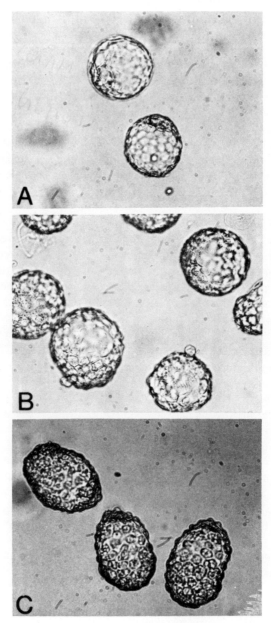

Fig. 1. Morphology of blastocysts developing in normal and delayed uteri and in culture. Blastocysts removed from uteri on the fourth day of pregnancy (A), cultured from the two-cell stage for three days in the medium of Whitten and Biggers (1968) (B), or removed from uteri seven days after ovariectomy delay (C) are pictured after one day of culture in supplemented NCTC-109 medium. Phase contrast microscopy, x122.

282

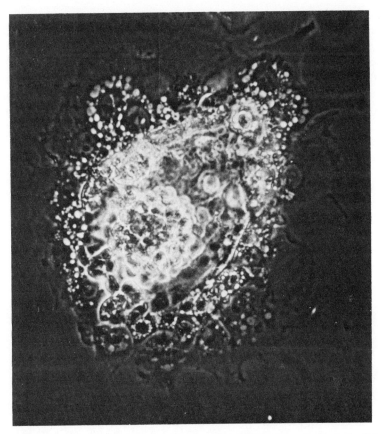

Fig. 2. Fourth day mouse blastocyst after three days culture in a plastic dish. The blastocyst was cultured in NCTC-109 medium supplemented with 10% heat-inactivated (56°, 30 min) fetal calf serum, penicillin (100 μg/ml), streptomycin (100 μg/ml) and kanamycin (100 μ/ml), at 37° in 5% CO_2-95% air (Sherman, 1974c). Phase contrast microscopy, x245.

the granules is known, so it is not possible to determine what role, if any, these structures play in the process of implantation or in other trophoblast functions. The subsequent development of secondary trophoblast cells from the ectoplacental cone and the further proliferation of ICM cells in culture is described elsewhere (Sherman, 1974c).

Primary monolayer cultures of uterine cells are generated by trypsinizing entire uterine horns (see Salomon and Sherman, 1974). The resultant mono-layers contain morphologically uniform cells (Fig. 3). These cells do not resemble epithelial or muscle cells; it is therefore assumed that the majority of them are derived from the uterine stroma. These cells are capable of proliferating in the medium normally used for blastocyst culture (NCTC-109 or NCTC-135, supplemented with heat inactivated fetal calf serum and antibiotics; see Sherman, 1974c, for details) and, if seeded heavily in culture dishes, they form a confluent monolayer within 24 hours. This monolayer can be maintained with occasional changes of medium for at least 2-3 weeks.

Blastocysts taken directly from the uterus of a normally pregnant or ovariectomized mouse, or cultured from the two-cell stage, are all capable of eventually attaching to a monolayer of uterine cells. After attachment, blastocysts behave in the same way regardless of their prior developmental schedule. Within a day of attachment, trophoblast cells begin to grow out, just as they do on a plastic surface (Fig. 4A). In the process, the underlying monolayer cells are displaced and a plaque is formed. Cole and Paul (1965) noted a similar effect using feeder layers. They proposed that this displace-ment was purely mechanical, i.e., feeder layer cells were pushed aside as the trophoblast cells migrated out along the dish. In many cases, however, we have noted a clear halo *ahead* of the trophoblast cells as they migrated into the monolayer (e.g., Fig. 4A). This may suggest secretion of cytolytic enzymes by the trophoblast cells; Kirby (1965) has made a similar proposal on the basis of his studies of blastocysts implanted under the testis capsule.

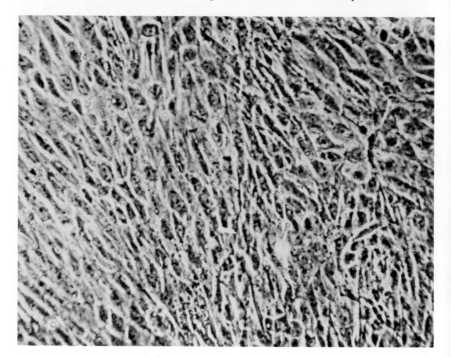

Fig. 3. Morphology of mouse uterine monolayers after one day in culture. The monolayer was generated from uteri dissected from previously superovulated and mated females on the fourth day of pregnancy (Salomon and Sherman, 1974). Conditions were the same as those used for blastocyst cultures. Phase contrast microscopy, x198.

The increase in plaque size continues for a few days as trophoblast cells displace uterine cells. After 5 to 6 days in culture, however, this process ends as the trophoblast appears to lose invasiveness (Fig. 4B). The trophoblast cells

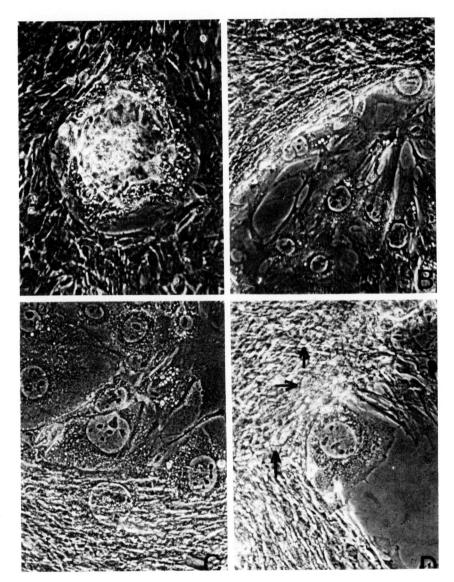

Fig. 4. Implantation of blastocysts on uterine monolayers. Time in culture: (A) 3 days; (B) 7 days; (C) 10 days; (D) 12 days. Note increase in size of trophoblast nuclei. Arrows in D locate giant nuclei of totally engulfed trophoblast cells. In all cases, magnification = 130x. Phase contrast microscopy.

at the periphery of the growth now intermingle with uterine cells (Fig. 4C), and engulfment of the giant cells eventually takes place (Fig. 4D). Primary and secondary trophoblast cells not at the periphery of the growth survive for long periods of time (up to 35 days in culture) as they do in the absence of uterine monolayers (Sherman, 1974c), but these cells do not show any signs of invasiveness. For some reason, the uterine cells fail to grow into the plaque; therefore, after 6 days in culture, plaque size remains constant for days or even weeks. The eventual loss of trophoblast invasiveness parallels that observed *in utero* (see Billington, 1971). This loss of invasiveness by trophoblast *in vitro* along a time scale not unlike that *in vivo* may indicate that decidualization is not the only factor responsible for arresting trophoblast outgrowth (Kirby and Cowell, 1968). Alternatively, the uterine monolayer cells in culture may come to share the invasion-arresting properties of the decidua.

Cole and Paul (1965) found that blastocysts could attach to, and grow out along, HeLa and L cell feeder layers. We have tested blastocyst implantation on cells derived from human, rat, hamster, and mouse tissues, including four established cell lines derived from long term culture of mouse blastocysts (Sherman, 1974c). In every case, blastocysts attached to the monolayers, and trophoblast cells grew out in the same invasive manner observed with uterine monolayers (e.g., Fig. 5). This is not surprising, since Kirby has demonstrated trophoblast invasion of several organs (Kirby, 1965) and even tumors (Kirby, 1962). These results are consistent with Solomon's (1966) claim that rodent trophoblast cells are among the most invasive cell types known.

The development of the ICM in the presence of uterine monolayers is variable, although in no case is it as extensive as it is in the absence of the monolayer (Sherman, 1974c). For example, although the ICM may consist of a small ball of cells or a rather large clump (regardless of whether the blastocyst had been in delay, or had developed *in vitro* or *in utero*), vesicle formation does not occur. Before degenerating, however, the ICM can in some cases give rise to a single cell layer, presumably distal endoderm which overlies the trophoblast (Fig. 6). In other cases, (Fig. 7), the ICM can actually become separated from the trophoblast monolayer and come to rest on uterine cells; Bryson (1964, 1965) also observed this separation in peritoneal chambers. The reason for eventual failure of ICM cells in the presence of uterine monolayers is not clear. Perhaps trophoblast cells can compete with uterine cells for essential nutrients more successfully than ICM cells. It may also be that trophoblast cells can withstand the slightly acidic conditions generated by the confluent uterine monolayer better than ICM cells.

C. Effects of Substratum on Blastocyst Attachment

In the experiments to be described, the sample size usually varied between 20-30 blastocysts. Unlike earlier experiments with endometrial strips (e.g., Glenister, 1965) or intact uterine horns (Grant, 1973), where the

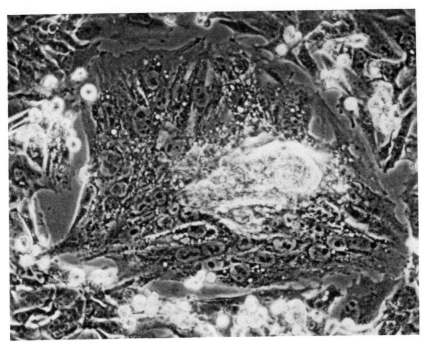

Fig. 5. Implantation of mouse blastocyst on MB21 cell monolayer. After 4 days in culture, the blastocyst has formed a plaque in a monolayer of cells derived from long term cultures of mouse blastocysts (Sherman, 1974c). Note MB21 cells raised off the surface of the dish just ahead of the periphery of the trophoblast cells. Phase contrast microscopy, x196.

frequency of blastocyst attachment was somewhat low, almost all of the blastocysts under observation in this study did attach, regardless of substratum used (Table 1). Virtually all blastocysts that attached to the monolayers outgrew as described in the previous section; there was no apparent difference in outgrowth whether the blastocysts were normal (4th day) or delayed for 7 days, or whether the monolayers were generated from uteri of 4th day pregnant or 7 day delayed mice.

Fig. 8 illustrates data from a typical experiment for the hatching and attachment times of blastocysts in the presence or absence of a uterine monolayer. Since 100% of the blastocysts do not always hatch and attach, we have found it convenient in analyzing our data to consider the time at which 50% of the blastocysts have attached or, where indicated, have hatched. This will be referred to as the "T50." From the data in Fig. 8A, for example, blastocysts have a T50 (hatching) of 32 hrs and a T50 (attaching) of 41.5 hours in the absence of a uterine monolayer, whereas, in the presence of a monolayer (Fig. 8B), the T50 values are 36 hrs and 49 hrs, respectively. In this experiment, then, there was a 7.5 hr difference in the T50 for attachment with and without a uterine monolayer. As we have indicated previously

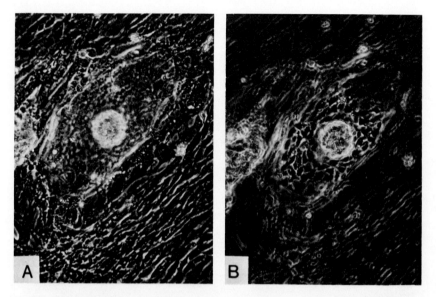

Fig. 6. Development of parietal endoderm cells in a blastocyst implanting on a uterine monolayer after three days in culture. (A) Focussed on the trophoblast cells; (B) focussed on the clump of ICM cells and the layer of parietal endoderm overlying the trophoblast. Phase contrast microscopy, x165.

(Salomon and Sherman, 1974), in some experiments (4/19 to date) there is no difference in the attachment rates on plastic and uterine monolayer, i.e., the T50's are within 3 hours of each other. However, on average, the T50 for blastocyst attachment on uterine monolayers is about 8 hr greater than that for attachment on plastic (Table 2). Delayed blastocysts show the same effect. Attachment time is not markedly affected by the source of the uterine monolayer (i.e., uteri from pregnant or delayed mice). The results confirm that blastocysts from ovariectomized hosts are released from delay upon introduction into culture (Gwatkin, 1966b). It is also apparent that monolayers consisting of cells from delayed uteri do not impede attachment more than cells derived from the uteri of pregnant animals.

The differences between attachment rates on cells vs. plastic, when observed, are not due to differences in the rate of hatching since: (1) although the T50 for hatching is occasionally greater in dishes with monolayers than in those without, the time differences are never as great as those for attachment; (2) premature removal of the zona by treatment with pronase (Mintz, 1962) does not alter the attachment rate; and (3) the great majority of delayed blastocysts have already hatched *in utero*, and yet they show the same retardation of attachment on uterine monolayers as do normal blastocysts.

Gwatkin (1966b) and Spindle and Pedersen (1973) have pointed out that a number of amino acids are required for blastocyst attachment to a plastic or a collagen substratum. We have considered the possibility that blastocyst

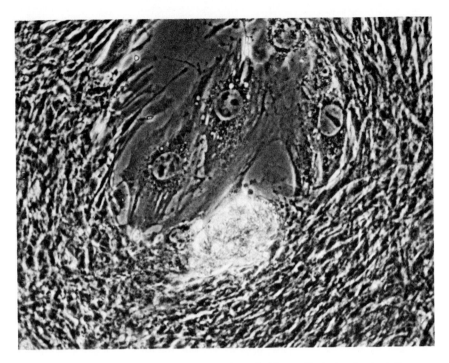

Fig. 7. Separation of inner cell mass and trophoblast cells during implantation on a uterine monolayer. After 7 days of culture, the round clump of ICM cells has come to lie over uterine monolayer cells. Phase contrast microscopy, x198.

TABLE 1

Attachment Frequency of Normal and Delayed[a] Blastocysts

Type of blastocyst	Type of substratum	Number of experiments	Number attached / Total	Percent attached
Normal	Plastic	8	470/516	91
	Uterine monolayer – pregnant mice[b]	11	530/590	90
	Uterine monolayer – delayed mice[c]	4	192/212	91
Delayed	Plastic	2	64/71	82
	Uterine monolayer – pregnant mice[b]	4	104/116	90
	Uterine monolayer – delayed mice[c]	4	109/124	88

[a]Blastocysts collected 7 days after ovariectomy.

[b]Prepared from uteri of previously superovulated mice on the fourth day of pregnancy.

[c]Prepared from uteri of mice 7 days after ovariectomy.

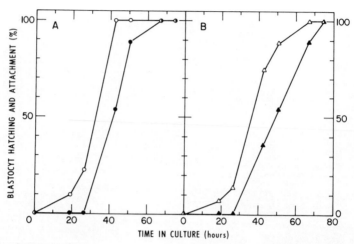

Fig. 8. Rates of blastocyst hatching and attachment in cultures without and with uterine monolayers. After 24 hours in culture, blastocysts were split into two groups of 30; one group was placed in a new culture dish (A), while the second was placed in a new dish containing a uterine monolayer (B). The times of hatching and attachment were determined by periodical inspection under a dissecting microscope. Attachment was considered to have taken place when blastocysts failed to become dislodged from the substratum during swirling of the culture dish. The abscissa indicates the total time spent by blastocysts in culture. O and Δ, percent of blastocysts hatched; ● and ▲, percent of blastocysts attached.

attachment on uterine monolayers is somewhat retarded because the large numbers of uterine cells present reduce the levels of these amino acids, or some other essential molecules, in the medium. To test this, we carried out split-dish experiments: uterine cells were seeded on a petri dish and subsequently the cells in half the dish were scraped off the surface and a low barrier was placed along the center of the dish. Blastocysts were then placed on both sides of the barrier. In three such experiments, even though the blastocysts shared the same medium, those on the side of the dish containing the uterine monolayer had T50's for attachment, 4, 9, and 12.5 hours greater than those exposed to the plastic surface. It therefore appears as though it is actually the cells themselves acting as a substratum which affect blastocyst attachment rate.

D. Effect of Dextran-Norit Treated Serum on Blastocyst Attachment

In his studies, Glenister (1965) was unable to discern any effect of added progesterone and estrogen upon invasiveness of rabbit blastocysts into endometrial strips. Similarly, Grant (1973) could demonstrate no dependence upon added estrogen for mouse blastocyst implantation on uterine horns *in vitro*,

TABLE 2

*Effect of Substratum on Attachment Time of Normal
and Delayed Blastocysts*

Type of blastocyst	Type of substratum	Number of experiments	T50[a]± standard deviation
Normal	Plastic	17	43.2 ± 7.5
	Uterine monolayers – pregnant mice	19	51.6 ± 8.1
	Uterine monolayers – delayed mice	7	53.5 ± 6.9
Delayed	Plastic	4	40.6 ± 8.1
	Uterine monolayers – pregnant mice	6	51.9 ± 12.1
	Uterine monolayers – delayed mice	6	47.1 ± 16.7

[a]Time in hours required for attachment of 50% of the blastocysts.

although he reported that unphysiologically high levels of progesterone increased the invasiveness of trophoblast after blastocyst attachment. Since these workers used serum in their studies and made no effort to remove endogenous steroids, the significance of their observations is questionable. We have shown (Salomon and Sherman, 1974) that extraction of serum with dextran-coated norit at 45°C adsorbs out progesterone and estrogen such that these steroids can no longer be detected by sensitive radioimmunoassay procedures (Abraham *et al.*, 1971; Tulchinsky and Abraham, 1971). From our radioimmunoassay studies, we calculate that progesterone levels in medium containing heat-inactivated (56°, 30 min) fetal calf serum (HIFCS) are about 50 pg/ml. Medium containing dextran-norit treated serum (DN-HIFCS) contains, *at most*, one-fifth that amount. Values of estrogen in medium containing HIFCS and DN-HIFCS are respectively about 75 pg/ml, and, *at most*, 2 pg/ml. The total amount of estrogen in DN-HIFCS-containing medium is therefore at least 4-5 orders of magnitude lower than the dose required to overcome implantation delay in ovariectomized mice by subcutaneous injection (Smith and Biggers, 1968), and at least three orders of magnitude lower than that required to reverse implantation delay in the rat by local administration of the steroid to the uterus (Psychoyos, 1973). Furthermore, in these and other studies carried out involving steroid hormone therapy (e.g., McLaren, 1971) several hundred micrograms, or even milligram amounts, of progesterone must be administered along with estrogen in order for implantation to occur. Judging from these studies, we feel it is likely that if any estrogen or progesterone is in fact present in our cultures containing DN-HIFCS, the concentrations are very much lower than those required to render the uterus receptive *in vivo*.

As indicated previously (Salomon and Sherman, 1974), extraction of steroids from the serum does not impede blastocyst attachment either to plastic surfaces or uterine monolayers. Pooled data from a number of experiments illustrates that attachment rates in medium containing HIFCS or DN-HIFCS are similar in the case of normal blastocysts (Table 3). However, Smith (1968) has suggested that blastocysts are capable of taking up estrogen from their milieu; therefore, blastocysts in these studies might already have been "sensitized" with estrogen before being removed from the uterus. Consequently, these studies were repeated using blastocysts obtained from two-cell embryos cultured in defined medium (Whitten and Biggers, 1968) and with delayed blastocysts maintained in ovariectomized hosts for seven days in the absence of any hormone therapy. In these cases as well, (Salomon and Sherman, 1974; Table 3) the rates of implantation in medium supplemented with DN-HIFCS were not markedly different from those in medium containing HIFCS.

In studies with delayed blastocysts, two interesting observations were made. First, blastocysts are extremely sticky upon removal from the delayed uterus. Typically, 3 hours after placement in culture, 50-90% of the blastocysts have attached either to plastic or to monolayers. However, this phase is only temporary; by 20-30 hrs of culture, all, or almost all, of the blastocysts have become free again (see, e.g., Fig. 11) and permanent attachment subsequently begins.

Two media have been used in these studies: NCTC-109 and NCTC-135 (McQuilkin *et al.*, 1957; Evans *et al.*, 1964). These media differ only in the content of cysteine (260 mg/1 in NCTC-109, but absent in NCTC-135). Although normal blastocysts may attach slightly more quickly in NCTC-109 medium compared to NCTC-135, even when the same lot of serum is used, delayed blastocysts attach substantially more quickly in NCTC-109. This explains the large standard deviation in the T50's for delayed blastocysts in

TABLE 3

*Effect of Steroid Extraction of Serum on
Attachment Time of Normal and Delayed Blastocysts*

Type of blastocyst	Type of serum	Number of experiments	T50 ± standard deviation
Normal	HIFCS	36	47.1 ± 8.8
	DN-HIFCS	19	50.4 ± 7.8
Delayed	HIFCS	10	45.8 ± 13.6
	DN-HIFCS	10	47.7 ± 14.1

Tables 2 and 3. Gwatkin (1966b) and Spindle and Pedersen (1973) noted an absolute requirement for cystine for blastocyst attachment in their system, but they did not report on the effects of cysteine addition. It must, however, be mentioned that the two media used in our experiments were purchased from different suppliers. Controlled studies are therefore necessary before any conclusions can be drawn.

E. Factors Controlling Blastocyst Attachment in Culture

1. Time in Culture. In the absence of steroid hormone regulation of blastocyst attachment in culture, other factors were considered which might be responsible for the determination of the time of attachment. For instance, blastocyst attachment might merely require that a certain fixed period of time be spent in culture. This is unlikely since the T50 for attachment of blastocysts removed from uteri early on the fifth day of gestation is about 15 hours, approximately 24 hours less than the T50 of fourth day blastocysts. Furthermore, the development of blastocysts from two-cell embryos in culture is retarded by one half to one day under our culture conditions. Accordingly, such blastocysts, although equivalent in age to fifth day blastocysts freshly removed from the uterus, require 30-35 hours to implant upon placement into NCTC medium; this T50 is midway between the T50's of blastocysts which have developed *in vivo* for four and five days.

Another possible explanation for attachment time *in vitro* is that a fixed contact time between the zona-free blastocyst and the substratum is required. However, this is also unlikely since normal blastocysts treated with pronase to remove the zona pellucida as well as the majority of delayed blastocysts, which have hatched *in utero*, are exposed to the substratum for about a day longer than normal blastocysts which are in their zonas when placed in culture. Yet, T50's are similar in all three cases.

All the above observations are, however, consistent with the proposal that the trophectoderm cells must undergo some change which renders the blastocyst capable of attaching. This proposal has previously been suggested (Mintz, 1971; Sherman, 1971). In support of this proposal are the observations by Dickson (1969) that a number of morphological changes occur in trophoblast cells just prior to implantation.

The fact that the average hatching and attachment time is about one half day longer for blastocysts *in vitro* than *in vivo* might suggest that our culture conditions are still not optimal for these events. On the other hand, this retardation may be due to the unavoidable trauma involved in the operation of removing blastocysts from the animal. In this laboratory, exposure to room temperature for up to a few hours occurs during collection of blastocysts, and this may further retard blastocyst maturation. Preliminary observations suggest that blastocyst cells have an extremely low mitotic index within twelve hours of being placed in culture (unpublished observations). We have not as yet eliminated the possibility that cell number plays a role in the timing of blastocyst attachment.

2. *Serum Concentration*. As observed by others (Gwatkin, 1966a,b; Spindle and Pedersen, 1973; Jenkinson and Wilson, 1973) blastocyst attachment is suboptimal in the absence of serum when a plastic substratum is used (Fig. 9A). Even a small amount of serum added to the medium increases the rate of attachment and, in some cases, the final percentage of blastocysts attached (Fig. 9A). Although a number of blastocysts can attach to the culture dish in the absence of serum, outgrowth of trophoblast does not occur under these conditions.

In the presence of uterine monolayers serum was not required for blastocyst attachment (Fig. 9B) and subsequent invasion of the monolayer (Fig. 10). This observation is not unexpected, since Jenkinson and Wilson (1973) have reported that blastocysts could implant on a collagen substratum in the absence of serum. In their studies albumin was present in the medium while we added no macromolecules. Undoubtedly, however, some high molecular weight material must have been in the medium as a result of secretion by, or breakdown of, uterine monolayer cells.

3. *Uterine Fluid*. Steroid hormones may mediate blastocyst implantation *in vivo* by neutralizing or inactivating an implantation inhibitor in the uterine fluid (see McLaren, 1973 and Psychoyos, 1973). If this were the case, then it would be expected that blastocyst attachment *in vitro*, in the absence of uterine fluid, would not require steroid hormone action. We have therefore begun a study of the effects of uterine fluid on attachment of blastocysts *in vitro* in the presence of serum depleted of steroids (i.e., DN-HIFCS). In the presence of uterine monolayers, either from pregnant or delayed mice, the addition of as much as one milligram uterine fluid protein per milliliter medium has no effect on the T50 of attachment of normal or delayed

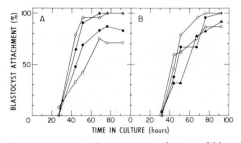

Fig. 9. Effect of serum concentration upon attachment of blastocysts on plastic and on uterine monolayers. Blastocyst attachment rates were measured on plastic (A) and on uterine monolayers (B) as described in the legend to Fig. 8. Serum concentrations were as follows: O, none; ●, 2%, Δ, 5%; ▲, 10%.

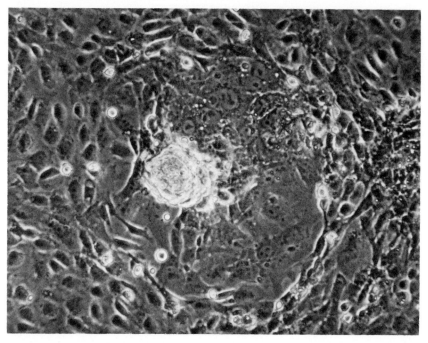

Fig. 10. Blastocyst implantation on a uterine monolayer in the absence of serum. Confluent monolayers were generated from uterine cells in the presence of medium supplemented with 10% fetal calf serum. Medium and serum were subsequently removed, and after several washes of the monolayer, blastocysts were added in medium lacking serum. Although the uterine cells did not survive well and the monolayer soon became sparse, trophoblast cells grew out into the surviving uterine cells in a normal manner. Photograph taken after 3 days in culture. Phase contrast microscopy, x198.

blastocysts, whether the source of the fluid is non-pregnant, delayed, fourth day, or fifth day uteri (see, e.g., Fig. 11). On the other hand, in the absence of a monolayer, concentrations of 400 micrograms or more of uterine fluid protein per milliliter medium were toxic to some blastocysts and retarded the attachment rate of others, regardless of the state of the uteri which were the sources of the fluid. A toxic effect of uterine fluid has also been observed by Psychoyos (1973) in rats. Curiously, concentrations of uterine fluid protein of 300 micrograms protein per milliliter medium or less appeared in preliminary studies to have a positive effect upon ICM proliferation from blastocysts cultured in the absence of a monolayer. Overall, it appears as though uterine fluid, regardless of source, does not have the ability to inhibit implantation *in vitro* in the absence of steroids, at least in the concentration range employed. The toxic effects of high concentrations of uterine fluid on blastocysts in the absence, but not presence, of uterine monolayers merits further study.

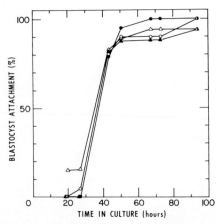

Fig. 11. Effect of uterine fluid on the attachment of blastocysts to uterine monolayers. Both blastocysts and uterine monolayers were obtained from ovariectomy delayed hosts. Addition of 300 μg of uterine fluid protein was made to the cultures as follows: O, none; ●, uterine fluid from previously superovulated, fourth day pregnant mice; Δ, uterine fluid from randomly selected, non-pregnant females; ▲, uterine fluid from 7 day ovariectomy delayed females.

However, this observation must be viewed cautiously, since our uterine fluid preparations, although filtered prior to use, undoubtedly contain contaminants from uterine cells disrupted during collection of the fluid, as well as from maternal serum. Mouse serum has already been found to be toxic to blastocysts in culture (Sherman and Chew, 1972).

F. Assessment of the System

The above studies indicate that upon removal from the maternal environment, both normal and delayed blastocysts are "activated" (McLaren, 1973), i.e., they can attach to, and implant upon, a monolayer of uterine cells, even in the virtual absence of steroid hormones. Neither cells nor fluid from delayed uteri nor both factors together are capable of impeding blastocyst activation *in vitro* (Fig. 11). Although our system offers the advantage of studying the interaction between the blastocyst and uterine cells, we have clearly failed to reproduce the maternal control of implantation *in vitro*. In fact, we are aware of only one reported method of delaying blastocyst activation in culture without permanently damaging the blastocyst, namely, that of omitting certain essential amino acids from the medium (Gwatkin, 1966b). However, Gwatkin (1969) later showed that these amino acids were not in short supply in the fluid of delayed uteri. Sherman and Barlow (1972) further showed that blastocysts delayed *in vitro* through amino acid deprivation were not necessarily arrested in the same part of the cell cycle as blastocysts in ovariectomy delay *in utero*.

A number of possibilities remain which may explain the failure to date, on our part and on the part of others, to control implantation *in vitro*. Formally, we must consider that exceedingly low levels of estrogen and progesterone may be adequate to induce implantation if localized at the attachment site. We are presently carrying out experiments designed to reduce even further the possibility that the levels of steroids in our cultures are adequate to induce implantation. It should be mentioned that Jenkinson and Wilson (1973) have succeeded in demonstrating blastocyst attachment and trophoblast outgrowth on a collagen substratum in the absence of serum. However, the objection could once again be leveled that trace amounts of steroids were present in their collagen preparations. In our studies to date, oxygen concentrations and pH have been relatively well regulated. These parameters both appear to change *in utero* during implantation (Yochim, 1971; McLaren, 1970b; Beier, 1974) and may be under hormonal control. Such changes can be reproduced for study in culture. Unlike those of Glenister (1961) and Grant (1973), our *in vitro* system suffers from the fact that blastocysts do not encounter the uterine epithelium as they do during attachment *in utero* and in organ culture. These cells may play a definite role in the control of implantation *in vivo*.

There are a number of other obvious differences between implantation *in utero* and on monolayers in culture: implantation *in vivo* is three-dimensional, while in our culture system there are only two dimensions; the blastocyst loses its spherical shape, and trophoblast cells lose their continuity about the embryo; the ICM does not develop well in the presence of uterine monolayers; finally, decidualization does not take place in culture. Nevertheless, our system may prove to be a model one for studying cell mobility and invasiveness. The acquisition of invasive properties by trophoblast cells and their subsequent loss of this ability (but not of viability) after a number of days in culture can be conveniently monitored at the gross morphological, histological, and biochemical levels.

III. BIOCHEMISTRY OF TROPHOBLAST DIFFERENTIATION *IN VIVO* AND *IN VITRO*

A. Previous Studies

We have mentioned above that blastocysts are capable of developing *in vitro* such that their morphological properties are similar to those observed *in vivo*. Superficially, then, it would appear as though maternal factors are not necessary for the decision-making process of the early embryo during its development. On the other hand, morphological characteristics are not necessarily the best or most faithful indicators of the developmental stage of a cell. We have therefore endeavored to characterize early embryonic cell types by their specific biochemical properties so that we could better assess the progress of differentiation *in vitro*. We began these studies by analyzing midgestation trophoblast, yolk sac, and embryo proper. Most of our studies have been aimed at describing markers in trophoblast and yolk sac rather than in the embryo, since the former have a more homogeneous cell population and develop best under our culture conditions. In some cases, e.g., DNA polymerases (Sherman and Kang, 1973), no qualitative differences were observed between the three fractions. In other instances, e.g., esterases (Sherman, 1972a, 1974b) and mucopolysaccharides (Shapiro and Sherman, 1974), qualitative differences between embryo, yolk sac, and trophoblast could be established. In still other studies, e.g., N-acetyl-βD-hexosaminidase (Bell and Sherman, 1973), quantitative differences were so large that the enzyme could be used as a marker of differentiation. These studies have been reviewed elsewhere (Sherman, 1974a); a list of markers appears in Table 4.

Having assembled a group of useful biochemical markers, we next ascertained the earliest time in development at which each was detectable

TABLE 4

Markers of Trophoblast and Yolk Sac Differentiation in Ectopic Sites and In Vitro

Cell type	Marker	Time of appearance *in utero* (gestation age)	Detected in ectopic sites	Detected in blastocyst cultures	Reference
Trophoblast	Polyploidy	5	+	+	Barlow and Sherman, 1972
	Alkaline phosphatase	5–8	+	+[a]	Sherman, 1972a
	Esterase A	5–8	+	+[a]	Sherman, 1972b, 1974b
	Esterase F uptake	5–8	+	+[a]	Sherman and Chew, 1972; Sherman, 1974b
	Δ^5,3β-Hydroxy-steroid dehy-drogenase	9	+	+[a]	Chew and Sherman, 1973, 1974
Yolk sac	Esterase D	9–11		+	Sherman, 1974b
	Esterase G	10–11		+	Sherman, 1974b
	N-Acetyl-β,D-hexosaminidase	9		+	Bell and Sherman, 1973

[a]Present in very small amounts.

(Table 4). In a number of cases, activity was observed in eighth day tissues, but not in fourth day blastocysts. The time of appearance could not be established more closely because adequate amounts of tissue could not be obtained for assay between the fifth and seventh days of pregnancy. We then determined whether these properties could be detected when blastocysts developed in the absence of the uterine environment, either in ectopic sites (under the kidney or testis capsule of an adult male), in the case of trophoblast, or in culture, for trophoblast and yolk sac markers. Trophoblast markers were always expressed in ectopic sites and, in all instances but one (see below), expression of the marker even agreed quantitatively with that *in vivo*. When blastocysts developed in culture, the differentiated properties of trophoblast and yolk sac were all expressed and often appeared at the same time *in vitro* as *in vivo*. The properties were restricted either to the trophoblast-containing monolayers or to the yolk sac vesicles, just as expected. Characteristically, trophoblast markers were only present at very low levels in culture.

B. *Expression of Trophoblast Δ^5,3β-Hydroxysteroid Dehydrogenase Activity In Vivo and In Vitro*

While a progesterone-primed uterus is necessary for blastocyst implantation in the mouse, progesterone is also required after implantation for the maintenance of pregnancy throughout most of the gestation period (Robson, 1938; Hall and Newton, 1947). Although the ovaries are a well recognized source of this hormone, histochemical studies suggested that rodent trophoblast also possesses Δ^5,3β-hydroxysteroid dehydrogenase (3β-HSD), the enzyme responsible for conversion of pregnenolone to progesterone (Deane *et al.*, 1962; Ferguson and Christie, 1967; Botte *et al.*, 1968). Utilizing a sensitive radioimmunoassay procedure for detecting progesterone generated by the enzyme, we have confirmed that mouse trophoblast homogenates possess 3β-HSD activity (Chew and Sherman, 1973, 1974). Activity was first detected in trophoblast homogenates on the ninth day of gestation, peaked on the eleventh day, and fell on the twelfth (Fig. 12). Recent studies suggest that the enzyme appears in cultured mouse blastocysts as early as the seventh day of gestation (Sherman, 1975). However, we have been unable to detect 3β-HSD in preimplantation mouse blastocysts, contrary to the report of Dickmann and Dey (1974) that rat blastocysts possess 3β-HSD activity.

The production of progesterone by the gonads appears to be mediated by gonadotropin or estrogen in various mammals (see Savard, 1973); it has been reported that luteinizing hormone (LH) increases specific 3β-HSD activity in the mouse testis (Hafiez *et al.*, 1971) and the rabbit ovary (Rubin *et al.*, 1965). We sought to determine whether the production of progesterone by trophoblast, as well as the levels of 3β-HSD, might be under the same type of

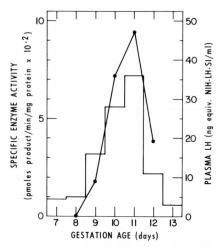

Fig. 12. Mouse trophoblast Δ^5,3β-hydroxysteroid dehydrogenase activity and maternal plasma luteinizing hormone levels during pregnancy. ●, 3β-HSD activity, taken from the data of Chew and Sherman (1974); histogram, LH content in maternal plasma, taken from the data of Murr *et al*. (1974).

control. Initially, we approached the question by comparing the levels of 3β-HSD activity in homogenates of trophoblast developing from blastocysts *in utero*, under the kidney capsule of a male host, or in culture (Fig. 13). It was not particularly surprising that the 3β-HSD activity of trophoblast in culture was only a fraction of that of uterine trophoblast, since this was also observed for other trophoblast markers (Table 4); however, it was notable that the 3β-HSD specific activity in ectopic trophoblast was also substantially lower. Other biochemical properties of trophoblast appear to be equally well expressed whether the tissue is obtained from the uterus or from ectopic sites (Sherman, 1974a). The results of the experiment decribed in Fig. 13 were therefore not inconsistent with the proposal that the levels of trophoblast 3β-HSD might be stimulated by LH or some other maternal factor.

Murr *et al*. (1974) have measured the plasma levels of LH, follicle stimulating hormone, and prolactin during pregnancy in the mouse. Comparison of 3β-HSD activity with maternal plasma gonadotropin levels showed an excellent correlation of plasma LH with trophoblast specific enzyme activity between the eighth and twelfth days of pregnancy (Fig. 12). If LH were involved in the regulation of trophoblast 3β-HSD levels, it might be expected that the temporal pattern of enzyme activity *in vitro* would be quite different from that *in vivo*, since the LH levels in the medium would be restricted to that contained in the added serum. To test this, trophoblast layers were dissected out from pregnant females on the eighth through twelfth days of gestation. The cells were then trypsinized and placed in culture

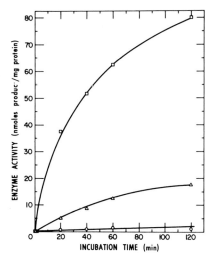

Fig. 13. Δ^5,3β-hydroxysteroid dehydrogenase activity in mouse trophoblast cells from normal pregnancy, ectopic growths and cultured blastocysts. Crude homogenates were made from trophoblast cells dissected from uteri on the eleventh day of gestation (□), from growths developing from fourth day blastocysts cultured under kidney capsules for seven days (Δ), or from fourth day blastocysts cultured in supplemented NCTC-109 medium for seven days (O). 3β-HSD activity in the homogenates was determined by measuring the production of progesterone from pregnenolone by radioimmunoassay (Chew and Sherman, 1974).

overnight. Thereafter, the cultures were incubated for 24 hour intervals with pregnenolone. Since most of the progesterone formed soon appears outside the cells (Chew and Sherman, 1974), formation of product was monitored by collecting the medium after each 24 hour incubation period and assaying it for progesterone content by radioimmunoassay. These cultures contained a mixture of giant polyploid cells and diploid ectoplacental cone cells in different proportions. Because enzyme activity resides mainly in the polyploid cell fraction, both in uterine trophoblast (Deane *et al.*, 1962; Ferguson and Christie, 1967; Botte *et al.*, 1968; Chew and Sherman, 1973, 1974) and in trophoblast cultures (Salomon and Sherman, unpublished results), only the relative changes in enzyme activity from day to day were compared between cultures.

Cultures of eighth day trophoblast produced optimal amounts of progesterone between the eleventh and twelfth days of culture; thereafter, enzyme activity fell rapidly (Fig. 14). Similar results were observed with ninth and tenth day cultures. Eleventh and twelfth day trophoblast produced very small amounts of progesterone and activity fell sharply after each day in culture. In other words, despite the gestation age at which trophoblast was removed from the uterus, progesterone production peaked between the

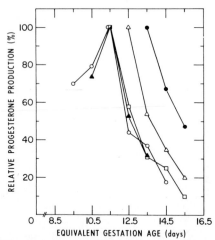

Fig. 14. Production of progesterone by mouse trophoblast cells in culture. Trophoblast cultures were generated from tissue dissected from the uterus on the 8th (O), 9th
(▲), 10th (□), 11th (△) and 12th (●) days of pregnancy. Half days are used in the
abscissa to indicate the relative progesterone production over the 24 hr period studied,
e.g., the activities for 10.5 days represents the progesterone production between the 10th
and 11th equivalent gestation day.

eleventh and twelfth day; if, on the other hand, the trophoblast cells were
dissected out after optimal enzyme levels had been reached *in utero*, activity
continued to fall in culture. The characteristic *in vivo* pattern of 3β-HSD
activity therefore appears to be reproduced *in vitro* outside the influence of
changing gonadotropic hormone levels. These results are preliminary in the
sense that we cannot yet be certain that we are not just monitoring the
programmed death of giant trophoblast cells in culture, or that trophoblast
cultures of later gestation age are capable of producing large amounts of
progesterone, but also have a newly acquired ability to further metabolize the
steroid. (Trophoblast cultures from earlier gestation ages do not appear to be
capable of further metabolizing the progesterone formed; in ninth day
trophoblast cultures at the peak of their activity, for example, more than 90%
of the added pregnenolone is detected as progesterone after the 24 hour
incubation period). In support of the data in Fig. 14 it will be shown
elsewhere that blastocyst cultures containing only trophoblast cells possess an
identical 3β-HSD profile, i.e., enzyme activity rises to a peak between the
eleventh and twelfth days after fertilization and subsequently falls (Sherman,
1975).

The levels of gonadotropic hormones in rat plasma during pregnancy have
also been determined (Linkie and Niswender, 1974; Marushige *et al.*, 1973).
Although the patterns in the two studies do not agree with each other, in

both cases the rat plasma LH levels are about one order of magnitude lower than those in the mouse. We have also determined levels of trophoblast 3β-HSD in rat trophoblast throughout most of the gestation period (Marcal *et al.*, 1975). This is possible in the rat but not in the mouse because trophoblast and decidua in the latter fuse so tightly after the twelfth day of gestation that they cannot be separated. The bulk of the placental 3β-HSD activity in fact resides with the trophoblast and not the decidua (Fig. 15). In the rat, the specific enzyme activity is highest in the trophoblast on the thirteenth day of gestation (Marcal *et al.*, 1975) and this peak of activity does not correlate well with either of the two plasma LH profiles given for the rat (Linkie and Niswender, 1972; Marushige *et al.*, 1973). We therefore considered the possibility that some other maternal factor might control 3β-HSD both in the ovary and trophoblast. Consequently, we have compared enzyme patterns in the two tissues between the ninth and twenty-first days of gestation. The highest specific 3β-HSD activity during this time in the ovary occurs on the nineteenth day, not in good agreement with that of trophoblast. Since the mass of the ovary is relatively constant after the fifteenth day of gestation, whereas that of trophoblast rises throughout pregnancy (Marcel *et al.*, 1975), we also compared the *total* enzyme activities in trophoblast and ovary corrected in each case for the number of conceptuses (Fig. 15). It can be seen

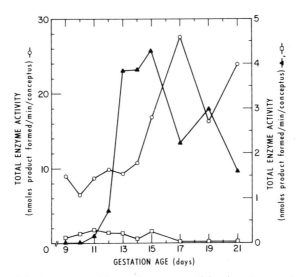

Fig. 15. Δ⁵,3β-hydroxysteroid dehydrogenase activity in rat ovary trophoblast and decidua homogenates through pregnancy. Crude homogenates were assayed for 3β-HSD activity as described in Marcal *et al.* (1975). The activity is expressed as the total enzyme capacity of the entire homogenate divided by the number of conceptuses. (*V*), ovary; (▲), trophoblast; (□) decidua.

that the activities in trophoblast and ovary do not follow the same pattern, suggesting that 3β-HSD levels in the two tissues are not coordinately controlled.

C. Significance

Our studies with 3β-HSD and other trophoblast markers demonstrate clearly that the levels of enzyme activity are much higher in uterine trophoblast preparations than they are in trophoblast derived from blastocyst cultures. The reason for this difference may be a trivial one, i.e., our culture conditions for trophoblast development may be suboptimal. However, at the morphological level, trophoblast cells appear to be well-developed, and strongly resemble those *in utero*. Furthermore, trophoblast cells characteristically survive longer under our culture conditions than they do *in utero* (Sherman, 1974c). Contact with tissue or, alternatively, some serum factors not present in fetal calf serum might be essential for the complete realization of trophoblast differentiation at the molecular level. Both possibilities would be consistent with our observations with ectopic implants, wherein most trophoblast markers are fully expressed; biochemical studies on trophoblast during interaction with uterine monolayers might distinguish between the two alternatives.

With respect specifically to trophoblast 3β-HSD activity, we are faced with a number of new and interesting observations which are, however, not easy to reconcile with each other. For example, although the timing of the patterns of trophoblast 3β-HSD activity *in utero* and in culture are strikingly similar, the enzyme activity is much lower in cultured than in uterine trophoblast; it would therefore appear as though the enzyme is being correctly programmed *in vitro* in one sense (temporally) but not in another (quantitatively). This could be explained if the expression of the 3β-HSD gene were activated at the same time *in vitro* as *in vivo* and if activation were programmed to last for a relatively short time in both cases (i.e., until the eleventh day of gestation). After that time, the observed drop in activity would reflect the rate of decay of the 3βHSD messenger RNA and of the enzyme itself. Consistent with this proposal is the observation that specific enzyme activity falls sharply and at about the same rate from peak levels, both *in vivo* (Fig. 12) and *in vitro* (Fig. 14); in all cases, specific enzyme activity has decreased by about 50% on the day following the peak activity day. According to this hypothesis, then, the failure of cultured trophoblast to produce levels of 3β-HSD equivalent to those observed *in utero* would not affect the temporal pattern of enzyme activity. The reduced levels of 3β-HSD activity *in vitro* could be due either to the production of less messenger RNA or to a less effective translation of the message.

Whether or not the above hypothesis is correct, the mechanism of activation of the 3β-HSD gene remains unexplained. Because of the correlation between plasma LH levels and trophoblast 3β-HSD activity in the mouse, LH might be considered a candidate as an activator of the 3β-HSD gene. On the other hand, the lack of relationship between plasma LH and trophoblast 3β-HSD levels in the rat weakens the argument. Furthermore, to date, we have been unable to demonstrate any effect of LH on 3β-HSD activity in cultured trophoblast cells (Salomon and Sherman, unpublished results). For the time being, then, we assume that the similarity between the patterns of plasma LH and trophoblast 3β-HSD levels is coincidental.

The temporal patterns of 3β-HSD activity in the trophoblast and ovary in the rat do not coincide (Fig. 15). Consequently if there is a maternal factor controlling 3β-HSD in trophoblast, it may not be controlling the ovary enzyme in the same way. This could be explained if the enzyme in the two tissues were not identical, and such a possibility is being investigated.

Finally, it must be recognized that while our culture conditions interrupt fetal-maternal interactions, they also disrupt fetal-fetal interactions in the sense that ICM derivatives do not develop normally. It may therefore be that the absence of the maternal environment is responsible in large part for the reduced expression of the biochemical markers of trophoblast cells, but the failure of ICM to develop fully may also be important. The role of early embryonic cell interactions in development will be considered elsewhere (Sherman, 1975).

IV. CONCLUSION

With recent improvements in blastocyst culture conditions it has become apparent that the early mouse embryo can at least initiate processes of differentiation in the absence of direct maternal influence. On the other hand, the results presented above clearly illustrate that the mother can regulate the progress of embryonic development *in utero* in ways that cannot yet be reproduced in culture. Furthermore, as we have shown, while early embryonic cells in culture may bear a striking resemblance to their counterparts *in vivo*, the quantitative aspects of some biochemical properties may still be under extrinsic control.

ACKNOWLEDGMENT

We wish to thank Ms. Nancy J. Chew for technical assistance.

REFERENCES

Abraham, G. E., Swerdloff, R. S., Tulchinsky, D., and Odell, W. D. (1971). Radio-immunoassay of plasma progesterone. *J. Clin. Endocrin.* **32,** 619–624.

Anderson, J. W. (1959). The placental barrier to gamma-globulins in the rat. *Amer. J. Anat.* **104,** 403–430.

Barlow, P. W., and Sherman, M. I. (1972). The biochemistry of differentiation of mouse trophoblast: studies on polyploidy. *J. Embryol. Exp. Morphol.* **27,** 447-465.

Beier, H. M. (1974). Oviducal and uterine fluids. *J. Reprod. Fertil.* **37,** 221–237.

Bell, K. E., and Sherman, M. I. (1973). Enzyme markers of mouse yolk sac differentia-tion. *Develop. Biol.* **33**, 38–47.

Billington, W. D. (1971). Biology of the trophoblast. *Advan. Reprod. Physiol.* **5**, 27–66.

Billington, D. W., Graham, C. F., and McLaren, A. (1968). Extra-uterine development of mouse blastocysts cultured *in vitro* from early cleavage stages. *J. Embryol. Exp. Morphol.* **20**, 391–400.

Botte, V., Tramontana, S., and Chieffi, G. (1968). Histochemical distribution of some hydroxysteroid dehydrogenases in the placenta, foetal membranes and uterine mucosa of the mouse. *J. Endocrin.* **40**, 189–194.

Bowman, P., and McLaren, A. (1970). Cleavage rate of mouse ova *in vivo* and *in vitro. J. Embryol. Exp. Morphol.* **24**, 203–207.

Brambell, F. W. R. (1958). The passive immunity of the young mammal. *Biol. Rev.* **33**, 488–531.

Brambell, F. W. R. (1970). *In*: "The Transmission of Passive Immunity from Mother to Young" North Holland Publishing Co., London.

Brinster, R. L. (1969). *In Vitro* cultivation of mammalian ova. *Advan. Biosci.* **4**, 199–233.

Brunschweig, A. E. (1927). Notes on experiments in placental permeability. *Anat. Rec.* **34**, 237–244.

Bryson, D. L. (1964). Development of mouse eggs in diffusion chambers. *Science* **144**, 1351–1353.

Bryson, D. L. (1965). Discussion after paper by Cole and Paul. *In* "Preimplantation Stages of Pregnancy" (G. E. W. Wolstenholme and M. O'Connor, eds.), pp. 113–118. Little, Brown, Boston, Massachusetts.

Chambon, Y. (1949). Realization de retard de l'implantation par les faibles doses de progesterone chez la ratte. *C. R. Soc. Biol.* **143**, 756–758.

Chew, N. J., and Sherman, M. I. (1973). $\Delta^5,3\beta$-Hydroxysteroid dehydrogenase activity in mouse giant trophoblast cells *in vivo* and *in vitro. Biol. Reprod.* **9**, 79.

Chew, N. J., and Sherman, M. I. (1974). Biochemistry of differentiation of mouse trophoblast: $\Delta^5,3\beta$-hydroxysteroid dehydrogenase. Submitted for publication.

Cole, R. J. (1967). Cinemicrographic observations on the trophoblast and zona pellucida of the mouse blastocyst. *J. Embryol. Exp. Morphol.* **17**, 481–490.

Cole, R. J., and Paul, J. (1965). Properties of cultured preimplantation mouse and rabbit embryos, and cell strains derived from them. *In* "Preimplantation Stages of Pregnancy" (G. E. W. Wolstenholme and M. O'Connor, eds.), pp. 82–112. Little, Brown, Boston, Massachusetts.

Daniel, J. C., Jr. (1964). Some effects of steroids on cleavage of rabbit eggs *in vitro. Endocrin.* **75**, 706–710.

Deane, H. W., Rubin, B. L., Driks, E. C., Lobel, B. L., and Leipsner, G. (1962). Trophoblastic giant cells in placentas of rats and mice and their probable role in steroid-hormone production. *Endocrin.* **70**, 407–419.

Dickmann, Z., and Dey, S. K. (1974). Steroidogenesis in the preimplantation rat embryo and its possible influence on morula-blastocyst transformation and implantation. *J. Reprod. Fertil.* **37**, 91–93.

Dickson, A. D. (1969). Cytoplasmic changes during the trophoblastic giant cell transfor-mation of blastocysts from normal and ovariectomized mice. *J. Anat.* **105**, 371–380.

Evans, V. J., Bryant, J. C., Kerr, H. A., and Schilling, E. L. (1964). Chemically defined media for cultivation of long-term cell strains from four mammalian species. *Exp. Cell Res.* **36**, 439–474.

Ferguson, M. M., and Christie, G. A. (1967). Distribution of hydroxysteroid dehydro-genases in the placentae and foetal membranes of various mammals. *J. Endocrin.* **38**, 291–306.

Glass, L. E. (1970). Transmission of maternal proteins into oocytes. *Advan. Biosci.* **6,** 29–58.

Glenister, T. W. (1961). Observations on the behaviour in organ culture of rabbit trophoblast from implanting blastocysts and early placentae. *J. Anat.* **95,** 474–484.

Glenister, T. W. (1965). The Behaviour of trophoblast when blastocysts effect nidation in culture. *In* "The Early Conceptus, Normal and Abnormal" (W. W. Park, ed.), pp. 24–26. University of St. Andrews Press, Edinburgh.

Graham, C. F. (1971). Virus assisted fusion of embryonic cells. *In* "Third Karolinska Symposium on Research Methods in Reproductive Biology" (E. Diczfalusy, ed.), pp. 154–167. Bogtrykkeriet Forum, Copenhagen.

Grant, P. S. (1973). The effect of progesterone and oestradiol on blastocysts cultured within the lumina of immature mouse uteri. *J. Embryol. Exp. Morphol.* **29,** 617–638.

Gwatkin, R. B. L. (1966a). Defined media and development of mammalian eggs *in vitro.* *Ann. N. Y. Acad. Sci.* **137,** 79–90.

Gwatkin, R. B. L. (1966b). Amino acid requirements for attachment and outgrowth of the mouse blastocyst *in vitro.* *J. Cell Physiol.* **68,** 335–344.

Gwatkin, R. B. L. (1969). Nutritional requirements for post-blastocyst development in the mouse. *Int. J. Fertil.* **14,** 101–105.

Hafiez, A. A., Philpott, J. E., and Bartke, A. (1971). The role of prolactin in the regulation of testicular function: the effect of prolactin and luteinizing hormone on 3β-hydroxysteroid dehydrogenase activity in the testes of mice and rats. *J. Endocrinol.* **50,** 619–623.

Hall, K., and Newton, W. H. (1947). The effect of oestrone and relaxin on the X-ray appearance of the pelvis of the mouse. *J. Physiol.* **106,** 18–27.

Hsu, Y.–C. (1971). Post-blastocyst differentiation *in vitro.* *Nature* **231,** 100–102.

Hsu, Y.–C. (1973). Differentiation *in vitro* of mouse embryos to the stage of early somite. *Develop. Biol.* **33,** 403–411.

Hsu, Y.–C., Baskar, J., Stevens, L. C. and Rash, J. E. (1974). Development *in vitro* of mouse embryos from the two-cell stage to the early somite stage. *J. Embryol. Exp. Morphol.* **31,** 235–245.

Jenkinson, E. J., and Wilson, I. B. (1973). *In vitro* studies on the control of trophoblast outgrowth in the mouse. *J. Embryol. Exp. Morphol.* **30,** 21–30.

Kirby, D. R. S. (1962). Ability of trophoblast to destroy cancer tissue. *Nature* **194,** 696–697.

Kirby, D. R. S. (1965). The "invasiveness" of the trophoblast. *In* "The Early Conceptus, Normal and Abnormal" (W. W. Park, ed.), pp. 68–74. University of St. Andrews Press, Edinburgh.

Kirby, D. R. S., and Cowell, T. P. (1968). Trophoblast-host interactions. *In* "Epithelial-Mesenchymal Interactions" (R. Fleischmajer and R. E. Billingham, eds.), pp. 64–77. Williams and Wilkins, Baltimore, Maryland.

Kirkpatrick, J. F. (1971). Differential sensitivity of preimplantation mouse embryos *in vitro* to oestradiol and progesterone. *J. Reprod. Fertil.* **27,** 283–285.

Koch, C., Boesman, M., and Gitlin, D. (1967). Maternofoetal transfer of γG immunoglobulins. *Nature* **216,** 1116.

Linkie, D. M., and Niswender, G. D. (1972). Serum levels of prolactin, luteinizing hormone, and follicle stimulating hormone during pregnancy in the rat. *Endocrin.* **90,** 632–637.

McLaren, A. (1970a). The fate of the zona pellucida in mice. *J. Embryol. Exp. Morphol.* **23,** 1–19.

McLaren, A. (1970b). Early embryo-endometrial relationships. *In* "Ovo-Implantation. Human Gonadotropins and Prolactin" (P. O. Hubinont, F. Leroy, C. Robyn and P. Leleux, eds.), pp. 18–37. Karger, Basel.

McLaren, A. (1971). Blastocysts in the mouse uterus: the effect of ovariectomy, progesterone and oestrogen. *J. Endocrin.* **50**, 515–526.

McLaren, A. (1973). Blastocyst activation. *In* "The Regulation of Mammalian Reproduction" (S. J. Segal, R. Crozier, P. A. Corfman, and P. G. Condliffe, eds.), pp. 321–334. Thomas, Springfield, Illinois.

McQuilkin, W. T., Evans, V. J., and Earle, W. R. (1957). The adaptation of additional lines of NCTC clone 929 (strain L) cells to chemically defined protein-free medium. *J. Nat. Cancer Inst.* **19**, 885–907.

Marcal, J. M., Chew, N. J., Salomon, D. S., and Sherman, M. I. (1975). Δ^5,3β-hydroxysteroid dehydrogenase activities in rat ovary and trophoblast during pregnancy. In preparation.

Marushige, W. K., Pepe, G. J., and Rothchild, I. (1973). Serum luteinizing hormone, prolactin and progesterone levels during pregnancy in the rat. *Endocrin.* **92**, 1527–1530.

Menke, T. M., and McLaren, A. (1970). Mouse blastocysts grown *in vivo* and *in vitro*: carbon dioxide production and trophoblast outgrowth. *J. Reprod. Fertil.* **23**, 117–127.

Mintz, B. (1962). Experimental study of the developing mammalian egg: removal of the zona pellucida. *Science* **138**, 594–595.

Mintz, B. (1964). Formation of genetically mosaic embryos and early development of lethal (t^{12}/t^{12})-normal mosaic. *J. Exp. Zool.* **157**, 273–292.

Mintz, B. (1971). Control of embryo implantation and survival. *Advan. Biosci.* **6**, 317–340.

Murr, S. M., Bradford, G. E., and Geschwind, I. I. (1974). Plasma luteinizing hormone, follicle-stimulating hormone and prolactin during pregnancy in the mouse. *Endocrin.* **94**, 112–116.

New, D. A. T. (1971). Culture of fetuses *in vitro*. *Advan. Biosci.* **6**, 367–378.

New, D. A. T., and Daniel, J. C., Jr. (1969). Cultivation of rat embryos explanted at 7.5 to 8.5 days of gestation. *Nature* **223**, 515–516.

Nuttig, E. F., and Meyer, R. K. (1963). Implantation delay, nidation and embryonal survival in rats treated with ovarian hormones. *In* "Delayed Implantation" (A. C. Enders, ed.), pp. 233–251. University of Chicago Press, Chicago, Illinois.

Pinsker, M. C., Sacco, A. G., and Mintz, B. (1974). Implantation-associated proteinase in mouse uterine fluid. *Develop. Biol.* **38**, 285–290.

Psychoyos, A. (1973). Hormonal control of ovoimplantation. *Vitamins and Hormones* **31**, 201–256.

Psychoyos, A., and Alloiteau, J. J. (1962). Castration precoce et nidation de l'oeuf chez la ratte. *C. R. Soc. Biol.* **156**, 46–54.

Robson, J. J. (1938). Quantitative data of the inhibition of oestrus by testosterone, progesterone and certain other compounds. *J. Physiol.* **92**, 371–382.

Rubin, B. L., Hilliard, J., Hayward, J. N., and Deane, H. W. (1965). Acute effects of gonadotrophic hormones on rat and rabbit ovarian Δ^5-3β-hydroxysteroid dehydrogenase activities. *Steroids, Suppl.* **1**, 121–130.

Salomon, D. S., and Sherman, M. I. (1974). Implantation and invasiveness of mouse blastocysts on uterine monolayers. Submitted for publication.

Savard, K. (1973). The biochemistry of the corpus luteum. *Biol. Reprod.* **8**, 183–202.

Shapiro, S. S., and Sherman, M. I. (1974). Sulfated mucopolysaccharides of midgestation embryonic and extraembryonic tissues of the mouse. *Arch. Biochem. Biophys.* **162**, 272–280.

Sherman, M. I. (1971). Discussion following paper by Mintz. *Advances Biosci.* **6**, 341.

Sherman, M. I. (1972a). The biochemistry of differentiation of mouse trophoblast: alkaline phosphatase. *Develop. Biol.* **27**, 337–350.

Sherman, M. I. (1972b). Biochemistry of differentiation of mouse trophoblast: esterase. *Exp. Cell Res.* **75,** 449–459.

Sherman, M. I. (1974a). *In vivo* and *in vitro* differentiation during early mammalian embryogenesis. *Front. Rad. Therapy* **9,** 122–134.

Sherman, M. I. (1974b). Esterase isozymes during mouse embryonic development *in vivo* and *in vitro*. *In* "Third International Conference on Isozymes" (C. L. Markert, ed.) Academic Press, New York. In press.

Sherman, M. I. (1974c). Long term culture of cells derived from mouse blastocysts. Submitted for publication.

Sherman, M. I. (1975). The role of cell-cell interaction in the development of the early mouse embryo. *In* "The Early Development of Mammals." (M. Balls and E. A. Wild, eds.) Cambridge University Press, Cambridge, U.K. In press.

Sherman, M. I., and Barlow, P. W. (1972). Deoxyribonucleic acid content in delayed mouse blastocysts. *J. Reprod. Fertil.* **29,** 123–126.

Sherman, M. I., and Chew, N. J. (1972). Detection of maternal esterase in mouse embryonic tissues. *Proc. Nat. Acad. Sci.* U.S. **69,** 2551–2555.

Sherman, M. I., and Kang, H. S. (1973). DNA polymerases in midgestation mouse embryo, trophoblast and decidua. *Develop. Biol.* **34,** 200–210.

Smith, D. M. (1968). The effect on implantation of treating cultured mouse blastocysts with oestrogen *in vitro* and the uptake of [^{3}H] oestradiol by blastocysts. *J. Endocrin.* **41,** 17–29.

Smith, D. M., and Biggers, J. D. (1968). The oestrogen requirement for implantation and the effect of its dose on the implantation response in the mouse. *J. Endocrin.* **41,** 1–9.

Solomon, J. B. (1966). Relative growth of trophoblast and tumour cells co-implanted into isogenic mouse testes and the inhibitory action of methotrexate. *Nature* **210,** 716–718.

Spindle, A. I., and Pedersen, R. A. (1973). Hatching, attachment, and outgrowth of mouse blastocysts *in vitro*: fixed nitrogen requirements. *J. Exp. Zool.* **186,** 305–318.

Sugiwara, S., and Hafez, E. S. E. (1965). Electrophoretic patterns of proteins in the blastocoelic fluid of the rabbit following ovariectomy. *Anat. Rec.* **158,** 115–120.

Tulchinsky, D., and Abraham, G. E. (1971). Radioimmunoassay of plasma estriol. *J. Clin. Endocrin.* **33,** 775–782.

Whitten, W. K. (1957). The effect of progesterone on the development of mouse embryos *in vitro*. *J. Endocrin.* **16,** 80–85.

Whitten, W. K. (1971). Nutrient requirements for the culture of preimplantation embryos *in vitro*. *Advan. Biosci.* **6,** 129–139.

Whitten, W. K., and Biggers, J. D. (1968). Complete development *in vitro* of the preimplantation stages of the mouse in a simple chemically defined medium. *J. Reprod. Fertil.* **17,** 399–401.

Yochim, J. M. (1971). Intrauterine oxygen tension and metabolism of the endometrium during the preimplantation period. *In* "The Biology of the Blastocyst" (R. J. Blandau, ed.), pp. 363–382. University of Chicago Press, Chicago, Illinois.

A Determinant Role for Progesterone in the Development of Uterine Sensitivity to Decidualization and Ovo-Implantation

Stanley R. Glasser and James H. Clark

*Department of Cell Biology and
Center for Population Research
Baylor College of Medicine
Houston, Texas 77025*

I. Introduction 311
II. Synchrony 312
III. The Decidual Cell Response 314
IV. Uterine Sensitivity and Receptivity 319
V. DCR as a Model for the Mechanism of Progesterone Action 320
VI. Uterine Receptivity to Ovo-implantation 330
VII. The Nature of Pre-Nidatory Estrogen 332
VIII. The Role of Steroid Hormone Receptors in the Development
of Uterine Sensitivity and Receptivity 336
References 342

I. INTRODUCTION

Prior to 1950 the relationship between the embryo/fetus and its maternal host was described as either maternally or embryonically directed. In that year evidence provided by Fawcett (1950) forced the consensus that cooperativity exists between the fertilized ovum and the uterus. A quarter century later we still find ourselves addressing the question as to how this cooperativity is established.

Blandau (1949) was the first to distinguish the existence of pre-implantation contact between the rat blastocyst and the endometrium. It was proposed that decidualization, that structural and functional differentiation of the endometrial stromal cell prerequisite to implantation, was induced during this contact. Viable blastocysts appear to have inductive properties not characteristic of pre-implantation ova (Alden and Smith, 1959) or blastocysts with non-viable blastomeres (Segal and Nelson, 1958).

The communication between zygote and the maternal host is, in fact, more sophisticated and complex. This may be derived from the data of Chang

311

(1950) who was the first to demonstrate a precise synchrony between the stage of development of the ovum, the corpus luteum, and the host uterus. This communicative interlock has been reported to be established very early after copulation. There are changes in the secretory rates and patterns of progesterone and 20α-dihydroprogesterone which follow a fertile mating that are not noted in pseudopregnancy. In as yet unconfirmed papers the corpus luteum is thought to "recognize" the pre-implantation blastocyst, perhaps through the release, by the blastocyst, of an HCG-like material which exerts a luteotrophic influence (Haour and Saxena, 1974; Fuchs et al., 1974).

In extending the work of Chang (1950), Noyes and his associates (1963) provided a body of data concerning early embryonal-maternal relationships that despite its relevance remains neglected. Thus, although these studies have fixed the concept of synchronization as a standard in this area of research very few investigators have yet to enlarge the principals inherent in these experiments.

II. SYNCHRONY

The simplicity of the Noyes studies is depicted in Fig. 1. Recipient females were mated with vasectomized males so that females were pseudo-pregnant until fertilized ova were transferred from the donors at varying stages

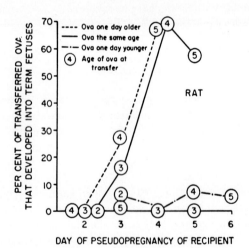

Fig. 1. The percentage of ova that survive to term following transfer, on different development days, into the uterine horns of recipient animals at different stages of pseudo-pregnancy. Included in these data are 60 recipients and 855 ova. Note the success of synchronous transfers (days 4 and 5) and asynchronous transfers in which ova were one day older than the uterus compared to other experimental groups. (From Noyes, R. W. et al., 1963)

of pregnancy into their uteri. The presence of a vaginal plug was taken as day 1 of development. It is evident that no day 2 ova and very few day 3 ova successfully survived transfer regardless of the age of the uterus. There was modest success following transfer of day 4 ova into day 3 recipients but the limits of the technique were reached on day 4 with either synchronous day 4 or asynchronous day 5 ova. Implantation in that colony occurred on day 5.

The success of transfer and implantation of ova of the same age or one day older, but not younger, into a uterus at least 4 days pseudopregnant elicited a series of questions, as to the relative developmental tolerance of the zygote compared to that of the endometrium, which remain unanswered in the practical sense. Only the rudiments of the developmental biology underlying the comparative success of synchronous and asynchronous transfers have been disclosed (Psychoyos, 1966, 1973) and then only in descriptive terms.

The conditions for interaction between ovum and endometrium, which result in successful implantation, are obviously not random. This may be counted among the costs of evolving to viviparity. These penalties also include the time dependent challenges of gamete transport and fertilization as well as the evolution of species specific protocols of cleavage and maturation which fill the interval between ovulation and entry into the uterus. Regardless of the duration of this pre-implantation interval the uterus must, in that period of time, undergo those sequential changes which lead first to the stage of uterine sensitivity precedent to uterine receptivity towards the blastocyst. The final series of synchronous interactions, culminating in the implantation of the blastocyst, must satisfy criteria on a multiplicity of organizational levels. They can only be initiated when the ovum has become a blastocyst and the uterus has achieved that summary stage of development, regulated by an ordered sense of hormone dependent events which is termed "receptive." In the natural course of events receptivity of the uterus occurs at a time coincident with blastulation.

Synchrony ensures embryonic development under optimally regulated conditions by protecting the component systems, i.e., zygote and uterus, against aberrations and alterations in substance and in time. With these safe-guards operational it is still estimated that the over-all pregnancy wastage during the period of pre-implantation and implantation in the human amounts to 30%. Because reproductive efficiency can be maintained at this level the uterus came to be considered a "favored or privileged" site for mammalian development (Glasser, 1972).

However, there seems to be a paradox because although studies of synchrony certify to the relative interdependence of the ovum and the uterus it is now well established that the ovum can differentiate *in vitro* and can grow and implant at ectopic, *in vivo* sites. Furthermore, the uterine responses which characterize sensitivity and receptivity to the developing embryo, i.e.

decidual cell responses, can be initiated independently of the blastocyst. That the ovum and uterus retain characteristics of developmental independence, while establishing the critical parameters of interdependence, does not diminish but accents the importance of the integrity of a functional reproductive system. Even if the uterus, in terms of these definitions, no longer qualifies as a "privileged" site in the immunological sense (Billingham, 1964), the processes and structures which regulate synchronous development of the blastocyst and uterus guarantee an order and efficiency to the mammalian reproductive process that promises evolutionary advantage.

III. THE DECIDUAL CELL RESPONSE

By virtue of their relative inaccessibility and the limits of the technical culture of the time, maternal-embryonal relationships were at first more easily defined by recording changes in the biology of the uterus. Loeb (1908 a, b) described an endometrial response (deciduoma) which could be induced by "indifferent" stimuli in a uterus conditioned by the proper endocrine environment. The morphological, temporal, and endocrine events occurring in the uterine stroma, induced by artificial means in the absence of fertilized ova, were essentially identical to those following natural uterine responses to embryonic stimuli (decidua). For these reasons, the decidual cell reaction (DCR) has acquired increasing credibility as a model for the events associated with implantation.

The DCR allows analysis of the separate roles of ovum and uterus because stimuli and information which may be contributed by the fertilized ovum can be excluded. It is thus feasible to investigate critically those factors which permit the uterus to differentiate structurally and functionally, so that it might respond to deciduogenic stimuli - including the blastocyst. Whether induced artificially (deciduoma) or naturally (decidua), the validity of decidualization as a model for the study of these unique uterine events rests on the corollary that the artificial stimuli which evoke deciduoma and the natural stimuli provided by the developing ovum involve a final common pathway. A new "organ," with specific functions, is formed within 2 to 3 days. The DCR may serve as a model for the study of the cell differentiation and the processes of intercellular communication. Most importantly, it is possible that the DCR may serve as an analog for those events involved in the implantation of the blastocyst.

Three determinants of this uterine response have attracted the major share of investigative attention: the endocrinology and chronology of uterine sensitivity to DCR and, to a lesser extent, studies of the nature of various inducing agents (DeFeo, 1963a, b; Yochim and DeFeo, 1963). The decidual cell response can be elicited in cycling rats, pregnant, pseudopregnant, and lactating rats as well as in ovariectomized rats maintained in certain hormonal regimens since these animals all fulfill the basic requirements first described by Loeb (1908a, b).

The state of uterine preparation for ovo-implantation has classically been assayed by testing the endometrium for a decidual response following some form of treatment. Of the various animal models available for the DCR, the ovariectomized rat, maintained on one of various hormonal regimens, proves generally to be most productive. After removal of endogenous controls, a discriminating and sensitive system is produced for qualitatively and quantitatively evaluating the various regulatory factors, including the gonadal steroids, that might influence the induction, differentiation, and growth of the DCR.

The generalized protocol used in our laboratory (Fig. 2) provides for ovariectomy of rats on the morning they present a sperm positive vaginal

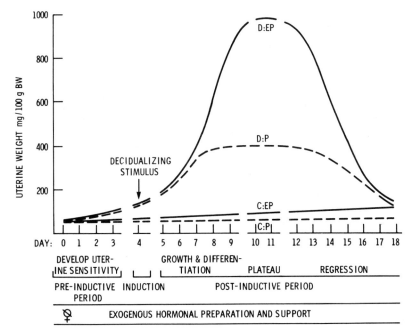

Fig. 2. The life history of a decidual cell response (DCR). Non-traumatized horn (C); traumatized horn (D); ovariectomized animal supported on 1 μg estrone and 2 mg progesterone per day (EP); ovariectomized animal supported on 2 mg progesterone per day only (P). Decidualizing stimulus (Needle scratch) given between 1030-1230 hours on day 4. Note lack of significant weight differences (D horn) when EP and P groups are compared during the first 72 hours. After growth plateaus, i.e., D:EP, day 9-11, regression proceeds in spite of continued hormonal support. (Glasser, S. R., 1975, in press)

smear. It is our custom, because it lends itself so much better to embryo-logical studies, to designate this as day "0". Protypically the castrate animal is treated daily with subcutaneous injections of 1 g estrone (E) and 2 mg of progesterone (P). In terms of producing the largest decidual mass this is an optimal ratio of the two hormones as determined on both an absolute and relative basis (Yochim and DeFeo, 1963).

The capacity of the uterus to initiate and sustain decidualization may be challenged by either physical (needle scratch, threads, etc.), chemical (any variety of solutions, media, buffers), or natural means such as provided by the blastocyst (DeFeo, 1967). As in the ova transfer experiments it can be shown that stimulation of the uterus on days, 0, 1, and 2 evokes no decidual response. Traumatization of the uterus on day 3 evokes only a marginal DCR (Fig. 9) which is dependent on the nature and magnitude of the stimulus. It is imperative that only one horn of the uterus be stimulated while the contralateral (control) horn serves as a monitor of each particular animal's internal environment.

Stimuli given on day 4 are deciduogenic. The response to such stimuli is precipitate growth (Fig. 2; days 4-9). The initial rate of growth of the stimulated horn is approximately 75-100% per day (DeFeo, 1967; Glasser, 1963, 1972). There are net increases in DNA (von Berswordt-Wallrabe, 1969l; Heald and O'Grady, 1969; Glasser, 1972), RNA (Shelesnyak and Tic, 1963a, b; Glasser, 1965a, 1972, 1975; Glasser and O'Malley, 1970; O'Grady et al., 1970), protein (Glasser, 1965b, 1972), glycogen (Glasser, 1957; Cecil et al., 1962), water (Glasser, 1957, 1963; Bitman et al., 1960), and those enzymes generally associated with growth phenomena (Lobel et al., 1965). Smooth endoplasmic reticulum increases in amount and complexity as stromal cells undergo hypertrophy and hyperplasia (Martin and Finn, 1969; Finn and Martin, 1970; Glassi, 1968; Tachi et al., 1972). Endoreduplication is another correlate of these changes (Ansell et al, 1974). A further increase in the rate of growth (125+%) occurs after the first 72 hours (Fig. 2). By day 9 (120 hours after the deciduogenic stimulus) the decidualized horn may exceed the weight of the contralateral control horn by more than 700%. The DCR evoked by natural stimuli is more modest and localized. The growth of the deciduoma peaks at day 9 in spite of continued hormonal support. Within 48-72 hours after attaining this plateau, although hormonal support is maintained, the DCR begins to regress steadily.

Ovariectomized rats maintained on 2 mg/day of progesterone alone will demonstrate a DCR growth curve that is qualitatively similar to that produced in uteri conditioned and maintained with both estrogen and progesterone (Fig. 2). Induction of decidualization is a progesterone determined response (Glasser, 1972) and the initial rate of DCR growth elicited under these conditions (0-72 hours, P only) is identical to that recorded with both hormones. Quantitative differences are not observed until 72 hours have elapsed since DCR was initiated (Fig. 4). At that time absolute and relative differences in the rates of DCR growth and regression may be noted between animals maintained in P only as compared with rats receiving both E and P.

Of particular importance, in any hormonal regimen, is the notation that the DCR may be elicited only during a sharply limited period of time, even on day 4. There are, of course, factors which modulate the response, e.g., strain of animal, specific hormone or hormones, nature of the inductive stimulus, etc. (DeFeo, 1967). However, irrespective of the animal model, sensitivity is attained after specific preparation, on only one day (day 4 in the rat) becoming maximal some time during that day. Unless there are rather specific alterations in the hormonal milieu regression of the sensitive state will begin soon thereafter. By day 5, in the rat, DCR can no longer be induced.

On the basis of studies in his laboratory, Shelesnyak (1960, 1962) assigned the major role in the initiation of decidualization to estrogen. Estrogen is given determinant influence at two separate time periods, e.g., during the proestrus preceeding the pregnancy of pseudopregnancy and again on day 3 preparatory to the inductive stimulus.

Estrogen does indeed play a role at these two time periods but the determinant role in terms of sensitivity to deciduogenic stimuli resides with progesterone (Glasser, 1972; Psychoyos, 1973). As we recognize it now, the role of estrogen proves to be delegated to augment and modulate progesterone induced changes.

A rational role has been suggested for estrogen as a stimulus of stromal mitosis during the pre-implantation period (Tachi et al., 1972; Finn, 1971). The earliest morphological correlates were provided by Krehbiel (1937) who defined the ctyologic basis for the hypertrophy and hyperplasia which follows induction of DCR. During early pregnancy or pseudopregnancy the proliferative pattern of the uterine cells of rat and mouse may be characterized by an apparent shift in mitotic activity from the luminal and glandular components to the stromal epithelium (Tachi et al., 1972). The interpretive analysis of these changes has been cast strongly in terms of the mitogenic action of estrogen, in such a way that the functional role of progesterone is difficult to define.

Estrogen provokes a marked increase in the mitotic index of the endometrium of ovariectomized-adrenalectomized rats which is more marked in the luminal rather than the glandular epithelium and absent in the stroma (Tachi et al., 1972). Pre-treatment of these animals with progesterone nominally suppresses the mitogenic action of estradiol on the luminal and glandular epithelium and subsequently redirects its action to the stroma.

The same general pattern may be seen in pregnant and pseudopregnant rats. Mitotic figures may be observed in luminal and glandular epithelium until midnight of day 2. It is of interest to note that estradiol titers are relatively low following ovulation until midnight of day 1 (Figs. 11, 12). By noon of day 3, in the face of rising plasma progesterone concentrations, mitotic activity has ceased in surface epithelia. A marked increase in stromal mitotic activity may be recorded by midnight of day 3 at which time the blastocysts are free in the uterine lumen. Mitoses are confined to subepithelial regions. Cell division ceases in pseudopregnant animals by day 7 but continues in pregnant rats with the involvement of deeper antimesometrial layers (Tachi *et al.*, 1972).

The mitogenic action of estradiol is so striking that the function of progesterone relative to these changes critical to ovo-implantation has not been fully appreciated. Progesterone, per se, has mitogenic activity in that it can condition and support stromal hypertrophy and hyperplasia in the absence of estrogen (Glasser, 1972). There is no question that estrogen is mitogenic with respect to the luminal and glandular epithelium but it is ineffective in the stroma in the absence of progesterone (Tachi *et al.*, 1972). Since decidualization cannot proceed in the absence of progesterone it is relevant to inquire into the determinant action of progesterone that has sensitized the stroma to the pronounced mitogenic action of estrogen. It is also germane to inquire if the inhibition of mitosis in surface epithelia is an important consequence in the reproductive biology of the animal.

It should no longer be required that the action of progesterone be explained in terms of estrogen. For these reasons it is no longer an exercise in esthetics, given the uniqueness of the DCR as a model and the techniques presently available to us, to insist on more precise knowledge and definition of the specific action of each hormone involved in the growth and differentiation of uterine endometrium. Failing that information we cannot constructively analyze the problems inherent in this process which is so pivotal to the evolutionary success of viviparity.

The difficulty in resolving the role of proestrus estrogen may reside in the fact that, in spite of a significant number of correlates, the DCR is really an imperfect analog of implantation (Glasser, 1963). Three, rather than two, hormonally directed systems may be operable in the pregnant animal, e.g., maturation of the blastocyst, maturation of the uterine endometrium, and the singular actions involved in the interrelationships of these two systems. Another complexity arises from the fact that most of the discriminating analytical methods that have been brought to bear on problems of growth have yet to be applied to the study of decidualization and implantation. Thus the analysis of cell differentiation in the stroma and the differential response of varied cell populations to the hormones (Tachi *et al.*, 1972; Finn and Martin, 1970; Marcus, 1970) requires more rigorous techniques for kinetic analysis of the cell cyle (Gelfant, 1963) than have previously been used.

IV. UTERINE SENSITIVITY AND RECEPTIVITY

There is a most promising period in which to define the specific actions of estrogen and progesterone in regulating the uterine responses incident to synchronous maturation of the developing embryo. It encompasses that time frame which immediately precedes and includes that period in which the uterus is sensitive to deciduogenic stimuli and to the egg. Thus an initial productive step might be the cataloging of the differences in the uterus that exist between day 3 (poor success of ovum transfer) and day 4 (day of maximum transfer success and time of natural implantation; Fig. 1). This is the transitional period that would be involved in the development of uterine sensitivity. A similar comparison between days 4 and 5 would, under most conditions, describe the loss of sensitivity and might be equally as instructive.

In a series of comprehensive studies that generated much interest DeFeo, (1963a, b) and Yochim and DeFeo (1963) developed and described the thesis of maximal sensitivity. Maximum sensitivity was defined as that period during which a deciduogenic stimulus could elicit its maximum response. The end point of this challenge was the weight of deciduoma measured 120 hours after induction. Maximum sensitivity was measured in intact pseudopregnant rats and in ovariectomized animals maintained on estrogen and/or progesterone in various concentrations. In their own work DeFeo and Yochim defined time dependent variables that impinged on the development of sensitivity, e.g., duration of action of the inducer or deciduogenic stimulus and thus the duration of the sensitive period. It was DeFeo's contention based on his thesis of maximal response that only traumatic stimuli (needle scratch, etc.) could evoke DCR in castrate animals maintained on progesterone alone whereas nontraumatic stimuli (balanced salt solutions) required the cooperative effect of estrogen for DCR to be induced. However, very respectable DCR could be elicited with Hanks solution instilled intraluminally in uteri of castrate rats maintained on progesterone alone (Glasser, 1965b, 1972). This was a satisfactory mode because we were concerned with the threshold for induction and the biochemical composition of the DCR rather than the maximal quantitative response.

The development of maximal sensitivity, based in large measure on treating appropriate animals with the proper ratio (absolute and relative) of estrogen and progesterone has generated much data. However, the quality of the data, i.e., maximum DCR weight 5 days after induction, obviates the basic question - what is sensitivity?

What is it that occurs between days 3 and 4 that allows nonspecific stimuli to dramatically alter the biochemical and structural organization of the uterus. It was our contention that measuring the greatest possible weight increase of DCR at 120 hours post-induction was too nonspecific and far removed in time to permit any definition of causal factors. The end-point of a maximal stimulus is, in terms of embryo-maternal relationships, almost unreal since under the conditions set forth by DeFeo the uterus can be driven to produce more decidual tissue in a single experiment than an animal might produce in a year of cyclic reproduction.

The problem as we saw it in terms of identifying the cause and effect relationships of hormonal determinants was to begin a consideration of threshold sensitivity. By seeking DCR specific parameters that could be measured during the period of induction and during the earliest stages of decidualization we could begin to sketch a rudimentary definition of sensitivity. Defined in such terms it would be possible to determine which hormone(s) made possible and directed metamorphosis from the insensitive to the sensitive uterus. Data concerning the rate and magnitude of DCR was only of secondary importance.

V. DCR AS A MODEL FOR MECHANISM OF PROGESTERONE ACTION

Transient acquisition of uterine sensitivity would seem to be most efficiently regulated by the induction of a limited number of enzymes. Thus the differentiation of the subepithelial cell to a decidual cell may require hormonally induced and modulated gene activation and transcription and translation of chromosomal information. What is required is a model, such as the chick oviduct system developed by O'Malley and his associates (1969) that will define the precise biochemical mechanism of progesterone and estrogen action. We believe that the decidual cell response system to be described below can serve as such a model. The study of such a model would be directed to the description of those early changes. We might possibly note the synthesis of new information or alterations in existing genetic capacity for transcription mediated by the hormone and expressed in terms of RNA and protein synthesis of new decidual tissue.

The protocol for developing a mammalian model for the biochemical definition of hormonal induction of uterine sensitivity is depicted in Fig. 3. Bilateral ovariectomy is performed on rats during the morning that vaginal sperm are identified. This is designated day "0". Animals are then treated with either the sesame oil vehicle, 1 μg estrone, 2 mg progesterone, or a combination of both these hormones. Treatment continues throughout the course of the experiment. The decidual cell response is evoked by either physical (needle scratch) or chemical (0.1 ml Hanks solution) means in a single horn between 1000 and 1400 hours on day 4. This process of induction ends the pre-inductive period.

Deciduoma or control tissue from the contralateral unstimulated horn is taken for analysis at various time periods after induction. The process of

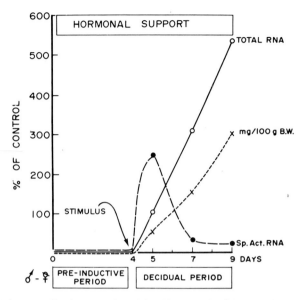

Fig. 3. The generalized protocol used by Glasser and O'Malley (1970) in their studies on the mechanism of progesterone action. These data represent the response of the traumatized uterus to 2 mg progesterone per day. They are prototypic in a qualitative sense for other active progestins with or without estrogen. It is evident from this and accompanying figures (4-8) that effective decidualization, based on gravimetric, histological and biochemical criteria, can be induced with progesterone alone. (Glasser, S. R., 1972)

decidualization is characterized by an immediate and rapid increase in the rate of RNA synthesis with an accumulation of total RNA and subsequently tissue mass. This increase in the absolute weight of the decidual horn should be expressed as a function of the change in the unstimulated horn (Fig. 4A). This is an important consideration in the evaluation of these data. The control horn is also subject to the endocrine environment imposed on the animal and will respond, in some manner, to that treatment. Thus, the endocrine milieu is different for each of the four groups described above. Measured on day 4 this reflects the character of the uterine substrate on which the deciduogenic stimulus acts; measured at any time thereafter this reflects the basic resources of the responding system. For each of the four groups in this experiment the weight and composition of the unstimulated uterine horn is significantly different from each other. For this reason, although the absolute weight increase in the animals treated with estrogen and progesterone is greater than that noted with progesterone alone, the relative weight increase, at 120 hours, is smaller (Fig. 4A). Failure to consider alterations in the unstimulated horn is a prominent cause of misinterpretation. Thus, the uterotrophic effect of

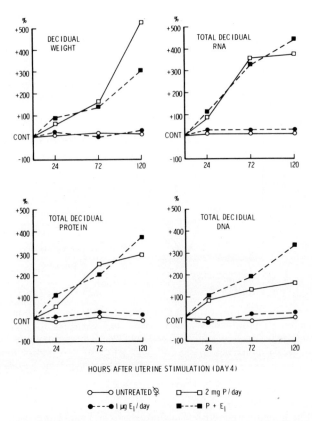

HOURS AFTER UTERINE STIMULATION (DAY 4)

○——○ UNTREATED ♀   □——□ 2 mg P / day

●---● I μg E₁ / day   ■---■ P + E₁

Fig. 4. Relative changes in some parameters of decidual growth following stimulation of ovariectomized rats maintained on either 1 mg estrone/day (E_1, ●), 2 mg progesterone/day (P, □) or a combination of these hormones (P + E, ■). One group of animals was left untreated (O). Uteri were traumatized on day 4. Decidual wet weight (upper left), total decidual RNA (upper right), total decidual protein (lower left), and total decidual DNA (lower right), all demonstrate that there are no significant differences in the decidual responses when P are compared with P + E animals during the first 72 post-inductive hours. There is a general growth response to E which occurs equally in both unstimulated and traumatized horns but no DCR occurs.

estrogen, if only the stimulated horn is assayed, could be categorized as a decidual response. However, the contralateral control horn is also stimulated by estrogen; thus no relative increase in mass is recorded. This was confirmed by examination of the histology of this tissue.

The increase in total decidual DNA (Fig. 4B) is representative of the hyperplasia and growth that occurs in this tissue. The relative increase in both progesterone treated groups is parallel through the 72 hours at which time there is a further increase in animals maintained on both estrogen and progesterone. Mitosis without cell division may also contribute to the polyploid character of mature deciduoma.

Total decidual RNA (Fig. 4B) rises immediately after stimulation. The pivotal influence of progesterone on these processes is emphasized by the fact that the role of estrogen is augmentive and is not expressed until after the 72 hours (Fig. 4B, C, D). The deciduogenic stimulus produces no relevant changes, compared with their own control horns, in animals treated with estrogen alone or left untreated. A concomitant increase in total decidual protein parallels (Fig. 4C) the changes noted in decidual RNA.

It is evident from the data obtained from this mode that all the changes that characterize the immediate post-inductive decidual response, biochemical as well as morphological, can be induced in the presence of progesterone alone. They cannot be induced experimentally in the presence of estrogen alone. In combination with progesterone the specific influence of estrogen is not demonstrable until the 72nd hour - well past the inductive period. Thus the DCR mammalian model is different from the chick oviduct model (O'Malley et al., 1969). In the reproductively competent rat the role of estrogen should not be considered permissive but rather as a modulating agent which would qualitatively and quantitatively augment progesterone initiated changes.

Though there are apparent differences in the two models, the chick oviduct system provides for a rather precise definition of the biochemical mechanism by which progesterone plays a regulatory role in growth and differentiation. The details of this system are widely published and were presented to a symposium of this Society (O'Malley et al., 1974). Thus, although all hormones do not necessarily influence target tissue protein synthesis in an identical manner we thought that for a first approximation the procedures that were so successful in the elucidation of the biochemical mechanism of hormone action in the chick oviduct could be fruitfully adapted to the study of the decidualizing rat uterus.

Hormone directed growth and differentiation implies the potential synthesis of new and perhaps unique regulatory proteins. Although the uterus, unlike the chick oviduct, does not possess marker proteins, i.e., ovalbumin and avidin, we predicted the appearance of new populations of uterine proteins. The synthesis of these products would be an indication of prior synthesis of new RNA species, including messenger RNA, in response to progesterone.

Protein synthesis was studied by labeling with amino acids during a 3 hour in vitro incubation. "Decidual" tissue from untreated rats labeled with [^{3}H] valine, was combined with deciduoma from progesterone treated rats labeled with [^{14}C] valine. Soluble protein extracts (105,000 x g) of the combined tissues were then subjected to polyacrylamide SDS gel electrophoresis. No stimulation was observed in amino acid incorporation into proteins of decidual tissue of untreated rats whereas incubation of decidual tissue from progesterone treated rats yielded a 60% increase in amino acid incorporation.

Gels were sliced, solubilized, and counted and ^{3}H-^{14}C ratios were calculated from the corrected ^{3}H counts of untreated animals and ^{14}C counts from the soluble protein patterns from progesterone animals. Gel ratio analysis of double labeled gels has great utility in these long and rather complex experiments with 4 experimental groups. This procedure serves to eliminate errors in technique and differential protein degradation. It also accents the differences in the rate of synthesis between groups and sub-groups (control vs. DCR tissue) and pinpoints the loci at which protein synthesis is being stimulated or arrested by a specific hormonal (regulatory) regimen.

A qualitative change in the pattern of synthesized proteins is evident from the variance in ^{3}H-^{14}C ratios depicted in Fig. 5. These type of data

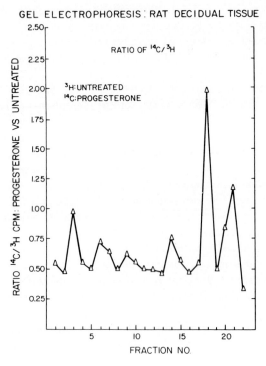

Fig. 5. ^{14}C/^{3}H ratio analysis of double label polyacrylamide SDS gels of rat uterine proteins. Traumatized uteri from untreated rats (labeled with [^{3}H] valine) and progesterone treated rats ([^{14}C] valine; ^{3}H/^{14}C = 10:1) were combined after 3 hours of incubation *in vitro*. Soluble protein extracts (105,000 x g) were applied to 7.5% gels. ^{14}C/^{3}H ratios were calculated from corrected counts.

substantiate the appearance of new protein populations as a result of progesterone action of decidual tissue which are unique to progesterone.

The changing patterns of protein synthesis suggest that progesterone may have a prior influence on the rate and character of nuclear RNA synthesis as has been described for estrogen in the chick oviduct (O'Malley et al., 1969, 1974) and the castrate rat uterus (Knowler and Smellie, 1971; Glasser et al., 1972a). To test this hypothesis each uterine horn was injected intraluminally with labeled cytidine at various intervals following DCR induction on day 4. In the data depicted in Fig. 6 we are comparing untreated and progesterone treated groups 24 hours after stimulation. Untreated animals received [^{3}H] cytidine in each horn and progesterone treated rats received [^{14}C] cytidine. After 60 minutes the decidual tissue from both groups was harvested and

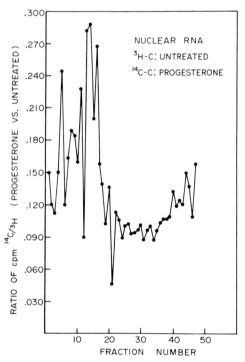

Fig. 6. Gel ratio analysis of rat uterine nuclear RNA. Untreated animals received 100 μC [^{3}H] cytidine in each horn and progesterone treated rats received 10 μC [^{14}C] cytidine 24 hours after decidirogenic stimulation. After one hour traumatized uteri from each group were combined and nuclear RNA was rapidly extracted by a hot phenol method. Double labelled purified nuclear RNA was analyzed by acrylamide agarose gel electrophoresis. Ratio analysis was derived from the corrected counts.

combined for extraction of nuclear RNA by a hot phenol method. Aliquots of the purified, extracted double labeled nuclear RNA were analyzed by acrylamide-agarose gel electrophoresis. Ratio analysis (Fig. 6) of these gels suggests that progesterone does indeed induce a qualitative change in transcription by eliciting the synthesis of different species of RNA's.

Deciduogenic stimuli evoke marked and immediate changes in total RNA. This results in a progressive shift in RNA/DNA from approximately 1.1/1.0 to 3.0/1.0 in 24-48 hours. Thus, the assay of specific RNA activity of decidual nuclear RNA within 24 hours of decidualization suggested itself as a more specific quantifiable evaluation of sensitivity as opposed to maximum decidual growth at 120 hours. Measurement of specific activity of decidual RNA at earlier time periods after induction (1 or 4 hours) requires further technical modification to attain the accuracy and precision required for an assay. It is a very promising and probable possibility.

This type of assay is done with castrate female rats placed in each of the four treatment groups. On day 4 one uterine horn is stimulated and 24 hours later each horn was injected with [^{3}H] cytidine. One hour later control and decidual tissue is removed, weighed, and processed. The incorporation of [^{3}H] cytidine into the acid insoluble fraction was counted. The specific activity of decidual RNA is expressed as a function of the specific activity of RNA of the contralateral control horn (Fig. 7). It is clearly demonstrable that DCR induced in progesterone treated animals is characterized by the specific synthesis of RNA. The incorporation of [^{3}H] cytidine is significantly greater (250-275%) than in DCR of animals receiving both progesterone and estrogen (60-80%) when compared to the incorporation into tissue from the appropriate control horn. Other than a general growth promoting influence reflected in an increase in total RNA multiple estrogen injections do not enhance the specific activity of RNA of the stimulated horn relative to its own control horn. While these data do strongly suggest a unique role for progesterone the use of the same experimental approach to specify the uniqueness of day 4 versus day 3, e.g., the provision of a molecular basis for uterine sensitivity, will not be reported at this time.

Growth and differentiation as manifested in the decidual response represents alterations in transcription of the genome. The transcriptive capacity of chromatin extracted from nuclei isolated from uteri was assayed by incubation with bacterial (E. coli) RNA polymerase. The template capacity of chromatin from each of the four experimental groups is depicted in Fig. 8. Also included is data from pregnant animals whose fallopian tubes were ligated on day 0 (day of vaginal sperm). These uteri were not stimulated on day 4. Within the limitations inherent in the use of bacterial polymerase it is evident that the extent of the transcriptive capacity of chromatin isolated from either the pregnant uterus or that maintained on progesterone alone increases daily. A

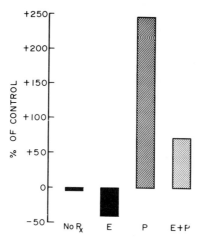

SPECIFIC ACTIVITY (^{3}H-CYTIDINE/µg RNA) IN UTERI
OF CASTRATE RATS 24 HOURS AFTER DECIDUALIZATION

Fig. 7. Specific activity of uterine RNA measured 24 hours after stimulation of one horn on day 4. The protocol for these experiments is depicted in Fig. 3. On day 5 each uterine horn is injected intraluminally with 10 µC [^{3}H] cytidine in 0.05 ml; one hour later the tissue is removed and processed. It is evident that progesterone (P = 2 mg/day) induces decidualization characterized by the specific synthesis of RNA. The rate of incorporation into uterine RNA is significantly greater than that noted in animals receiving both estrogen and progesterone. These differences are related to the appropriate control and are not attributable to alterations in total decidual RNA. The uterotrophic action of multiple doses of E is reflected in a net increase in total RNA but traumatization produces no DCR and no change in relative specific RNA synthesis (Glasser, S. R., 1972).

marked increase is noted between days 3 and 4 after which the template capacity of chromatin from the pregnant uterus is severely restricted whereas the template of progesterone treated animals remains open (Glasser *et al.*, 1972b). The pattern of transcriptive activity of uteri from animals maintained on both progesterone and estrogen signifies two important processes. Firstly, that the exposure of animals to constant and elevated titers of both estrogen and progesterone from day 0 is not analogous to the hormonal environment seen during pre-implantation stages of pregnancy. Secondly, these data suggest the role that estrogen may play, in these circumstances, of restricting gene expression and altering and eventually terminating uterine sensitivity. This facet of estrogen action will be discussed below.

If the transcriptional events we ascribe to progesterone, i.e., synthesis of new RNA species and subsequently new proteins, are really related to the induction of deciduoma, it follows that interference with RNA synthesis

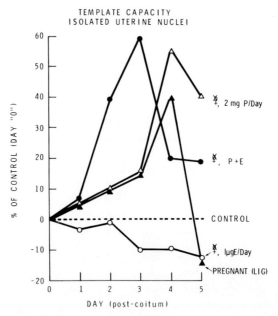

Fig. 8. Template capacity of uterine chromatin isolated from untreated, estrone (1 µg/day) treated, progesterone (2 mg/day) and estrone and progesterone treated rats ovariectomized on day 'O' of pregnancy. Intact, tubally ligated, pregnant rats are also included. Template capacity was assayed daily with purified E.coli RNA polymerase. Under the conditions of assay the template (chromatin DNA) is rate-limiting and the incorporation of radioactivity into an acid-insoluble product is linear. (See Glasser *et al.*, 1972a)

should result in blocking the progesterone supported decidual response and the biochemical processes which characterize it. The obvious approach to this question is to attempt to block DCR related processes with an inhibitor such as actinomycin D. The first and more general attempt (Fig. 9) demonstrates that whereas induction of DCR with Hanks balanced medium in either progesterone or progesterone and estrogen treated rats is predictably successful, instilling actinomycin D (1 - 5 µg dissolved in Hanks) into the uterine horn completely inhibits decidualization. Without producing any cytotoxic effects actinomycin D inhibits the DCR by blocking new RNA synthesis and thus RNA accumulation and sequential increases in decidual protein and decidual mass (Table 1). We have thus demonstrated the specific action of progesterone in stimulating RNA and protein synthesis, both qualitatively and quantitatively, and we have proven with the use of actinomycin D that RNA synthesis and its sequelae are critical processes in the evolution of uterine sensitivity to the induction of deciduoma (Glasser, 1956b, 1972; Heald *et al.*, 1972).

One pervasive reservation to the non-dependent action of progesterone requires further resolution. The animals used in these experiments were cyclic.

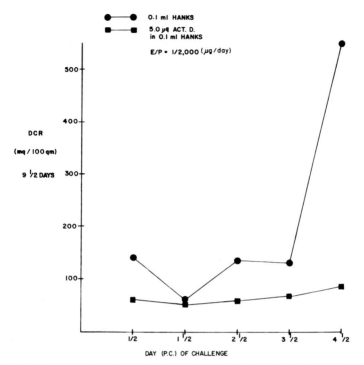

Fig. 9. Influence of actinomycin D on decidualization. Animals were bilaterally ovariectomized on the morning that sperm were present in the vagina (day ½) and maintained on 1 μg estrone and 2 mg progesterone/day thereafter. On each succeeding day different groups of animals were stimulated by instillation of 0.1 ml Hanks medium in the left uterine horn (●) and 5 mg actinomycin D in Hanks into the right uterine horn (■). Note that the deciduogenic stimulus of Hanks is not elicited to any significant degree until day 4½ and that the inclusion of actinomycin D blocks the action of Hanks on DCR.

Thus, 24-36 hours prior to their introduction into the protocol they had been exposed, during their proestrus, to elevated estrogen titers. Since the spectrum of responses to a single injection of estradiol are manifested during the ensuing 24-48 hours (Glasser et al., 1972a; Clark et al., 1973) it is possible that animals used in these studies are still under the influence of their previous proestrus estrogen. It is impossible to produce any animals completely free of estrogen exposure. A reasonable approach to resolving this problem was to use immature animals. These rats were ovariectomized at 21 days of age and held, without treatment, for seven days in an estrogen-free environment. Progesterone injections were begun on day 28 and uteri were stimulated at the time of the fifth injection. Under these rather stringent conditions it is evident (Table 2) that progesterone produces all the same responses that characterize decidualization in the mature, ovariectomized animal.

TABLE 1

Effect of Actinomycin D on Decidual Response

(5 µg/horn at 60 minutes pre-stimulation on day 4)

Day 5	no AMD		AMD	
	Control	DCR	Control	DCR
mg/100 g B.W.	57	101	72	77
	(+ 77%)		(+ 2%)	
cpm [³H] cytidine* per µg RNA	48	173	39	43
	(+260%)		(+10%)	
Total RNA (µg)	742	1241	654	732
	(+ 67%)		(+12%)	
Total protein (µg)	5420	8670	6484	6381
	(+ 60%)		(- 2%)	

* See Figure 7

The specificity of the responses described above suggest that active progestational compounds regulate uterine sensitivity to deciduogenic factors by stimulating events at the level of nuclear RNA transcription. This yields the synthesis of new RNA species and sequentially new populations of peptides and proteins. It is also evident that some of the products synthesized or induced by progesterone, are responsible for the evolution of uterine sensitivity. Although estrogen stimulates the synthesis of many uterine proteins, including some unique species, these molecules do not produce those factors which sensitize the uterus.

VI. UTERINE RECEPTIVITY TO OVO-IMPLANTATION

The state of uterine preparation for implantation of the blastocyst has classically been assayed by testing the endometrium for a decidual response following some form of stimulation. However, if the progesterone sensitized uterus were to be used as a recipient for ovum transfer, rather than for

TABLE 2

Comparison of Progesterone Induced RNA Changes[1]
(Mature vs. Immature Rats) 24 Hours after Decidualization

	DCR Weight	Total RNA	Sp. Act. RNA[2]
Mature	+ 53.0	+ 81.1	+ 243.0
Immature	+ 78.0	+ 40.9	+ 167.8

[1] All data expressed as % of control
[2] cpm [³H] cytidine/µg RNA/60 minutes (see Figure 7)

traumatic decidualization, successful implantation of synchronous blastocysts could not be demonstrated (Psychoyos, 1969, 1973a). Blastocysts would remain free in the uterus in a state of developmental diapause for periods extended by the daily administration of progesterone.

Certain "inducing agents" were found not to be capable of producing maximal DCR responses in castrate females maintained on progesterone alone. Shelesnyak (1960) suggested that the sensitivity of the uterus to these high threshold inducers was dependent on the presence of estrogen. In fact, minute amounts of estrogen prove to be required to terminate the period of embryonic diapause (delayed implantation) and to initiate implantation (Yoshinaga, 1961; Zeilmaker, 1963; Nutting and Meyer, 1964; Meyer, 1970). Within 24 hours after exposure to the estrogen the blastocysts will have implanted.

The hormonal basis by which the sensitive uterus evolves to one receptive to the implanting blastocyst is understood in only very general terms. The details are rather obscure. In the pregnant rat, ovariectomized on day 3 and maintained on daily progesterone, estrogen is required if the operation is performed early in the morning but not if the ovaries are removed in the afternoon (Chambon, 1949; Cochrane and Meyer, 1957; Zeilmaker, 1963; Psychoyos, 1973a). These type of data became the keystone of a vast library of indirect data in support of the idea that a pre-nidatory release of estrogen occurred sometime during mid-day of day 3. Although these experiments are open to alternative explanations, i.e., estrogen has to be secreted continuously until mid-day of day 3, the pre-nidatory release of estrogen was presumed to be transient rather than continuous and was termed the "estrogen surge." Controversy in the absence of direct measurement of ovarian estrogen secretion was never resolved but a time-worn consensus in favor of a "surge" colored the thinking of many investigators.

Though pre-nidatory estrogen release is generally believed to be limited to a short, discrete secretory period on day 3, it may play a more significant role than the relatively large amount of estrogen secreted during proestrus. Psychoyos (1973a, b) instructs us that the timing of estrogen release is one of the two factors which signal the time at which events leading to implantation are to begin. The other factor is the moment at which endometrial maturation by progesterone is considered complete. Only when the basic uterine preparation by progesterone, which requires a minimum of 48 hours, is complete can estrogen act to modify uterine sensitivity to a stage of receptivity to stimuli of low intensity such as the blastocyst. At the same time processes are initiated by estrogen which complete the requisite progesterone-estrogen sequence and terminate the period of sensitivity.

If the completion of the progesterone-estrogen sequence is postponed by the continued administration of progesterone in the absence of estrogen (Fig.

10, lines 3, 4) the endometrium remains sensitized to decidualization (Meyers, 1970; Psychoyos, 1973b). Estrogen, as noted above, completes this sequence and the uterus must be stimulated with 24 hours with either an artificial or natural (blastocyst) inducer (Fig. 10, line 1). A refractory state soon develops and neither stimulus can evoke the characteristic changes (Fig. 10, lines 2, 5). A second period of sensitivity cannot be originated (Fig. 10, line 6) until 48 hours have elapsed since the completion of the original progesterone-estrogen sequence (Fig. 10, lines 7, 8). In the rat the basic sequence would be completed, following a pre-nidatory estrogen "surge" during the latter hours of day 3 of pregnancy or pseudopregnancy (Psychoyos, 1973a, b). The altered threshold required for ovo-implantation occurs during the afternoon of day 4. The precipitous decrease in the transcriptive capacity of uterine chromatin from pregnant rats (Fig. 8) compared with castrate animals receiving progesterone might be viewed as confirming the view that this pre-nidatory estrogen acutely modulates the nature of progesterone induced gene expression.

VII. THE NATURE OF PRE-NIDATORY ESTROGEN

The source of pre-nidatory estrogen is thought to be the luteinized ovary which can, in fact, synthesize estrogens during the early stages of pregnancy (Zmigrod and Lindner, 1972). There is a growing library of evidence that these steroids are released into the blood (Shaikh and Abraham, 1969; Yoshinaga et al., 1969). However, there is still notable controversy regarding

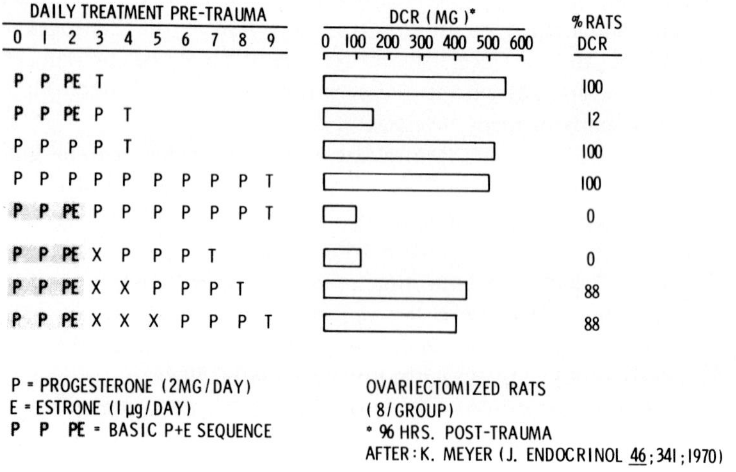

DAILY TREATMENT PRE-TRAUMA 0 1 2 3 4 5 6 7 8 9	DCR (MG)* 0 100 200 300 400 500 600	% RATS DCR
P P PE T	(bar)	100
P P PE P T	(bar)	12
P P P P T	(bar)	100
P P P P P P P P P T	(bar)	100
P P PE P P P P P P T	(bar)	0
P P PE X P P P T	(bar)	0
P P PE X X P P P T	(bar)	88
P P PE X X X P P P T	(bar)	88

P = PROGESTERONE (2MG/DAY)
E = ESTRONE (1 μg/DAY)
P P PE = BASIC P+E SEQUENCE

OVARIECTOMIZED RATS
(8/GROUP)
* 96 HRS. POST-TRAUMA
AFTER: K. MEYER (J. ENDOCRINOL 46;341;1970)

Fig. 10. The effect of a single injection of estradiol-17β or of an interruption in the course of progesterone treatment on the sensitivity of the uterus to traumatic decidualization. The rats used in these experiments were bilaterally ovariectomized and treated with progesterone (adapted from Meyers, K., 1970).

the precise onset of ovarian estrogen secretion in the pregnant rat with particular reference to the chronology of its relationship to uterine receptivity to nidation.

The concept of an estrogen "surge" as the *sine qua non* of early pregnancy has dominated the literature on hormonal regulation of decidualization and implantation (Shelesnyak *et al.*, 1963, 1970). For almost a decade this concept has received consensus acceptance and usage in studies relating to the transient nature and precise timing of uterine sensitivity, and in the results of precisely timed ovariectomy and replacement therapy in endoprivic mated and pseudopregnant rats. In spite of its utility the concept of estrogen "surge" has not been universally accepted. DeFeo (1967) argued that chronic estrogen and progesterone output would be equally as effective. Glasser (1965b, 1972) demonstrated that estrogen was not obligatory for the development of uterine sensitivity to decidualization. This idea has been confirmed and refined by Psychoyos (1969, 1973b) in his description of the progesterone-estrogen sequence. Finn (1971) has summarized the issue by noting that these studies show only how implantation could be accomplished. The mechanisms operating in the cyclic female remain to be revealed.

The primary difficulty in assigning importance to the "surge" has been the difficulty in proving its existence. Assessment of estrogen secretion into ovarian venous blood of pregnant rats has been the subject of a large number of contradictory studies. Yoshinaga *et al.* (1969) used a one point bioassay based on vaginal cornification to estimate total estrogen secretion (Fig. 11). The observed increase in the late morning of day 1 became maximal on day 3 before falling to baseline before the morning of day 4. The authors interpret these data as endorsing the "surge" hypothesis; however, it is our judgement that the titers occurring between days 2 and 3 do not really vary significantly. Estrogen values, elevated over a 36 hour period, do not constitute a "surge."

Shaikh and Abraham (1969) also provided data supporting the "surge" hypothesis. Radioimmunoassay of crude ether extracts of blood did not detect estradiol until titers rose markedly between 0800 and 1000 hours on day 3 (Fig. 11). There was almost as remarkable a fall in estradiol levels. These data were so consistent with the definition of the "surge" that Shelesnyak and Marcus (1971) were prompted to propose that the quotation marks be deleted when the expression estrogen "surge" was used.

The most careful evidence was provided by the analysis of ovarian venous effluents by gas-liquid chromatography (Nimrod *et al.*, 1972). These experiments offer direct evidence of ovarian secretion of estradiol and estrone during the early stages of pregnancy (Fig. 11). Secretion rates, even at peak post-conception days 2-7, were 15-20 times less than values measured during the morning of proestrus. The secretory pattern certainly does not describe anything like a "surge" or a transient rise. Rather estradiol secretion rises

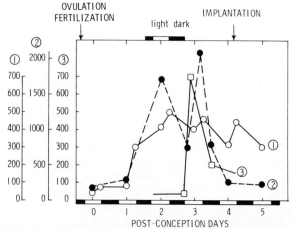

Fig. 11. A comparison of data derived by three different methods used to evaluate estrogen secretion during the pre-implantation stages of pregnancy in the rat.

during the afternoon of day 1 and remains elevated at 4-5 times that level through the period of implantation. The estradiol output during the pre-implantation period could be extrapolated to values of 0.05 μg/rat/day which is close to the dose required for induction of implantation (0.06-0.08 μg; Perel and Lindner, 1970). Thus, it is not unreasonable to conclude that the use of exogenous estradiol to induce implantation in lactating or spayed rats maintained on progesterone is a physiological and not a pharmacological maneuver (Cochran and Meyer, 1957; Psychoyos and Alloiteau, 1962).

On the basis of these data we feel that the inferences derived from the indirect evidence regarding event timing and the duration of estrogen secretion in early pregnancy (Shelesnyak et al., 1963; Shelesnyak and Marcos, 1971) are in need of revision. Rather than a transient "surge" of estrogen secretion on day 3 a sustained plateau of estrogen secretion, initiated on day 1, is to be observed (Yoshinaga et al., 1969; Nimrod et al., 1972). Ovarian venous titers of estradiol and estrone are notably elevated on day 1. By day 2 the levels are further elevated and the plasma values through day 5 are not distinguishable from each other on a daily basis. These data cover the time frame which includes day 3 - the day of the putative "surge." The precise timing of uterine receptivity to the blastocyst, which is confined to a short period on day 4, cannot be determined solely by a transient surge of estrogen which is assumed to precede nidation by 12-20 hours. The actual situation is obviously more complex and involves continuous secretion throughout the pre-implantation period of both progesterone and estrogen.

Concerning endometrial sensitivity Psychoyos minimizes his own contributions when he insists that the literature of the 1950's and early 1960's recognized that the action of estrogen was on a uterus primed by progesterone. This is a generous interpretation because the sense of that literature was not to dwell on or even give notation to the regulatory actions of progesterone but rather to define the predominant role of estrogen. In defining the precise time course of progesterone dependent uterine sensitivity and its evolution to a state of receptivity by estrogen it has now become possible to reevaluate the modulatory action of estrogen and the primary mechanisms of progesterone.

Evaluation of plasma levels of estradiol and progesterone, derived from the literature and from our analyses, is shown in Fig. 12. Estradiol titers during the morning of proestrus exceed those measured on day '0' by 15 to 20-fold. No "surge" can be demonstrated on day 3 (Nimrod *et al.*, 1972). Although there is good agreement regarding the patterns during days 0-3, there is ambivalence regarding the drop in titers on day 4. Progesterone secretion also experiences a short pre-ovulatory rise. Plasma values fall and then gradually rise again through day 1 (Fig. 12). They increase sharply (almost 3-fold) by day 2 and gradually thereafter (McCormack *et al.*, 1974).

In deciding to interpret these data as representing a sustained plateau of estrogen secretion (vs. "surge") during the pre-implantation period requires that we suggest some role for this estrogen. Inspection of Fig. 12 shows that

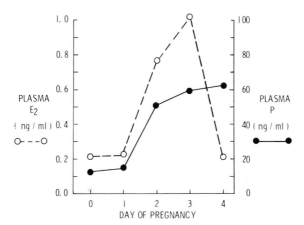

Fig. 12. Peripheral plasma estradiol titers (Nimrod *et al.*, 1972) and peripheral plasma progesterone titers (McCormack *et al.*, 1974) during various pre-implantation stages of the pregnant rat.

relatively high pre-implantation levels are attained before the animal completes 48 hours of sustained progesterone action. Yet the basic sequence, as described by Psychoyos (1973a, b) is not completed in discrete physiological terms, until day 4. The problem approaches solution if we apply certain of the principles derived from studies of the mechanism of hormone action and the dynamics of specific macromolecular target cell receptors for the steroid hormones to its resolution.

VIII. THE ROLE OF STEROID HORMONE RECEPTORS IN THE DEVELOPMENT OF UTERINE SENSITIVITY AND RECEPTIVITY

Thinking with regard to the action of estrogen on the pre-implantation uterus has become increasingly unilateral. The action of estrogen is not directed solely towards nidation but rather serves to modulate the sequential processes, many initiated by progesterone which direct the estrus uterus through the consecutive and inter-dependent stages of sensitivity and receptivity. In this context it seems imperative that the secretion of estrogen be chronic rather than transient.

Estrogen responsive tissues contain unique macromolecules which bind estrogenic hormones in a stereospecific manner (Gorski et al., 1968; Jensen and DeSombre, 1972). It is generally accepted that when estrogen, E, enters uterine cells it is rapidly bound by a high affinity cytoplasmic receptor protein, R, and the receptor-estrogen complex, $R \cdot E$, is then transported from the cytoplasm to the nucleus of uterine cells (Gorski et al., 1968; Jensen and DeSombre, 1972) where it is probably bound to acceptor sites on chromatin (O'Malley and Means, 1974). The interaction of $R \cdot E$ with acceptor sites in the nucleus is considered to be important in the stimulation of RNA synthesis which ultimately results in estrogen-induced responses (O'Malley and Means, 1974). Furthermore, we have shown that the accumulation of the receptor complex by uterine nuclei is under the control of ovarian estrogens (Clark et al., 1972) and that the quantity of $R \cdot E$ in the nucleus is closely correlated with uterotrophic responses (Anderson et al., 1973). Thus, the $R \cdot E$ complex in the nucleus of uterine cells is probably of fundamental importance in the mechanism of action of the hormone (Clark et al., 1973). The ability of the $R \cdot E$ complex to stimulate growth of the uterus appears to depend not only on the initial accumulation of the complex by the nucleus, but more importantly, on the retention of the complex by the nucleus. We have suggested that the failure of the estrogen, estriol, to cause a significant increase in the uterine weight 24 hours after treatment was related to the inability of estriol to cause long-term retention of the estrogen receptor by the nucleus (Anderson et al., 1972a). In order to examine this hypothesis and to describe these relationships in more detail the amount of nuclear R·E and the extent of uterine growth were examined as a function of time following injection of estradiol or estriol.

Immature rats were injected with 1.0 μg of estradiol (E_2) or estriol (E_3) and the concentration of nuclear R·E was determined at various times following injection by the [^{3}H] estradiol exchange assay (Anderson et al., 1972b; Fig. 13). The uterine weight was also measured at various times after injection (Fig. 13). The concentrations of nuclear receptor estrogen complex are equivalent at 3 hours after an injection of either estradiol or estriol (Fig. 13). Uterine responses are also equivalent at three hours (Fig. 13; Table 3). However, by 6 hours the concentration of nuclear R·E which is elicited by

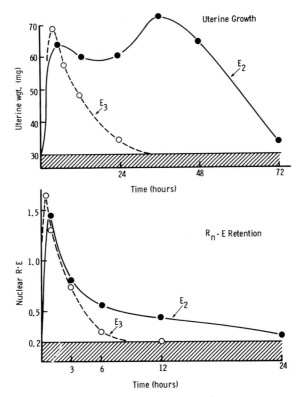

Fig. 13. The concentration-time parameters of nuclear receptor-estrogen binding and uterine response following an injection of estradiol of estriol. Immature rats were injected with 1.0 μg of either estradiol (closed circles) or estriol (open circles). At various times after treatment the uteri were weighed to the nearest 0.1 mg (upper) and the concentration of the nuclear RE complex (lower) was determined by the [^{3}H] estradiol exchange assay. The cross hatched portion in the lower figure represents the control level of nuclear RE in saline injected controls. (Anderson, Clark and Peck, 1972a)

TABLE 3

Effect of Estradiol and Estriol on Short Term Uterine Responses

Rats were killed at 3 hours after the injection of saline, estradiol (1 μg) or estriol (1 μg). The mean ± S.E.M. conversion of (^{14}C) glucose to (^{14}C) lipid (A), (^{14}C) protein (B) and (^{14}C) RNA (C) was calculated from at least 4 determinations with 2 uteri per determination. RNA polymerase activity, expressed as dpm of (^{3}H) UTP incorporated into RNA per 100 μg of RNA was determined by procedures described by Gorski. The means ± S.E.M. were determined from at least 4 determinations with three uteri per determination (Anderson, Peck and Clark, 1974, Endocrinology, in press).

Response	Saline	Estradiol	Estriol
A. (^{14}C) lipid (dpm/uterus)	2,001 ± 218	6,183 ± 604	7,272 ± 985
B. (^{14}C) protein (dpm/mg protein)	174 ± 4	320 ± 15	346 ± 25
C. (^{14}C) RNA (dpm/100 μg RNA)	2,227 ± 332	3,395 ± 319	3,570 ± 239
D. RNA polymerase (dpm/100 μg RNA)	803 ± 104	1,631 ± 219	1,850 ± 94

estriol has declined to near control levels while that by estradiol remains well above control. This rapid decline in the nuclear R·E$_3$ complex when compared to the R·E$_2$ complex is paralleled by a corresponding inability of estriol to maintain uterine wet weight (Fig. 13) and to elicit long-term growth response (Fig. 13). These experiments support the idea that the time required for the stimulation by the receptor estrogen complex of nuclear events which result in long term uterine growth responses is at least 6 hours, and that estriol is a weak estrogen in this regard because it does not promote long term retention of the receptor within the nuclear compartment.

The above experiment suggests that the accumulation and retention of the receptor by the nucleus is an important physiological function. If such were the case these binding interactions should manifest themselves during various reproductive states. The cyclic fluctuations of estrogens in the blood of the rat have been well described (Yoshinaga *et al.*, 1969; Shaikh, 1971) and would be expected to have profound influences on the translocation of the R·E complex to the nucleus in estrogen sensitive tissues. In order to test this proposal the amount of nuclear R·E was examined in the uterus during the various stages of the estrous cycle (Clark *et al.*, 1972). The number of R·E complexes per mg DNA is at a minimum in estrus and diestrus (approximately 1000 sites per nucleus; Fig. 14). The concentration increases between metestrus and diestrus (3500 sites/nucleus) peaking on the day of proestrus (5000 sites/nucleus). These cyclic fluctuations in the concentration of nuclear R·E during the estrus cycle closely parallel the rate of ovarian estrogen secretion (Hori *et al.*, 1968). The low concentration of nuclear R·E during estrus and metestrus parallel a minimum secretory rate. The rate of estrogen

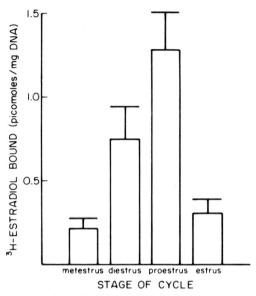

Fig. 14. The concentration of nuclear receptor estrogen complex in the rat uterus during the estrous cycle. The quantity of nuclear receptor estrogen complex was determined by the [³H] estradiol exchange assay for each day of the estrous cycle. Each value represents the mean ± S.E.M. of 5-6 determinations. (Clark, Anderson and Peck, 1972)

secretion increases in diestrus together with an increase in uterine nuclear R·E. The maximum concentration of nuclear R·E coincides with the proestrus peak in secretory rate.

These results indicate that the accumulation, via translocation, of R·E complexes by uterine nuclei is highly correlated with the blood levels of estrogen, and thus can be used as an indicator of estrogen fluctuations in the blood. This prompted us to measure the quantities of uterine nuclear R·E during the time of implantation in order to examine the important estrogen stimulated events that are taking place at this time.

The quantity of nuclear R·E increases significantly on days 2 and 3 during the implantation period (Fig. 15). This peak is followed by a decline on day 4 to levels that are not significantly different from those of days 0 and 1. These data support the proposal that estrogen blood levels are elevated on days 2 and 3 as suggested by Nimrod et al. (1972) and Yoshinaga et al. (1969) and these elevated levels probably cause the increased nuclear R·E that we have observed during this time. This accumulation of nuclear R·E is probably involved in the initiation of nuclear events that are important to the stimulation of implantation. The decline in nuclear R·E that is observed between day 3 and 4 may be the result of reduced blood levels of estrogen; however, this observation is open to other interpretations. We have recently demonstrated that progesterone reduces the quantity of cytoplasmic estrogen

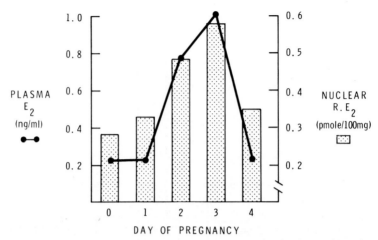

Fig. 15. The concentration of receptor estrogen complex in the nucleus of uterine cells during early pregnancy. The amount of nuclear RE was measured by RE-[^{3}H] estradiol exchange assay and the quantities are compared to the estrogen blood levels. The quantities of estrogen in the blood were adapted from the data of Yoshinaga, *et al.*, 1969.

receptor and this is accompanied by a reduced capacity of the uterus to respond to estrogen (Hsueh *et al.*, in press). Therefore, the decrease in nuclear R·E. observed on day 4 may be due to rising progesterone titers that would result in a decreased availability of receptor, hence decreased nuclear R·E. Regardless of the cause, the accumulation of R·E by the nucleus during this period is probably associated with the stimulation of important nuclear events that are required for blastocyst implantation.

One of the problems in trying to define the regulatory system of the pre-implantation uterus derives from the complexity of the uterus as an organ. We do not know where the available receptors are at each stage of development. Different cells, in differentially changing cell populations, may have different patterns of receptor availability for each hormone. Thus, the glandular epithelium may always have a population of active receptor-acceptor sites whereas other regions of the endometrium may have varying concentration of steroid hormone receptors with each change of the hormonal environment. As the plasma concentration of P increases with time the stromal cells become increasingly sensitive to the mitogenic influence of E (Tachi *et al.*, 1972). The glandular and surface epithelium decrease their responsiveness. Concomitantly a full spectrum of growth related biochemical responses, inducible with P but augmented by E can be measured in stromal cells. These changes may be the result of a differential redistribution of available receptor sites.

The full explanation of the conjoint actions of progesterone and estrogen during the pre-implantation period awaits a more extensive disclosure of the action of each hormone on gene expression. The uterus provides an additional complexity because of the diversity of target cells, their rather specific regional distribution and the fact that both hormones act on the same genetic apparatus during the same period of time.

Proestrus titers of estrogen and progesterone provide the endocrine preparation for mating behavior and ovulation. The effect of proestrus estrogen may be to enhance the available number of plasmic receptors for progesterone and to play a role, together with progesterone, in determining the sensitivity of a differential population of cells to the biochemical action of progestrone.

Rising progesterone levels (Fig. 12) evoke a serial synthesis of RNA, protein and other metabolites from an enlarging population of endometrial cells. Some of the regulatory proteins thus synthesized (Fig. 16) may be secreted into the uterine lumen where, in time, they interact with the blastocyst. Other structural and regulatory proteins contribute to the altera-

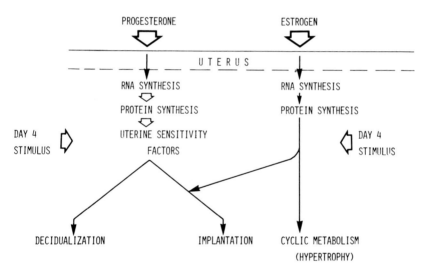

Fig. 16. A scheme representing the influence of estrogen and progesterone on uterine sensitivity to decidualization and receptivity to implantation. Progesterone stimulates the synthesis of specific RNA's and subsequently specific proteins. The products of these synthesis are responsible for the competence of the free blastocyst to differentiate embryonic derivatives after nidation, the developmental inhibition of the blastocyst prior to implantation and after a minimum of 48 hours, and the development of uterine sensitivity to deciduogenic stimuli. This uterine environment will not sustain ovum-endometrial interaction. Estrogen also stimulates the synthesis of specific RNA and proteins. These metabolites are directed towards tissue growth and are not compatible with decidualization or nidation. Estrogen modulates the synthetic processes of the progestational environment (uterine sensitivity factor) activating and altering the threshold for the process of implantation. This completes the hormonal sequence required for synthesis thereby restricting uterine sensitivity. (Glasser, S. R. 1972)

tions in the biology of the stromal cell. These changes are directed by progesterone and require 48 hours in order that the stromal cell acquire the sensitivity to respond to deciduogenic stimuli. For this reason, progesterone may be considered the major determinant for the uterine sensitivity. The sensitized uterus will not respond to stimuli of the nature provided by the blastocyst; thus, implantation in such uteri is not successful.

However, these changes are progressively modulated by increasing estrogen titers (Fig. 11) which elicit their own responses (Fig. 16). Hypothetically, among the regulatory proteins synthesized because of estrogen stimulation are those which act on the uterus as it passes the stage of sensitivity causing it to lower the stimulatory threshold. The modulation of the progesterone induced cellular biology by the action of estrogen converts the sensitive uterus to one receptive to interactions with natural stimuli such as the blastocyst (Fig. 16). Further synthetic events, expressions of the interactions of progesterone and estrogen with the acceptor site on chromatin, as progesterone titers continue to rise and estrogen titers fall, terminate these recognition events.

REFERENCES

Alden, R. H. and Smith, J. (1959). Implantation of the rat egg IV. Some effects of artificial ova in the rat uterus. *J. Exp. Zool.* **142**, 215–225.

Anderson, J. N., Clark, J. H. and Peck, E. J., Jr. (1972a). The relationship between nuclear receptor estrogen binding and uterotrophic responses. *Biochem. Biophys. Res. Commun.* **48**, 1460–1468.

Anderson, J., Clark, J. H. and Peck, E. J., Jr. (1972b). Oestrogen and nuclear binding sites: Determination of specific sites by [^{3}H] oestradiol exchange. *Biochem. J.* **126**, 561–567.

Anderson, J. N., Peck, E. J., Jr. and Clark, J. H. (1973). Nuclear receptor estrogen complex: Relationship between concentration and early uterotrophic responses. *Endocrinology* **92**, 1488–1495.

Billingham, R. E. (1964). Transplantation immunity and the maternal-fetal relationship. *New Engl. J. Med.* **270**, 667–672, 720–725.

Bitman, J., Wrenn, T. R., Cecil, H. C. and Sykes, J. F. (1960). Water and electrolyte composition of decidual tissue of the rat uterus. *Acta Endocrinol.* **34**, 465–472.

Blandau, R. J. (1949). Embryo-endometrial interrelationship in the rat and guinea pig. *Anat. Rec.* **104**, 331–360.

Cecil, H. C., Bitman, J., Connolly, M. R. and Wrenn, T. R. (1962). Glycogen in the decidual tissue of the rat uterus. *J. Endocr.* **25**, 69–76, 1962.

Chambon, Y. (1949). Realisation du retard de l'implantation par des faibles doses de progesterone chez la ratte. *Compt. Rend Soc. Biol.* **143**, 753–756.

Chang, M. C. (1950). Development and fate of transferred rabbit ova or blastocyst in relation to the ovulation time of recipients. *J. Exp. Zool.* **114**, 197–225.

Clark, J. H., Anderson, J. N. and Peck, E. J., Jr. (1973). Nuclear receptor estrogen complexes of rat uteri: Concentration-time-response parameters. *In*, "Receptors for Reproductive Hormones" (B. W. O'Malley and A. P. Means, eds.), pp. 15–59. Plenum Press, New York.

Clark, J. H., Anderson, J. N. and Peck, E. J., Jr. (1972). Receptor-estrogen complex in the nuclear fraction of rat uterine cell during the estrous cycle. *Science* **176**, 528–530.

Cochrane, R. L. and Meyer, R. K. (1957). Delayed implantation in the rat induced by progesterone. *Proc. Soc. Exptl. Biol. Med.* **96**, 155–159.

DeFeo, V. J. (1963a). Temporal aspect of uterine sensitivity in the pseudo-pregnant or pregnant rat. *Endocrinology* **72**, 305–316.

DeFeo, V J. (1963b). Determination of the sensitive period for the induction of deciduomata in the rat by different inducing procedures. *Endocrinology* **73**, 488–497.

DeFeo, V. J. (1967). Decidualization. *In*, "Cellular Biology of the Uterus" (R. M. Wynn, ed.), pp. 192–290. Appleton Century Crofts, N.Y.

Fawcett, D. W. (1950). Development of mouse ova under the capsule of the kidney. *Anat. Rec.* **108**, 71–92.

Finn, C. A. (1971). The biology of decidual cells *In*, "Advances in Reproductive Physiology" (M. W. H. Bishop, ed.), pp. 1–27. Academic Press, N.Y.

Finn, C. A. and Martin, L. (1970). The role of oestrogen secreted before oestrus in the preparation of the uterus for implantation in the mouse. *J. Endocr.* **47**, 431–438.

Fuchs, A. R., Beling, C. and Park, K. (1974). Corpus luteum recognition of blastocysts prior to implantation. Abst. 224 (p. A-167) Endocrine Society 56th Annual Meeting, Atlanta.

Galassi, L. (1968). Autoradiographic study of the decidual cell reaction in the rat. *Develop. Biol.* **17**, 75–84.

Gelfant, S. (1963). A new theory on the mechanism of cell division. *In*, "Cell Growth and Cell Division" (R. J. C. Harris, ed.), pp. 229–259. Academic Press, N.Y.

Glasser, S. R. (1957). Influence of goitrogens on induction of deciduoma in the rat uterus. *Fed. Proc.* **16**, 47.

Glasser, S. R. (1963). Placental dysfunction as the cause of prenatal radiation death: An endocrine oriented hypothesis. *In*, "International Symposium on the Effects of Ionizing Radiation in the Reproductive System" (W. Carlson and F. X. Gassner, eds.), pp. 361–376. Pergamon Press, N.Y.

Glasser, S. R. (1965a). Metabolic Requirements for Implantation. *In*, "Conference on Fertilization, Gamete Transport and Pre-Implantation Mechanism. Vanderbilt University.

Glasser, S. R. (1965b). Biochemical studies on a cellular receptor mechanism for uterine sensitivity. *Excerpta Medica, International Congress Series* **99**, 335.

Glasser, S. R. (1972). The uterine environment in implantation and decidualization. *In*, "Reproductive Biology" (H. A. Balin and S. R. Glasser, eds.), pp. 776–883. Excerpta Medica Fdn., Amsterdam.

Glasser, S. R. (1975). A molecular bioassay for progesterone and related compounds. *In*, "Methods in Enzymology" (B. W. O'Malley and J. Hardman, eds.), Vol. 36. Academic Press, N.Y.

Glasser, S. R. and O'Malley, B. W. (1970). Mechanism of progesterone action: A 'new' mammalian model system. 52nd Meeting, Endocrine Society, #120.

Glasser, S. R., Chytil, F. and Spelsberg, T. C. (1972a). Early effects of estradiol 17-β on the chromatin and activity of DNA dependent RNA polymerases (I and II) of the rat uterus. *Biochem. J.* **130**, 947–959.

Glasser, S. R., Spelsberg, T. C., Chytil, F. and O'Malley, B. W. (1972b). Progesterone induced changes in template capacity of uterine chromatin related to uterine sensitivity. Society for the Study of Reproduction (5th Meeting).

Gorski, J., Toft, D., Shyamala, G., Smith, D. and Notides, A. (1968). Hormone receptors: Studies on the interaction of estrogen with the uterus. *Recent Progr. Hormone Res.* **24**, 45–80.

Haour, F. and Saxena, B. B. (1974). Detection of a gonadotropin in rabbit blastocyst before implantation. *Science* **185**, 444–445.

Heald, P. J. and O'Grady, J. F. (1969). The uptake of [^{3}H] uridine into the nucleic acids of the rat uterus during early pregnancy. *Biochem. J.* **114**, 36–37p.

Heald, P. J., O'Grady, J. E., O'Hare, A. and Vass, M. (1972). Changes in uterine RNA during early pregnancy in the rat. *Biochim. Biophys. Acta* **262**, 66–74.

Hsueh, A. J. W., Peck, E. J., Jr. and Clark, J. H. Progesterone antagonism of the estrogen receptor and estrogen induced uterine growth. *Science* (in press).

Jensen, E. V. and DeSombre, E. R. (1972). Mechanism of action of the female sex hormones. *Annual Rev. Biochem.* **41**, 203–230.

Knowler, J. T. and Smellie, R. M. S. (1971). The synthesis of ribonucleic acid in immature rat uterus responding to oestradiol 17-β. *Biochem. J.* **125**, 605–614.

Krehbiel, R. H. (1937). Cytological studies of the decidual reaction in the rat during early pregnancy and in the production of deciduomata. *Physiol. Zool.* **10**, 212–233.

Lobel, B. L., Tic, L. and Shelesnyak, M. C. (1965). Studies on the Mechanism of Nidation XVII (Pt 1-5). *Acta Endocrinol.* **50**, 452-583.

Loeb, L. (1908a). The experimental production of the maternal part of the placenta in the rabbit. *Proc. Soc. Exp. Biol. Med.* (N.Y.) **5**, 102–105.

Loeb, L. (1908b). The production of deciduomata and the relation between the ovaries and the formation of decidua. *J. Amer. Med. Assoc.* **50**, 1897–1901.

Marcus, G. J. (1970). A cellular basis for implantation failure due to massive decidualization in the rat. 3rd Meeting, Society for the Study of Reproduction (#46).

Martin, L. and Finn, C. A. (1969). Duration of progesterone treatment required for a stromal response to oestradiol 17β in the uterus of the mouse. *J. Endocrinol.* **44**, 279–280.

Meyers, K. (1970). Hormonal requirements for the maintenance of oestradiol-induced inhibition of uterus sensitivity in the ovariectomized rat. *J. Endocrinol.* **46**, 341–346.

McCormack, S. A., Glasser, S. R. and Clark, J. H. (1974). Unpublished Data.

Nimrod, A., Ladany, S. and Lindner, H. R. (1972). Perinidatory Ovarian Oestrogen Secretion in the Pregnant Rat, determined by Gas Chromatography with Electron Capture Detection. *J. Endocrinol.* **53**, 249–260.

Noyes, R. W., Dickmann, Z., Doyle, L. L. and Gates, A. H. (1963). Ovum transfers, synchronous and asynchrous, in the study of implantation. *In,* "Delayed Implantation" (A. C. Enders, ed.), Univ. of Chicago Press, Chicago, Ill.

Nutting, E. F. and Meyer, R. K. (1964). Effect of Oestrone on delay of Nidation, Implantation and Foetal Survival in Ovariectomized Rats. *J. Endocrinol.* **29**, 235–242.

O'Grady, J. E., Heald, P. J. and O'Hara, A. (1970). Incorporation of [^{3}H] uridine into the ribonucleic acids of rat uterus during pseudo-pregnancy and in the presence of I.C.I. 46474 (trans-1-Co-B dimethyl-amino-ethoxyphenyl-1, 2-diphenyl-but-l-ene). *Biochem. J.* **119**, 609–613.

O'Malley, B. W., McGuire, W. L., Kohler, P. O. and Korenman, S. G. (1969). Studies on the Mechanism Steroid Hormone Regulation of Synthesis of Specific Proteins. *Rec. Progr. Hormone Res.* **25**, 105–160.

O'Malley, B. W., Means, A. R., Socher, S. H., Spelsberg, T. C., Chytil, F., Comstock, J. P. and Mitchell, W. M. (1974). Hormonal Control of Oviduct Growth and Differentiation. *In,* "Macromolecules in Growth and Development" (E. D. Hay, T. J. King and J. Papaconstantinou, eds.), pp. 53–80. Academic Press, N.Y.

O'Malley, B. W. and Means, A. R. (1974b). Female Steroid Hormones and target cell Nuclei. *Science* **183**, 610–620.

Perel, E. and Lindner, H. R. (1970). Dissociation of Uterotrophic Action from Implantation Inducing Activity in two Non-steroidal Oestrogens (Coumestrol and Genestrin). *J. Reprod. Fert.* **21**, 171–175.

Psychoyos, A. (1966). Recent research on egg implantation. *In,* "Ciba Foundation study group on egg implantation" (E. W. Wolstenholme and M. O'Connor, eds.), pp. 4–28. Churchill, London.

Psychoyos, A. (1969). Hormonal requirements for egg implantation. *In*, "Advances in Bio-sciences IV. Mechanisms Involved in Conception" (G. Raspe, ed.), pp 275–290. Pergamon Press, London.

Psychoyos, A. (1973). Hormonal control of ovo-implantation. *Vitamins and Hormones* **31**, 201–256.

Psychoyos, A. (1973b). Endocrine Control of Egg Implantation. *In*, "Handbook of Physiology, Section 7, Endocrinology, Vol. II, Part 2" (R. O. Greep and E. B. Astwood, eds.), pp. 187–215. American Physiological Society, Washington, D.C.

Psychoyos, A. and Alloiteau, J. J. (1962). Castration précoce et nidation de l'oeuf chez la ratte. *C. R. Seanc Soc. Biol.* **156**, 46–49.

Segal, S. J. and Nelson, W. D. (1958). An orally active compound with antifertility effects in rats. *Proc. Soc. Exp. Biol. Med.* **98**, 431.

Shaikh, A. (1971). Estrone and esteradiol in the ovarian venous blood from rats during esterous cycle and pregnancy. *Biol, Reprod.* **5**, 297–307.

Shaikh, A. A. and Abraham, G. E. (1969). Measurement of the estrogen-surge during pseudo-pregnancy in rats by radioimmunoassay. *Biol. Reprod.* **1**, 378–380.

Shelesnyak, M. C. (1960). Nidation of the Fertilized Ovum. *Endeavour* **19**, 81–86.

Shelesnyak, M. C. (1962). DECIDUALIZATION: The Decidua and The Deciduoma. *Perspect. Biol. Med.* **5**, 503–518.

Shelesnyak, M. C. and Tic, L. (1963a). Studies on the mechanism of decidualization: IV. Synthetic processes in the Decidualizing Uterus. *Acta Endocrinol.* **42**, 465–472.

Shelesnyak, M. C. and Tic, L. (1963b). Studies on the mechanism of decidualization: V. Suppression of synthetic processes of the uterus following inhibition of decidualization by an anti-oestrogen, ethanoxytriphetol (Mer-25). *Acta Endocrinol.* **43**, 462–468.

Shelesnyak, M. C., Krarcer, P. F. and Zeilmaker, G. H. (1963). Studies on the mechanism of decidualization. I. the oestrogen surge of pseudo-pregnancy and progravidity and its role in the process of decidualization. *Acta Endocrinol.* **42**, 225–232.

Shelesnyak, M. C., Marcus, G. J. and Lindner, H. R. (1970). Determinants of the decidual reaction. *In*, "Ovo-Implantation, Human Gonodotropine and Prolactin" (P. O. Hubimont, F. Leroy, C. Robyn, and P. LeReux, eds.), pp. 129. Krager, Barel.

Shelesnyak, M. C. and Marcus, G. J. (1971). Steroidal conditioning of the endometrium for nidation. *In*, "Advances in Bio-sciences VI: Schering Symposium on Intrinsic and Extrinsic Factors in Early Mammalian Development" (G. Raspe, ed.), pp. 303–316. Pergamon Press, N.Y.

Tachi, C., Tachi, S. and Lindner, H. R. (1972). Modification by progesterone of oestradiol-induced cell proliferation. RNA Synthesis and Oestradiol Distribution in the rat Uterus. *J. Reprod. Fert.* **31**, 59–76.

Von Berdswordt, R., Walbabe, R. and Turner, C. W. (1961). Influence of graded levels of progesterone in ovariectomized rats on placentomata formation measured by total DNA. *Proc. Soc. Exp. Biol. Med.* **107**, 469–471.

Yochin, J. M. and DeFeo, V. J. (1963). Hormonal control of the onset magnitude and duration of uterine sensitivity in the rat by steroid hormones of the ovary. *Endocrinology* **73**, 317–326.

Yoshinaga, K. (1961). Effect of local application of ovarian hormones on the delay in implantation in lactating rats. *J. Reprod. Fert.* **2**, 35–41.

Yoshinaga, K., Hawkins, R. A. and Stocker, J. F. (1969). Estrogen secretion by the rat ovary *in vivo* during the estrous cycle and pregnancy. *Endocrinology* **85**, 103–112.

Zeilmaher, G. H. (1963). Experimental studies on the effects of ovariectomy and hypophysectomy on blastocyst implantation in the rat. *Acta Endocrinol.* **44**, 355–366.

Zmigrad, A. and Lindner, H. R. (1972). Oestrogen biosynthesis by the rat ovary during early pregnancy. *Acta Endocrinol.* **69**, 127–140.

Subject Index

A

Abscisic acid, effect on, 167–169
  carboxypeptidase, 167–169
  translation, 167
Actinomycin D
  effect on cotton embryo, 168–170
  effect on decidualization, 328
Anti-androgen, 46
Androgen sterilization, effect of
  on estrogen binding proteins, 246
  on preoptic LH-trigger, 245–246
Anterior hypothalamic lesions, 245–246

B

Blastocyst, morphological
  differentiation, 210–211
Blastocyst attachment, in vitro, 286–296
  amino acid deprivation, 296
  in absence of uterine monolayer, 186–290
  on collagen substrate, 296
  controlling factors, 293–297
    serum concentration, 294
    time in culture, 293–294
    uterine fluid, 294–296
  in presence of uterine nomolayer, 288–290
  in steroid free serum, 290–293
Blastocyst implantation
  function of trophoblast cells, 281–282
  in nomolayers of uterine cells, 281–284
  morphological changes, 280–286

C

Cordycipin (3′ - deoxyadenosine), effect on
  cotton seed germination
  carboxypeptidase appearance, 170–173
  mRNA synthesis, 169–173

Cotton embryo
  development of, 166–167
  enzyme synthesis in carboxypeptidase,
    isocitritase, 167
  germination, mRNA requirement for,
    166–167

D

Decidual Cell, formation of, 320–321
Decidual cell response
  mechanism of progesterone action, 320–330
  uterine sensitivity to, 314–315
Decidualization
  actinomycin D effect on, 328
  characterization
    hyperplasia, 322
    increase in DNA, 322
    increase in mass, 321
    increase in protein, 323, 327–328
    RNA synthesis, 321
  induction of
    by estrogen, 316–317
    by progesterone, 316–317
Deciduogenic stimuli, 314–320
3′ - deoxycytidine, See Cordycipin
Determination,
  in mouse
    blastocyst ICM cells, 212–224
    trophoblast cells, 211
  in interspecific mouse: rat chimeras,
    213–214, 223–224
  of germ cells, 3–21
  in volvox, gonidial cells, 66–67
Development
  of mouse embryos *in vitro*, 279–280
  of preimplantation embryos *in
    vitro*, 279–280

mouse ICM, 210–211
mouse trophoblast, 210–211

347

DNA–RNA hybridization, cotton embryo
mRNA, 176–177

E

Embryoid bodies in transplantable
teratoma, 97–101
Enzyme synthesis
in cotton embryos, 167
in cotyledons, 167
Estradiol, binding capacity by
hypothalamus, 246
uterus, 246
Estrogen
feedback control of LRF-producing
nerons, 245
receptors
in hypothalamus, 248–250
in pituitary, 248–250
regulation of
uterine response, 319–320
protein receptors in uterus, 336–342
uterine response to, 327, 328
Estrogen–sensitive neurons
hypothalamic, 246–247
in preoptic-anterior hypothalamic
area, 244–245
Estrogen stimulation
of endometrium, mitotic index, 317–318
of stromal mitosis, 317
Estrogen surge, 244

F

Follicle stimulating hormone (FSH), release
of, 244–245

G

Germ cell determinant
demonstration of, 19
nature and action of, 19
U.V. sensitivity to, 13–14
Germ plasm
destruction of by
surgical removal, 13–19

U.V. irradiation, 13–19
electron dense bodies, 4–7
role of in formation of germ cells, 13–19
Germinal granules
effect of U.V. irradiation on, 14–19
E.M. structure of, 16–17
germinal vesicle breakdown, 9–10
in Drosophila, 5
origin and continuity of, 7–13
in Rana, 5
in Xenopus, 5
Gonadotrophin secretion
brain mechanism, 239–240
control of
cycle mechanism, 245
hypothalamic, 245
neuronal, 244
neurohormonal estrogen feedback,
244–245
by testosterone, 258
tonic mechanisms, 239–240

H

Hermaphrodites
from aggregation chimeras, 191–192
fertility in, 197–203
gonadal morphology in, 196–197
hypospadia, 190
in mice, 189–204
sex reversal, 189–190
spontaneous, 192–196
testicular feminization, 190
Δ^5, 3 β-hydroxysteroid dehydrogenase
activity
with gonadotrophin, 300–301
in ovary, 303–305
in mouse trophoblast, 299–301
in rat trophoblast, 302–305
in utero, 299–305
in vitro development, 299–305
hormonal control of, 399–301
LH control of, 399–301
Hypophysiotropic releasing factors (HTA),
neuronal synthesis of, 241
Hypothalamus
arteriodorsal region, 241
electrolytic lesions of, 245

hypophysiotropic area, function of, 240
releasing factors, 241

I

ICM
development in uterine monolayers, 286
development *in vitro*, 279–280
isolation of, 210
of mouse blastocyst, 210
properties of, 210–211
transplantation to blastocyst, 219–223
Implantation
of blastocyst, 312–314
interaction between ovum and
endometrium, 313
requirement for progesterone, 299
Inner cell mass (*see* ICM)

L

LH–RH, *see* Luteinizing hormone-
releasing hormone
LRF, *see* Luteinizing hormone–releasing
hormone
LRF fibers, arrangement of, 242–243
Luteinizing hormone, concentration in
plasma, 259–263
release of, 243–244
in androgen sterilized rats, 267–269
response to electrochemical stimulus,
267–269
Luteinizing hormone–releasing hormone
adenylcyclase response to, 260
immunofluorescent localization, 242
pituitary response to
in adults, 259–261
in castrated rats, 263
response of androgen sterilized rats
to, 263–266

M

Male anti-fertility agents, 40
alkylating agents, 40–41
1-amino, 3-chloro, 2-propanol
hydrochloride, 47

bis (dichloroacetyl) diamine, 41
α-chlorohydrin, 46
cyproterone acetate, 46
Male germ cells
incomplete cytokinesis, 36
intercellular bridges, 36–40
synchronized differentiation, 36–40
syncytial nature of, 36–40
Maternal–fetal interactions, in mouse
embryo, 277–279
Maturation of spermatozoa, 46–47
Membranes, in spermatozoa, 47–48
Mitochondria, of spermatozoan
midpiece, 48
Mitogenic activity
in endometrium, 317–318
of estradiol, 317–318
of progesterone, 317–318
Morphological differentiation of cleaving
mouse egg, 210
Mouse embryo
blastocyst implantation, *in vitro*, 280–292
development *in vitro*, 279–280
early somite stage, *in vitro*, 279–280
mRNA
in developing cotton embryo, 173–179
poly (A) addition to, 169–173, 177–185
storage of in cotton seeds, 169, 173–174
mRNP, in cotton cotyledons, 173–174

N

Norepinephrine induction of ovulation
by, 248–250

O

Ovarian teratocarcinogenesis, 99
Ovulation blocking by
anti-estrogens, 246–248
chlomiphene, 246

P

Parthenogenesis
initiation of, 108–111

in LT mice, 93, 99, 101–104
  spontaneous, 101–104
Parthenogenetic activation, 111–120
  activating agent, 112–113, 118–119
  and genetic constitution, 113–120
  in induced ovulation, 113
  post-ovulatory age, 119
Parthenogenetic agents, 108–111
Parthenogenetic embryos development
    of, 121–126
    pronucleate stage, 122–124
  mortality of, causes, 125–127
  post-implantation development, 124–125
  pre-implantation development of, 121–124
Parthenote
  implantation of, 102–103
  from LT mice, 102–104
Parvicellular neurosecretoar system, 240
Polar granules
  in Amphibian eggs, 4–5
  in *Drosophila*, 4
  in fertilized egg, 5–7
  in mature oocyte, 5–7
  in *Xenopus*, 5
Polar trophoblast–blastocyst orientation,
    228–231
Poly (A) in RNP, 174–175
Polyadenylation, of mRNA during
    germination, 179–185
Polyfollicular ovary syndrome, 245–246
Preimplantation
  membrane function in, 152–153
  metabolism during, 151
  mitochondrial function, 153–155
  role of embryonic gene products in, 148–151
  sterioid metabolism, 155
  transcription of, 136–146
    mRNA, 143–146
    ribosomal RNA, 140–143
    tRNA, 137
  translation during, 146
Pre-nidatory estrogen, nature of, 330–336
Preoptic-anterior hypothalamic area
  electrolytic lesions in, 248
  estrogen sensitive neurons in, 244–245
Preoptic LH-trigger, 244
Primordial germ cells, 4
Progesterone
  changes in plasma concentration, 266–268
    in adrenalectomy, 266

    in ovariectomy, 266
  decidual response to, 323
  production of by trophoblast cells, 301–302
  uterine response to
    protein synthesis, 323–326
    transcription, 325–330

R

Radioimmunoassay for
  follicle stimulating hormone, 256–258
  3 β-hydroxysteroid dehydrogenase, 299
  luteinizing hormone, 256–258
  luteinizing hormone-releasing factor, 242
  progesterone, 299–300
Ribonucleoprotein
  in cotton cotyledons, 173–174
    base composition of, 173
    CsCl analysis of, 174
    RNA from, 174
Ribosomal RNA synthesis in volvox, 72
Ribosomes
  in gonidia of *Volvox*, 71–73
  in somatic cells of *Volvox*, 71–73

S

Seminiferous epithelium, organization
    of, 28–35
Seminiferous tubules
  cells of
    germ cells, 29
    Sertoli cells, 28
  compartmentation of the epithelium, 32–35
  mobility of germ cells, 30–32
Sexual inducers
  in *Volvox*
    characterization, 78–82
    mode of action, 83–86
    production of, 82–83
    purification of, 78–82
Sexual reproduction, in *Volvox*, 74–82
Spermatid development
  acrosomal, 40–46
  ultrastructural analysis of, 41–43
  effects of drugs on, bis (dichloroacetyl)
    diamine, 46

Spermatogenesis
  endocrine control of, 25–50
    role of androgen, 26–28
    role of FSH, 26–28
    role of LH, 26–28
Spermatozoa
  maturation of, 46–47
  membranes, 47–48
Stored mRNA, 173
Synchronous development of blastocyst
    and uterus, 312–314
Synchrony, during pre-implantation and
    implantation, 312–314

T

Teratocarcinogenesis, 93–104
  genetic influences on, 94–95
  in male primoridal germ cells, 94–99
Teratomas, 93–104
  development from primordial germ cells,
    95–97
  embryo derived, 99–101
  embryoid bodies from, 97–99
  experimental production of, 96–97
Testicular teratomas, 94–97
Testosterone, and polyfollicular ovary
    syndrome, 245–246
Testosterone propionate, effect of,
  on porduction of sterility, 256
  on reproductive endocrinology, 256–257
Thyrotrophin-releasing factor, 241
Transcription
  in cotton embryo, 169–173
  during preimplantation, 136–146
Translation
  inhibition of by abcisic acid, 167–169
  during preimplantation, 146
Trophoblast
  differentiation, 133–134, 209–211
    biochemistry of, 297–305, 225–228
    morphological, 225–228
    physiological, 225–228
  Δ^5, 3 β-hydroxysteroid dehydrogenase
    activity, 299, 302–303
Trophoblast cells
  growth *in vitro*, 279–286
  loss of invasiveness, 286
Tubero-infundibular tract

arrangement of fibers, 242
function, 241

U

Uterine receptivity
  to blastocyst, 313–314
  to ovoimplantation, 330–332
Uterine sensitivity
  hormonal induction of, 320
  steroid hormone receptors in, 336–342
U.V. irradiaton
  effect on germ cell formation, 13–18
  effect on germ plasm, 16–19

V

Volvox
  androgonia, 75–77
  asexual cycle
    DNA synthesis in, 70–71
    effect of actinomycin D on, 69–70
    nuclear activity in, 67–70
  asexual reproduction, 58–59
  chloroplast ribosome, 71–73
  cytoplasmic ribosomes, 71–73
  DNA
    content of somatic cell, 70
    in mature gonidia, 70–71
    nuclear, 70
    satellite, 70
  eggs, 74
  female spheroids, 74–78
  fertilization, effect of
    soybean trypsin inhibitor, 77–78
    cycloheximide, 77–78
  flagella in
    female spheroids, 77–78
    sperm packetts, 77–78
  gonidia, satellite DNA content of, 71
  gonadial cell determination, 66–67
  gonadial cell differentiation, 64
  germination, 78
  male spheroids, 74–75
  reproductive cells,morphology, 55–58
  ribosomal RNA in, 71–73
  sexual inducers, 74–75, 78–86
  sexual reproduction, 74–78

sexual spheroids, 74
sexuality
   Type I, 74
   Type II, 74
   Type III, 74
   Type IV, 75
   Type V, 75
sheath, chemical composition of, 56–57
somatic cells, 55–57

sperm packets, 74, 77–78
sperm development, 75–77
stages of spheroid formation, 58–61
zygote formation, 78

Y

Yolk sac, biochemical markers, 198–299

A
B
C
D
E
F
G
H
I
J